Surfactants from Renewable Resources

Wiley Series in Renewable Resources

Series Editor

Christian V. Stevens, Department of Organic Chemistry, Ghent University, Belgium

Titles in the Series

Wood Modification: Chemical, Thermal and Other Processes
Callum A. S. Hill

Renewables-Based Technology: Sustainability Assessment
Jo Dewulf & Herman Van Langenhove

Introduction to Chemicals from Biomass
James H. Clark & Fabien E. I. Deswarte

Biofuels
Wim Soetaert & Erick Vandamme

Handbook of Natural Colorants
Thomas Bechtold & Rita Mussak

Surfactants from Renewable Resources
Mikael Kjellin & Ingegärd Johansson

Forthcoming Titles

Industrial Application of Natural Fibres: Structure, Properties and Technical Applications
Jorg Müssig

Thermochemical Processing of Biomass
Robert C. Brown

Surfactants from Renewable Resources

Edited by

MIKAEL KJELLIN

YKI, Institute for Surface Chemistry, Stockholm, Sweden

INGEGÄRD JOHANSSON

AkzoNobel Surfactants, Stenungsund, Sweden

WILEY

A John Wiley and Sons, Ltd., Publication

Registered office
John Wiley & Sons Ltd, The Atrium, Southern Gate, Chichester, West Sussex, PO19 8SQ, United Kingdom

For details of our global editorial offices, for customer services and for information about how to apply for permission to reuse the copyright material in this book please see our website at www.wiley.com.

Library of Congress Cataloging-in-Publication Data

Surfactants from renewable resources / editors, Mikael Kjellin, Ingegard Johansson.
p. cm.
Includes bibliographical references and index.
ISBN 978-0-470-76041-3
1. Surface active agents. 2. Environmental chemistry – Industrial applications. I. Kjellin, Mikael. II. Johansson, Ingegärd.
TP994.S8735 2010
668′.1 – dc22

2009036215

A catalogue record for this book is available from the British Library.

ISBN 978-0-470-76041-3 (H/B)

Typeset in 10/12pt Times-Roman by Laserwords Private Limited, Chennai, India.

Contents

Series Preface

Renewable resources, their use and modification are involved in a multitude of important processes with a major influence on our everyday lives. Applications can be found in the energy sector, chemistry, pharmacy, the textile industry, paints and coatings, to name but a few.

The area interconnects several scientific disciplines (agriculture, biochemistry, chemistry, technology, environmental sciences, forestry, etc.), which makes it very difficult to have an expert view on the complicated interaction. Therefore, the idea to create a series of scientific books, focusing on specific topics concerning renewable resources, has been very opportune and can help to clarify some of the underlying connections in this area.

In a very fast changing world, trends are not only characteristic for fashion and political standpoints, also science is not free from hypes and buzzwords. The use of renewable resources is again more important nowadays; however it is not part of a hype or a fashion. As the lively discussions among scientists continue about how many years we will still be able to use fossil fuels, with opinions ranging from 50 to 500 years, they do agree that the reserve is limited and that it is essential not only to search for new energy carriers but also for new material sources.

In this respect, renewable resources are a crucial area in the search for alternatives for fossil-based raw materials and energy. In the field of energy supply, biomass and renewable-based resources will be part of the solution alongside other alternatives such as solar energy, wind energy, hydraulic power, hydrogen technology and nuclear energy.

In the field of material sciences, the impact of renewable resources will probably be even bigger. Integral utilization of crops and the use of waste streams in certain industries will grow in importance, leading to a more sustainable way of producing materials.

Although our society was much more (almost exclusively) based on renewable resources centuries ago, this disappeared in the Western world in the nineteenth century. Now it is time to focus again on this field of research. However, it should not mean a *retour à la nature*, but it should be a multidisciplinary effort on a highly technological level to perform research towards new opportunities and to develop new crops and products from renewable resources. This will be essential to guarantee a level of comfort for a growing number of people living on our planet. It is 'the' challenge for the coming generations of scientists to develop more sustainable ways to create prosperity and to fight poverty and hunger in the world. A global approach is certainly favoured.

This challenge can only be dealt with if scientists are attracted to this area and are recognized for their efforts in this interdisciplinary field. It is therefore also essential that consumers recognize the fate of renewable resources in a number of products.

Furthermore, scientists do need to communicate and discuss the relevance of their work. The use and modification of renewable resources may not follow the path of the genetic engineering concept in view of consumer acceptance in Europe. Related to this aspect, the series will certainly help to increase the visibility of the importance of renewable resources.

Being convinced of the value of the renewables approach for the industrial world, as well as for developing countries, I was myself delighted to collaborate on this series of books focusing on different aspects of renewable resources. I hope that readers become aware of the complexity, the interaction and interconnections, and the challenges of this field and that they will help to communicate the importance of renewable resources.

I certainly want to thank the people of Wiley from the Chichester office, especially David Hughes, Jenny Cossham and Lyn Roberts, in seeing the need for such a series of books on renewable resources, for initiating and supporting it and for helping to carry the project to the end.

Last, but not least, I want to thank my family, especially my wife Hilde and children Paulien and Pieter-Jan, for their patience and for giving me the time to work on the series when other activities seemed to be more inviting.

Christian V. Stevens,
Faculty of Bioscience Engineering
Ghent University, Belgium
Series Editor '*Renewable Resources*'
June 2005

Preface

Surfactants are molecules that consist of one hydrophilic (water-loving) part and one hydrophobic (water-hating or oil-loving) part. The production of a surfactant is essentially a question of joining different types of these two categories with one another. Renewability refers to the sources for the hydrophilic and the hydrophobic groups.

There has been a substantial development during the last century to construct molecules that are more efficient than the fatty acid soaps that have been produced for over 2000 years. As pointed out in the chapter on surfactants (Oleochemical and Petrochemical Surfactants: An Overall Assessment) in the first book in the series about renewable products (*Renewables-Based Technology*: *Sustainability Assessment*), most surfactants today are readily biodegradable and low-toxic to the aquatic environment, which are the two criteria for 'green surfactants'. The majority of these surfactants are, however, synthesized from petroleum, which of course is non-renewable. This book will focus on renewable sources for surfactants that are also readily biodegradable and how an increased use of renewable sources might be achieved.

When it comes to the hydrophobic part of a surfactant, the natural oleochemical source predominantly offers straight hydrophobic chains with even amounts of carbon atoms. These structures are not always optimal and it has been shown that some branching that does not destroy the biodegradability is preferable from a performance point of view in many applications like cleaning, wetting, etc. On the hydrophilic side, one of the most interesting structural elements that forms the non-ionic surfactants as well as some of the anionic surfactants is ethylene oxide, which at present is made from petroleum sources, i.e. ethylene.

In both cases there are ways of making building blocks from 'natural' sources, for instance from ethanol from fermentation processes using 'green chemistry'. There are activities reviving the processes that were used as late as in the 1950s to produce a whole range of small and larger building blocks from ethanol, starting with acetaldehyde and condensing that to larger branched aldehydes, as well as producing ethylene that could be polymerized to polyethene or oxidized to ethylene oxide.

One could argue that the high-tech surfactants that we use today offer much less burden for the environment than the less efficient, more primitive versions of renewable surfactants that were made earlier, e.g. from fatty acid. Developing the 'green routes' to these advanced surfactants via green building blocks is then an important task for the future and efforts in this direction are thus reported in this book.

Another stumbling block on the road to renewable sources for surfactants are the market issues that come as a consequence of an increasing use of oleochemicals as fuels such as biodiesel. Due to subsidies the market is very much influenced by the support from governments, resulting in an increasing price for the classical hydrophobe sources. Not only cost but also availability is influenced, which in the end might result in a decrease in possible raw material amounts. This development is illustrated in Figure 1 where the price level for fatty acids is followed through the years 2004–2008. There is an obvious dependence on the diesel price which makes the level vary in an unforeseeable way.

Yet another complication is the property demand on the structure of the hydrocarbon chain, which is totally different when the oleochemical is used as an energy source from when it is used as the hydrophobic part of a surfactant. To produce energy through combustion you just need a certain amount of carbon material, but for a surfactant the behaviour is mostly determined by the length and structure of the hydrophobe. This means that, for example, tallow oil cannot be easily substituted by, for instance, palm oil to get the same surfactant properties. Therefore, if a couple of major power plants choose to use tallow oil for their combustion, they could easily consume the total amount produced in Europe. This would be a rather attractive option for the tallow producers, having to deal with only a couple of large-scale customers prepared to pay premium

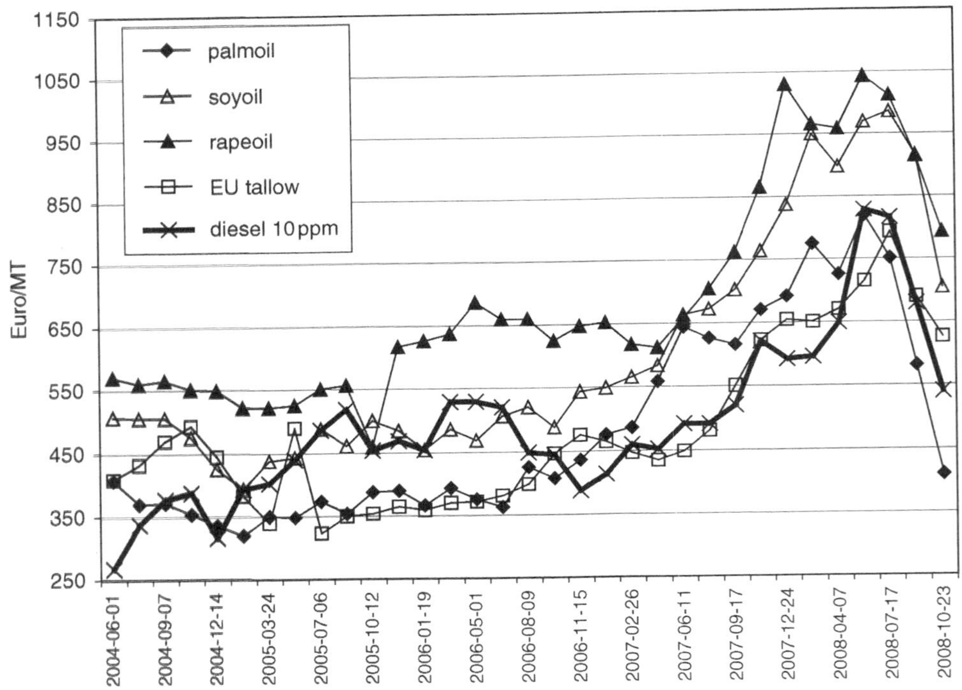

Figure 1 *Correlation between the prices for different raw materials between 2004 and 2008.*
Source: AkzoNobel

prices, having fewer delivery points, lower demands on quality and higher prices due to subsidies. The market might then be forced to go back to petroleum-based sources for surfactant production, i.e. synthetic fatty alcohols – a development that is not wished for by anyone.

It is thus important to create knowledge and awareness of the complicated issues involved in the raw material source uses when the market is driven by forces other than natural competition.

In this book you will find reviews treating both the traditional sources for hydrophobic as well as hydrophilic parts of surfactants, and some newer attempts. We have chosen to concentrate on issues that have an obvious potential for large-scale use and not the more academic investigations, however interesting they might be.

In the first part of the book we treat surfactant raw materials from different sources, crops, animals and wood, touching upon the biorefinery concept including carbohydrates and amino acids and short carboxylic acids like lactic acid, citric acid, etc.

The rest is devoted to different ways of creating new resources, i.e. green ethylene from green ethanol and complex mixtures from waste biomass. A high-flying concept like using algae as a new source is only mentioned very briefly since large-scale experience and knowledge is still lacking.

On top of that, green ways of using these raw materials, for instance in enzymatic processes or microorganism systems, are treated. An example of the use of living cells is the production of sophorolipids and rhamnolipids to be integrated in new 'green' detergents that have found their way to the market in the last 10–15 years and thus can be considered to be an established type of biosurfactant.

A few surface-active structures can be extracted directly from nature, such as lecithin and saponin. They are reviewed in separate chapters, showing that these historic types of surface-active materials are still in use in important areas like food and feed production and various cleaning applications.

Finally, the area is enlarged a bit by looking at larger surface-active molecules that one could describe as surface-active polymers or polymeric surfactants. Here mature types of products like cellulose derivatives and lignosulfonates, as well as the newer inulin products, are treated.

Mikael Kjellin
Ingegärd Johansson
Stockholm/Stenungsund, Sweden
2009

Acknowledgements

The editors for this book met in 1995 when the Centre for Surfactants Based on Natural Products (SNAP) started. Mikael was then a PhD student in surface chemistry at the Royal Institute of Technology and Ingegärd an industrial research leader at AkzoNobel Surface Chemistry in Stenungsund. In total, six academic and thirteen industrial partners collaborated within SNAP with the common goal to explore properties and applications of the next-generation environmentally friendly surfactants. The main outcome of the centre activities was 22 PhDs and over 200 scientific publications.

The networks between academic and industrial researchers created during the lifetime of SNAP also laid the foundation for future research collaborations. Two ongoing examples are the Controlled Delivery and Release Centre (CODIRECT) at the Institute for Surface Chemistry (YKI) and the Supramolecular Biomaterial Center (SuMo Biomaterials) at Chalmers University of Technology.

We thank our employers, the Institute for Surface Chemistry (YKI) and AkzoNobel Surface Chemistry, for giving us the opportunity to work with this book, which we feel covers an important topic for the future. We would also particularly like to thank all the authors for their contributions and for answering all our questions on top of all their other duties in their company or academic surroundings.

List of Contributors

Edgar Joel Acosta Department of Chemical Engineering & Applied Chemistry, University of Toronto, Canada

Patrick Adlercreutz Department of Biotechnology, Lund University, Sweden

David Grant Allen Department of Chemical Engineering & Applied Chemistry, University of Toronto, Canada

Kris Arvid Berglund Division Chemical Engineering, Luleå University of Technology, Sweden

Dirk W. G. Develter Ecover Belgium NV, Malle, Belgium

Steve J. J. Fleurackers Ecover Belgium NV, Malle, Belgium

Ralph Franklin AkzoNobel Surfactants, Brewster, USA

Flor Yunuen García-Becerra Department of Chemical Engineering & Applied Chemistry, University of Toronto, Canada

Stig Are Gundersen Borregaard LignoTech, Sarpsborg, Norway

Arafa Hamed Department of Botany, Aswan Faculty of Science, South Valley University, Aswan, Egypt

Rajni Hatti-Kaul Department of Biotechnology, Lund University, Sweden

Karlheinz Hill Cognis GmbH, Monheim, Germany

Thomas Hjertberg Borealis AB, Stenungsund, Sweden

David B. Hodge Division Chemical Engineering, Luleå University of Technology, Sweden

Bjarne Holmbom Process Chemistry Centre, Åbo Akademi University, Finland

Maria Rosa Infante Instituto de Química Avanzada de Cataluña, Barcelona, Spain

Leif Karlson AkzoNobel Functional Chemicals, Stenungsund, Sweden

Rolf Andreas Lauten Borregaard Corporate R&D, LignoTech Group, Sarpsborg, Norway

Bart Levecke Beneo Bio Based Chemicals, a Division of Beneo-Remy NV, Leuven-Wijgmaal, Belgium

Anna Lundgren Stiftelsen Chalmers Industriteknik, Gothenburg, Sweden

Carmen Moran Chemistry Department, Coimbra University, Coimbra, Portugal

Bernt O. Myrvold Borregaard Corporate R&D, LignoTech Group, Sarpsborg, Norway

Willem van Nieuwenhuyzen Lecipro Consulting, Limmen, The Netherlands

Wieslaw Oleszek Institute of Soil Science and Plant Cultivation, State Research Institute, Pulawy, Poland

Lourdes Perez Instituto de Química Avanzada de Cataluña, Barcelona, Spain

Aurora Pinazo Instituto de Química Avanzada de Cataluña, Barcelona, Spain

Ramon Pons Instituto de Química Avanzada de Cataluña, Barcelona, Spain

Ulrika Rova Division Chemical Engineering, Luleå University of Technology, Sweden

Anders Strand Process Chemistry Centre, Åbo Akademi University, Finland

Anna Sundberg Process Chemistry Centre, Åbo Akademi University, Finland

Martin Svensson Lantmännen Food R&D AB, Stockholm, Sweden

Tharwat Tadros Consultant, Wokingham, UK

Part 1

Renewable Hydrophobes

1
Surfactants Based on Natural Fatty Acids

Martin Svensson
Lantmännen Food R&D AB, Stockholm, Sweden

1.1 Introduction and History

Over the last 50 years or so consumer awareness and concern for the environmental impact of various household products has steadily increased, and contributed to consumer preferences in choosing, for example, soaps, detergents, cleaners and so on. Initially this concern was driven by the visible effects of certain products on the environment, for example river water. However, in recent years the interest has moved to the products' global effect on the environment and the 'total carbon load' has become an issue. In combination with the sharp increases in price and the competition for petroleum products, the economic importance of renewable or biological raw materials for the chemical industry has increased. This trend has been most visible in the energy and fuel sector, where the capacity for production of renewable products has increased dramatically. It has also manifested itself in the production of bioplastics. The detergent industry has also in the last decades increasingly turned its attention to natural raw materials to replace petrochemical products, either as hydrophilic or hydrophobic building blocks. Hydrophilic building blocks have been chosen from many different sources, for example sugars, amino acids, cellulose and other carbohydrates (as illustrated in many of the chapters of this book). Even though natural fats and their derivatives are common feed stocks of the detergent industry, efforts to find new hydrophobic materials have increased, mainly because of an awareness that natural hydrophobic compounds can yield properties that are not easily achieved through conventional synthesis from petrochemical products.

Surfactants from Renewable Resources Edited by Mikael Kjellin and Ingegärd Johansson

An interesting line of development is the use of unsaturated bonds in fatty acids for simple chemical modification to obtain bulkiness in the hydrophobic moiety of the surfactant [1].

Parallel to the growth of the petrochemical industry, the fats and oils industry has grown, and oleochemistry has become an important area of research and technology in several institutions and industries over the years. A large variety of products based on fats and oils have been developed since then, for different uses, such as low-fat spreads and drinks, emulsifiers and functional food ingredients and specialties for cosmetic and personal care applications [2]. These technological advances have also expanded the possibilities of using derivatives of fats and oils for surfactant synthesis.

The availability of oleochemicals has traditionally been dependent on the food and feed industry, where the oils and fats can be found as side-products (e.g. tallow, soya oil, fish oil) or main products (e.g. oils from rapeseed). The recent years' quest for alternative fuels based on fats and oils has led to an increased production and availability of high-quality oleochemicals for nonfood purposes, typically as methyl esters of fatty acids. The increasing demand, in combination with advances in genetics, biotechnology, process chemistry and engineering, are leading to a new or, rather, a return to an old manufacturing concept for converting renewable biomass to valuable fuels and products, generally known as the *biorefinery* concept. The gradual integration of crop-based materials and biorefinery manufacturing technologies offers a potential for new advances in sustainable biomaterial alternatives [3]. There is increased interest in reassessing and developing the biological materials in several fields of application, for example epoxidized oil as plasticizers and stabilizers for vinyl plastics [4], biobased materials [5, 6], reactive diluents [7, 8], surfactants [9], lubricants [10] and printing inks [11]. In this respect the interest has increased in developing new crops and varieties of old crops with higher yields and better performances in the production and final properties. Furthermore, it has become important to evaluate the environmental impact of bio-based products with respect to their entire life cycle, demonstrating that the choice of the raw material often turns out to be an important parameter influencing the life cycle performance [12].

This chapter will cover recent developments in the production, use and characterization of fatty acids and their derivatives as surface-active materials. However, the chapter will be limited to surfactants where the original, native, fatty acid plays an evident role in the properties of the surfactant and will not include the many surfactant classes in which the hydrocarbon backbone or carboxylic group have been modified (e.g. by epoxidation, hydrogenation, amidation) or where the surfactant properties are mostly decided by the variations in the polar head group (e.g. carbohydrate derivatives, amino acids).

1.2 Fats and Oils as Raw Materials

Most fatty acids are obtained by hydrolysis of oils from various oleochemical sources (animal, marine and plant) and the composition of fatty acids in the oil is determined by its origin and production method. An exception to this is the widely used tall oil fatty acid products, obtained as free fatty acids together with rosin acid from paper pulping. Animal sources, for example lard and tallow, are characterized by high concentrations of saturated fatty acids, while marine sources (fish oils) are characterized by long-chain and unsaturated acids. The fatty acid composition of oils from plant sources varies greatly

Table 1.1 Typical concentrations of different fatty acids in oils from commercially available variants of common oil crops

	Palmitic acid C16 : 0	Stearic acid C18 : 0	Oleic acid *cis*C18 : 1	Linoleic acid *cis,cis*C18 : 2	Linolenic acid C18 : 3	Other
'Normal' rapeseed	6		60	21	10	
High erucic rapeseed	4		11	12	9	
'Normal' linseed	10		18	14	58	
Tall oil (Scandinavian)	1	2	30	45		Pinolenic 9% Conjugated C18 : 3 5%
Conventional sunflower	12		19	68		
High oleic sunflower	7		83	10		
Conventional soya bean	15		23	54	8	
Palm oil	55	2.5	30	10		
Tallow	27	33	40	3		

Data collated from References [18] to [20].

depending on the plant origin and cultivar. Commercially exploited seeds such as soya, rape and sunflower have been the subject of many years of breeding programmes to obtain oils with particular fatty acid patterns. The fatty acid composition of a selection of fats and oils can be found in Table 1.1. In addition to breeding efforts on traditional oil crops, work is being done to domesticate alternative oil-rich plants that may yield new, potentially useful, fatty acids [16]. Furthermore, plants and organisms can also contain fatty acids with more unusual functionalities, such as conjugated alkenes, alkyne, epoxy and hydroxyl groups [17]. These unusual fatty acids have been classified by Spitzer [18], but the plants and organisms containing them are not domesticated and the oils and fats are only available in small quantities. However, the genes responsible for the synthesis of some of these have been identified and to some extent transferred to agriculturally useful crops [19, 20]. Modern crop development and genetic engineering approaches may, in the future, contribute to an even greater range of hydrophobic materials available for surfactant synthesis, and an increased need for basic studies of surface-active properties of fatty acids.

Traditionally, industrial oleochemistry has concentrated predominantly on exploiting synthetic methods applied to the carboxylic acid functionality of fatty acids, and less than 10% of the modifications have involved the hydrocarbon backbone of the fatty acid [21]. However, the continued development of oleochemistry opens up for several reaction routes involving selective transformation of the alkyl chain, for example epoxidation, sulfonation, with the potential of producing new highly-branched and charged hydrophobes from abundant natural material [15].

1.3 Fatty Acid Soaps

In the fat-splitting process, fats and oils are hydrolysed to glycerol and fatty acid. Prior to saponification the fatty acids can be purified by, for example, distillation in a specific fraction. Soaps of fatty acids are subsequently produced by the neutralization with various bases, resulting in a acid–soap salt with different positively charged counterions, for example Na, K, NH_4. In contrast to the fatty acids, the soaps are generally water soluble and display strong surfactant properties. The solubility and surface-active properties can be tuned by the nature and combination of fatty acids, counterions and the extent of polarization.

The surface activity and adsorption of fatty acids from a bulk solution to an interface is important in various applications, most importantly in personal cleansing applications where a small amount of the original fat is generally considered to have a beneficial effect on skin. The ability of fatty acid soaps to adsorb selectively to solid particles in aqueous solution is used in many applications, for example lubrication [22], flotation de-inking of paper [23] and purification of minerals [24]. The surface chemical aspects of the process of de-inking has been reviewed by Theander and Pugh [25]. The strong tendency of fatty acids to adsorb to liquid and solid surfaces is a topic of great interest for the more fundamental study of fatty acids. Their behaviour as a two-dimensional monolayer at the air–water interface (Langmuir films) or deposited on a substrate (Langmuir–Blodgett films) display a very rich phase transition behaviour and have been taken as potential models for biological membranes [26] or for fabrication of reliable electronic devices [27].

Many different techniques have been used and developed to study the phase behaviour and association at these monolayers [28, 29]. A large amount of studies have been carried out with various X-ray techniques, and the latest information on ordering and phase behaviour in monolayers using this and other methods have been reviewed by, among others, Schlossman and Tikhonov [30] and Duwez [31]. Dutta [32] surveyed some of the currently available experimental evidence regarding backbone ordering and order–disorder transitions in fatty acid monolayers. Ignes-Mullol *et al.* [33] discussed the rheological responses of the monolayer following various forcing processes.

When the straight-chain fatty acid structure is disturbed the ordering at the monolayer, and the properties, are also significantly altered. Several studies have also been published reporting the effect on the ordering as the fatty acid structure is disrupted by one or several alkyl groups [34], hydroxyl groups [35, 36] or unsaturations [37]. An example of this is the study by Siegel *et al.* [38] on the effect of the OH-group position of hydroxypalmitic acids on the monolayer characteristics. By coupling the results of surface pressure-area isotherm measurements and Brewster angle microscopy (BAM) they were able to demonstrate variations in the temperature dependence, as well as in the long-range orientational order. In the case of OH-substitution near the COOH head group ($n = 2$ or 3), irregular domain growth occurred while at OH-substitution in or near the mid-position ($n = 9$) of the alkyl chain, where regular patterning of the domains indicates high ordering. Alonso and Zasadzinski [39] measured the two-dimensional surface shear viscosity of fatty acid monolayers of different chain lengths. They demonstrated that the viscosity can increase by orders of magnitude at phase boundaries associated with tilted to untilted molecular order, providing that the underlying order is semicrystalline. Hence, untilted, long-range ordered phases are the most viscous films (see Figure 1.1). The association behaviour and adsorption to surfaces in liquids, both in pure water and organic solvents, have been studied by several workers [40–42]. Neys and Joos [43] performed very precise measurements of the surface adsorption of aqueous solutions of a homologous series of fatty acids. Additional information about the behaviour at

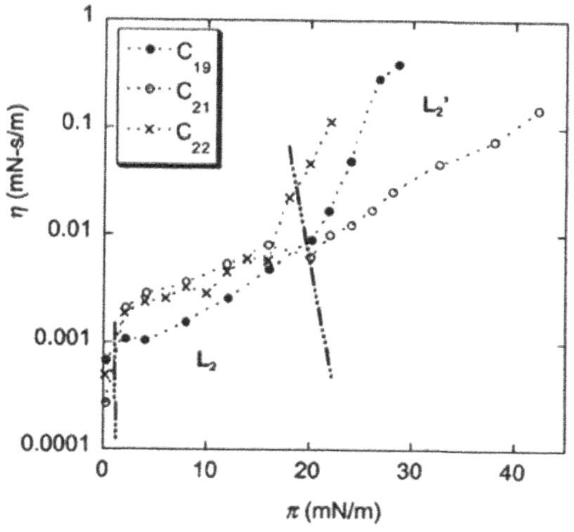

Figure 1.1 *Comparison of the surface shear viscosity η measured as a function of surface pressure for nonadecanoic (C19) at 30 °C, heneicosanoic (C21) at 25 °C and behenic acid (C22) at 20 °C. The temperature of each experiment was adjusted for the monolayers to undergo a transition from a tilted phase (L_2) to an untilted ($L_{2'}$) phase at approximately the same surface pressure. Dashed lines denote phase boundaries. In both the L_2 and $L_{2'}$ phases, the surface viscosity increases exponentially with surface pressure and, hence, with decreasing molecular tilt.*

Reprinted with permission from C Alonso and J A. Zasadzinski, A brief review of the relationships between monolayer viscosity , phase behaviour, surface pressure and temperature using a simple monolayer viscometer, *J. Phys. Chem. B*, **110**, 22185–22191. Copyright 2006 American Chemical Society.

the oil–water interface was obtained by Yehia [44], who found that the heat resistance through a monolayer of fatty acids/alcohols at an oil–water interface reaches a minimum at maximum packing of the species at the monolayer.

The relevance of these studies to the behaviour of other surfactants strengthens as the fatty acids become ionized and turn to soaps with an increasing pH. This transition, and its effect on surface-active properties, has consequently been subject to several studies. At low pH values, the predominant molecule is the undissociated fatty acid. At intermediate values (pH 4–8), undissociated acid, anionic carboxylates as well as so called acid soaps, $(RCOO)_2H^-$, coexist in the system. At alkaline pH, carboxylate anions and acid–soap salts, $(RCOO)_2HNa$, dominate the solution and the surface layer [45]. This change in chemical composition causes changes in the steric, electrostatic and bonding interactions between the molecules at the surface, which can be noticed as several phase transitions in the monolayer [46]. Miranda *et al.* [47] investigated the interactions between water and fatty acids as the monolayer changes from neutral to negatively charged soaps and concluded that the fatty acid monolayer is half-ionized at a pH as high as 10.5–12, as compared to the pK_a of acids in bulk water of 4.9. This was attributed to the locally higher pH at the interface, resulting from a higher concentration of protons at the surface, induced by the surface electric field. Wen and Lauterbach [48] measured the density, the molecular level structure and conformation of myristate or myristate/myristic acid monolayer at the air–water interface. At the intermediate pH (pH 9) it was concluded that the adsorbed monolayer contains not only myristate but also substantial amounts of myristic acid. By titrating a homologous series of C18 fatty acids with varying degrees of unsaturation, Kanicky and Shah [49] could conclude that the pK_a was related to the melting point of the fatty acid and area per molecule at the monolayer. The order of these pK_a values were in the same order as area per molecule values of the fatty acids in spread monolayers. This suggests that as area per molecule increases, the intermolecular distance increases and pK_a decreases due to reduced cooperation between adjacent carboxyl groups. Additionally, the same scientists [50] studied how the ionization of fatty acid varied with concentration. Below the critical micelle concentration (CMC), the value of pK_a was found to decrease as the solution was diluted to a lower concentration. Thus, it was concluded that this reduction in pK_a, even at concentrations well below the CMC, is attributed to the effect of submicellar aggregates on the ionization of the polar head group, leading to higher pK_a as compared to that of soap monomers. Mixing of soap molecules of unequal chain length decreases the pK_a of the solution as compared to that of the two individual components because of disorder produced by the unequal chain length. Kralchevsky *et al.* [51] studied how the natural pH and surface tension isotherms of sodium dodecanoate (laurate), $NaC1_2$, and sodium tetradecanoate (myristate), $NaC1_4$, solutions depend on the surfactant concentration at several fixed concentrations of NaCl. Depending on the surfactant concentration, the investigated solutions contain precipitates of definite stoichiometry of alkanoic acids and neutral soaps. The analysis reveals that the kinks in the surface tension isotherms of the investigated solutions correspond to some of the boundaries between the regions with different precipitates in the bulk. The information of the precipitation behaviour and equilibrium between different forms of the acid–soap complex in dilute and concentrated solutions is important for the understanding of bulk properties of soaps in various products, e.g. bars, detergents and liquid cleansing products.

The changing degree of ionization and packing behaviour of soaps as the pH in the solution varies can be observed in many properties of practical relevance. The pH- and pK_a-related phenomena of fatty acid behaviour and their technological applications were described by Kanicky *et al.* [52]. They found that optimum properties in various properties (foam height and stability, bubble lifetime, contact angle, water evaporation rate) were observed at a pH very near the pK_a of sodium laurate at concentrations below the CMC (Figure 1.2). Based on these observations, they proposed that at the pK_a a maximum ion dipole interaction takes place between ionized and unionized species, leading to a minimum in the area per molecule and an optimum in many properties. Similarly, Somasundaran and co-workers [53] found that the flotation of hematite with weakly anionic collectors, such as oleic acid, displays a distinct maximum at a pH of around 8. When the pH is decreased the presence of undissociated acid and acid–soap complexes increases significantly, leading to an increased surface activity of the oleate species and an improved flotation. If the pH is further decreased to the acidic region, the presence of the ionic soap and acid–soap complexes decreases while that of the undissociated acid remains the same, resulting in a decrease in the hematite flotation and an increase in surface tension. Therefore, the greatest number of surface-active species exists in the neutral pH range.

More recently, Novales *et al.* [54] reported the effect of organic counterions on dispersions of a fatty acid and hydroxyl-derivative salts in aqueous solutions that were

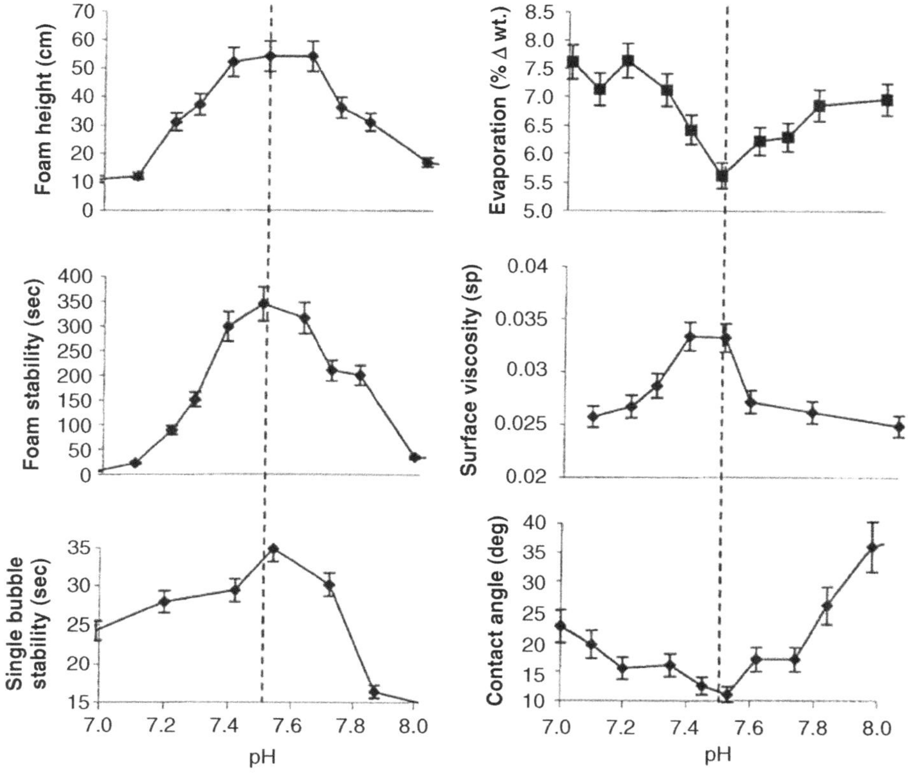

Figure 1.2 *Diagrams depicting maxima and minima in various interfacial properties, with respect to pH, of a sodium laurate solution.*

Reprinted with permission from J.R. Kanicky *et al.*, Cooperativity among molecules at interfaces in relation to various technological processes: Effect of chain length on the pK(a) of fatty acid salt solutions, *Langmuir*, **16**, 172–177. Copyright 2000 American Chemical Society.

further used to produce foams and emulsions. The tetrabutyl-ammonium salts of palmitic acid, 12-hydroxy stearic acid and 8-hydroxy palmitic acid formed isotropic solutions of micelles, whereas the ethanolamine salts of the same acids formed turbid, birefringent, lamellar solutions. This polymorphism demonstrated the effect of a hydroxyl group within the hydrophobic core layer. Foams and emulsions produced from ethanolamine salt solutions were more stable than those obtained from tetrabutyl-ammonium salt solutions. These results were explained in terms of counterion size, lipid molecular shape and the formation of hydrogen bonds between lipids in the core of the micelles.

Soap is generally not toxic to aquatic organisms. Reported EC50 values of laurates for algae, fish and *Daphnia* are 53.0, 11.0 and 10.2 mg/l, respectively [55]. As the solubility of soap is lower in environmentally relevant waters than well water, the bioavailability of soap is generally lower in environmentally relevant waters. Thus, it is generally accepted that soap is even less toxic in aquatic environments than under laboratory conditions using clean water [56].

1.4 Polyethylene Glycol Fatty Acid Esters

Direct ethoxylation of fatty acids and fats with conventional catalysts yields a complex mixture of mono- and diesters, as well as various polyethylene glycols as by-products, with a wide range in the number of polyethylene glycol units. Despite the inhomogeneity of the composition of the final product, they have found a wide use as emulsifiers in food, feed and technical applications and detailed studies of their emulsification and solubility/dispersibility properties have been carried out [57–59].

The disadvantages of direct ethoxylation have, from the late 1980s onwards, been resolved. Based on experiences of narrow-range ethoxylation catalysis, new catalysts have been developed that enable a direct ethoxylation of short-chain alkyl esters of fatty acids. Cox and Weerasooriya extensively described this alkoxylation technology and the properties of ethoxylated methyl esters of various fatty acids in a series of papers [60–62]. The difference in distribution of ethyleneoxide units between a fatty acid methyl ester ethoxylated with a conventional hydroxide catalyst or a more active Ca/Al catalyst was shown to be drastic (see Figure 1.3, from Reference [61]). The distribution is slightly peaked and the amount of unreacted fatty acid methyl ester significantly reduced. Thratnig [63] later described how a similar effect in the distribution of ethylene oxide (EO)units can be obtained with ethoxylation of fatty acids rather than the methyl ester. Furthermore, the technique has also been shown to be applicable to ethoxylation of several types of fatty acid esters like triglycerides or branched alkyl esters [62, 64], as well as for propoxylation of fatty acid esters [65]. Alejski *et al.* [66, 67] and Bialowas and Szymanowski [68] have contributed to the understanding of how the oxyethylation reaction of fatty acid methyl esters proceeds in stepwise incorporation of the ethylene oxide units. In particular, the ethoxylation of inexpensive methyl esters of common oils like that of rapeseed oil (rapeseed oil methyl ester, or RME) have been attractive, because of the increasing production of this ester as a biodiesel alternative in Europe [69, 70].

Several researchers have described properties of methyl ester ethoxylates [64, 70]. In general, polyoxyethylene esters of fatty acid methyl esters have been found to have good emulsifying, lubricating, dispersing and suspending power and these properties, combined with detergent and antistatic characteristics, provide a potential in a variety of textile processing applications. Good wetting, penetrating and dispersing properties have made them useful in adjuvants in agricultural products [71]. A comparison between fatty acid methyl ester ethoxylates and the corresponding range of alcohol ethoxylates, shows that the CMC is somewhat higher and surface tension at CMC lower for the methyl ester ethoxylates [72]. The methyl terminating EO chain leads to a lower foam profile and a lowering of the cloud points by approximately 10 °C [65]. However, the dishwashing capacity is not so good, due to the low solubilization ability of fats and the low foaming. Nonetheless, Renkin *et al.* [70] reported that the washing performance of rapeseed oil methyl ethoxylates with seven EO units in a laundry formulation could be considered as the equivalent of lauryl alcohol ethoxylates with the same number of EO units. Likewise, Littau and Miller [73] described the benefits of mixing the fatty acid methyl ester ethoxylate with conventional nonionic and anionic surfactants to achieve optimum performance in hard surface cleaning. Hama *et al.* [74] established structure–property

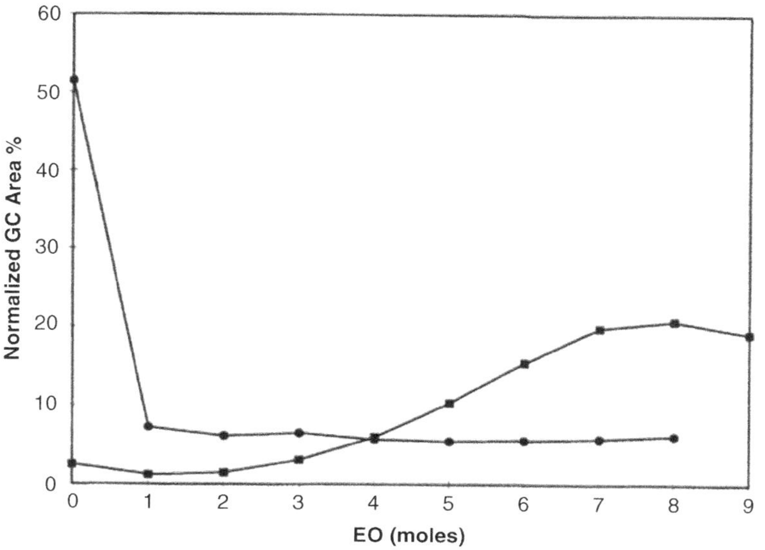

Figure 1.3 Distribution of ethylene oxide units for tetradecyl methyl ester ethoxylates prepared with conventional catalyst (NaOH) and proprietary catalyst: -●-, conventional catalyst, -■-, proprietary catalyst. Reproduced with kind permission from Springer Science + Business Media: *J Am Oil Chem Soc.*, Methyl ester ethoxylates, **74**, 1997, 847–859, MF Cox and U Weerasooriya.

relationships by varying fatty acid structure and amounts of EO. Ethoxylated methyllaurate with approximately 60–70% ethyleneoxide was found to be the most suitable as a base surfactant for household detergents.

Various tests have showed that methyl ester ethoxylates have an improved mildness to human skin compared to ordinary alcohol ethoxylates [62, 64]. From the standpoint of environmental properties, fatty acid methyl ester ethoxylates are readily biodegradable and an order of magnitude less toxic than alcohol ethoxylates [70, 75]. However, the beneficial environmental properties, such as rapid biodegradability, also have the drawback of a poorer hydrolytic stability, particularly in high alkaline or acid conditions. After 80 days at 40 °C there was a 4% decomposition at pH 7 and 13.5% at pH [61]. In a typical laundry detergent formulated in the range pH 8.5–10, the hydrolysis after two months' storage is insignificant [70].

1.5 Polyglycerol Fatty Acid Esters

An attractive alternative to ethoxylation, from an environmental point of view, is the possibility of using natural glycerol as the hydrophilic part of the fatty acid surfactant. Partially hydrolysed triglycerides, with one glycerol moiety, represent the most widely used surfactant of this kind, found as emulsifiers in many food and cosmetic products. In addition to these, polyglycerol fatty esters produced through a condensation reaction of fatty acids or partial glycerides with glycerol have been the attention of many studies. However, like the direct ethoxylation described above, this condensation reaction gives rise to a broad distribution in the hydrophilic head group, as well as a distribution of a number of fatty acids attached to the hydrophilic group and various glycerol oligomers as by-products. Hence, the product will consist of many different constituents.

Ishitobi and Kunieda [76] have investigated the effect of the oligoglycerol distribution on the phase behaviour, comparing one product with a broad distribution and one with a more narrow distribution. The phase diagram revealed a micellar region and a hexagonal phase at higher concentrations for both products. The more narrow-range product formed hexagonal phases at higher concentrations, has a higher cloud point, higher surface tension at corresponding concentrations and is a less efficient emulsifier. All effects are explained by the fact that the product with the broader distribution has a smaller effective cross-sectional area per hydrophobic chain and thus can pack more tightly in the interfaces. The challenge of studying these surfactants due to the variation in composition was also addressed by Duerr-Auster *et al.* [77]. They found that a commercial mixture of polyglycerol fatty acid esters (from palmitic and stearic acid) in water formed a lamellar morphology over the whole concentration range investigated. However, it was also found that the commercial mixture contained small amounts of unreacted fatty acid, in a dissociated, anionic, state. This small impurity had a pronounced stabilizing effect on the gel phase. In addition, the phase behaviour of commercial tetraglycerol [78], pentaglycerol [79] and decaglycerol fatty acid esters [80] have been reported.

In contrast, Kato *et al.* [81] prepared a series of purified polyglycerol monolaurates (C12Gn, with $n = 2, 3, 4, 5$). The phase behaviour and surfactant properties of these were compared with those of n-dodecyl polyoxyethylene monoethers (C12EOn) to examine the function of the hydrophilic part of these compounds. The surfactants followed similar trends in properties like the CMC, surface area at interface, detergency, foam height and stability. However, the foam heights of the glycerol-based surfactants were consistently higher and more stable than those of C12EOn. It was concluded that important surfactant properties, for example detergency, of polyglycerol monolaurates having few glycerol units (di- to tetraglycerol monolaurates) were on the same level as those of C12EOn having more oxyethylene units (hexa- and octaoxyethylene) (see the example in Figure 1.4). If the fatty acid chain is further increased to stearic acid (C18), the surfactant loses water solubility and forms a stable monolayer at the air–water interface [82].

Diglycerol esters of saturated fatty acids have recently been extensively studied by Shrestha and co-workers in both aqueous [83] and nonaqueous [84, 85] systems. The phase behaviour of caprate (C10) and laurate (C12) esters in water were found to be quite different from the solution behaviour of the myristate (C14) and palmitate (C16) esters (see Figure 1.5). In the former, a lamellar liquid crystal phase is present in the surfactant-rich region and it absorbs a substantial amount of water. The melting temperature of this phase is practically constant in a wide range of compositions. For the more hydrophobic surfactant the phases with solids and the extent of water solubilization are increased.

To conclude, polyglycerol fatty acid esters are edible nonionic surfactants, and, in combination with their low solubility in water and high surface activity, many of them are of interest as emulsifiers, dispersants, solubilizers, rheology modifiers in drugs, cosmetics or as specific food ingredients where controlled release is the goal (fragrances, flavourings) [86]. The capability of forming stable α-gel phases makes them useful as stabilizing foams and emulsions in food products [87]. However, like the previously

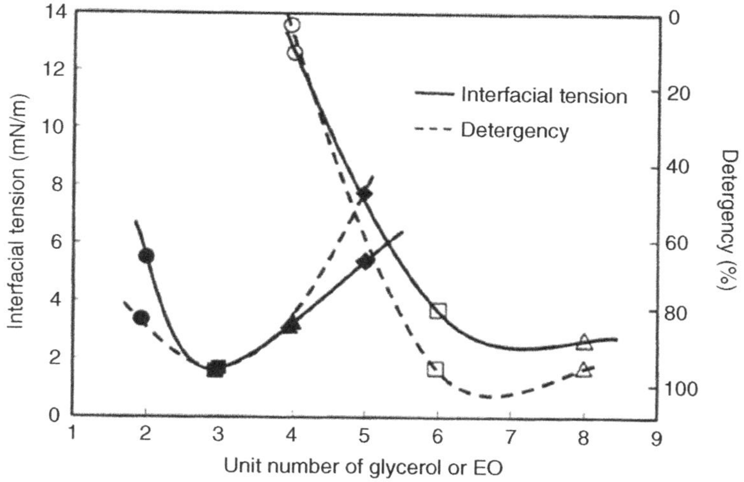

Figure 1.4 Plots of the interfacial tension of corn oil/surfactant solutions and detergency as a function of the number of glycerol or oxyethylene (EO) units of polyglycerol laurate (filled symbols) and polyoxyethylene lauryl ether (open symbols).

Reproduced with kind permission from Springer Science + Business Media: *J. Surfactants Deterg.*, Surfactant properties of purified polyglycerol monolaurates, **6**, 2003, 331–337, T Kato *et al.*

described esters of polyoxyethylene and fatty acid they are susceptible to hydrolysis in strong acid and alkaline environments.

1.6 Conclusions

As is evident from this review, the amount of work being done concerning fatty acids and their derivatives is immense. If the increasing interest in renewable sources for energy purposes in recent years can be combined with a sustainable cultivation and processing, it is expected that fatty acids will continue to grow as a widely available stock for detergents. To understand fully the behaviour and properties of surfactants derived from these fatty acids it is important to expand the studies also of more fundamental properties, such as the association behaviour of fatty acids in Langmuir films. These will, for example, show how basic information of the dissociation behaviour of soaps can be related to practical properties such as foaming and detergency. Another example is the formation of soap–acid complexes and precipitates at higher concentrations and the behaviour of common soap bars.

To overcome the problems with poor solubility in hard water or in the presence of salts and other ions, fatty acids have been used in a rich variety of reactions with polar compounds to produce many different types of surfactants. In this respect, an illustration of the slightest modification would be simple esterification with polar compounds to achieve surfactancy. The simplest of these esters are represented by the polyoxyethylene and glycerol esters. These are characterized by a good biodegradability, low toxicity and mildness to skin, making them useful in cleansing products, agriculture, food and feed

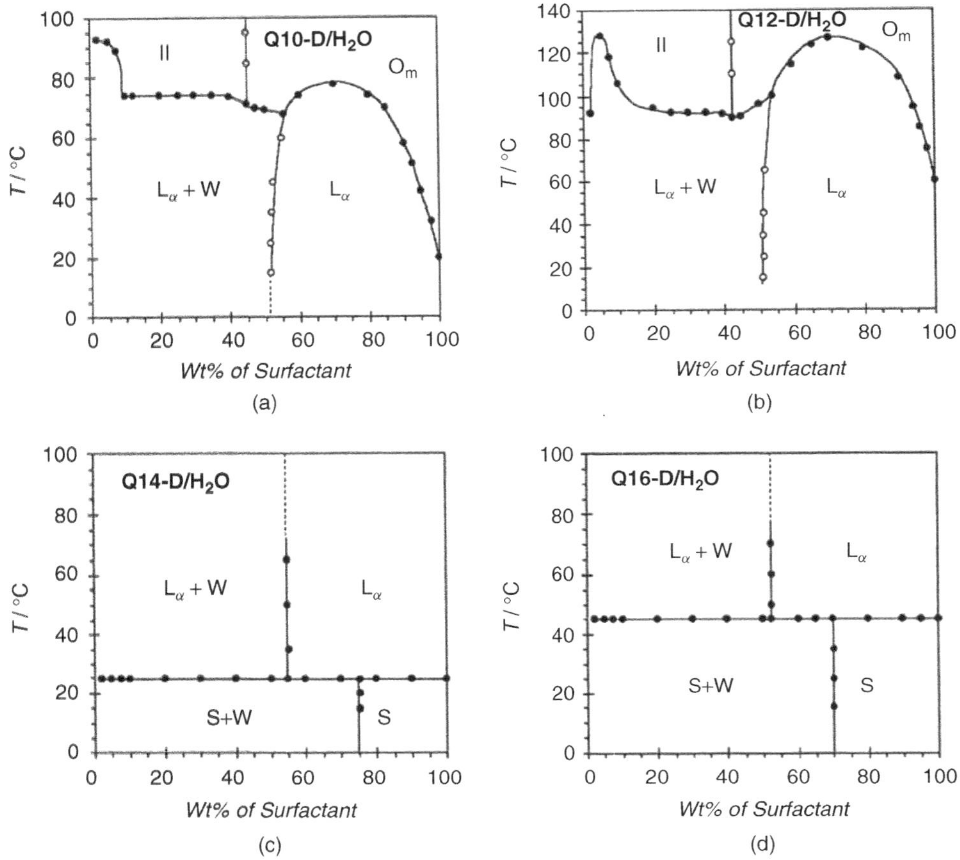

Figure 1.5 *Binary phase diagrams of diglycerol esters of fatty acids in water: (a) caprylate ester (C10), (b) laurate ester (C12), (c) myristate (C14), (d) palmitate (C16) (L_α = lamellar liquid crystals, W = excess water, II = two-liquid phase region, Om = isotropic reverse micellar solution, S = solid).*

L. K. Shrestha *et al.*, Aqueous phase behavior of diglycerol fatty acid esters, *Journal of Dispersion Science and Technology*, **28**, 2007, 883–891. Reproduced with permission from Taylor & Francis Group, http://www.informaworld.com.

formulations. However, the industrial synthesis of these has not been straightforward, yielding numerous side-products and a distribution of components. The recent years' discovery of a new catalyst for ethoxylation of fatty acid methyl ester has opened up the production of these types of products.

A drawback with ester-based surfactants are their susceptibility to hydrolysis if stored in aqueous formulations. The extent of this problem is not completely clear, but has to be kept in mind for any application of these surfactants. A way to overcome this is to convert the acid to an amide. The properties of this type of surfactant is the topic of another chapter in this book.

References

1. Holmberg, K. (1996) Unsaturated monoethanolamide ethoxylates as paint surfactants. *Prog. Colloid Polym. Sci.*, **101**, 69–74.
2. Hill, K. (2007) Industrial development and application of biobased oleochemicals, *Pure Appl. Chem.*, **79**, 1999–2011.
3. Vemula, P. K. and John, G. (2008) Crops: a green approach toward self-assembled soft materials. *Acc. Chem. Res.*, **41**, 769–782.
4. Bouchareb, B. and Benaniba, M. T. (2008) Effects of epoxidized sunflower oil on the mechanical and dynamical analysis of the plasticized poly(vinyl chloride). *J. Appl. Polym. Sci.*, **107**, 3442–3450.
5. Jin, F. L. and Park, S. J. (2008) Thermomechanical behaviour of epoxy resins modified with epoxidized vegetable oils. *Polym. Int.*, **57**, 577–583.
6. Hong, C. K. and Wool, R. P. (2005) Development of a bio-based composite material from soybean oil and keratin fibres. *J. Appl. Polym. Sci.*, **95**, 1524–1538.
7. Stenberg, C., Svensson, M., Wallström, E. and Johansson, M. (2005) *Surf. Coat. Int. B: Coat. Trans.*, **88**, 119–126.
8. Johansson, K. and Johansson, M. (2007) Fatty acid methyl esters as reactive diluents in coil-coatings. *Polym. Prepr.*, **48**, 857–858.
9. Piispanen, P. S., Persson, M., Claesson, P. and Norin, T. (2004) Surface properties of surfactants derived from natural products. *J. Surf. Detergents*, **7**, 147–159.
10. Sharma, B. K., Perez, J. M. and Erhan, S. Z. (2007) Soybean oil-based lubricants: a search for synergistic antioxidants. *Energy Fuels*, **21**, 2408–2414.
11. Hutchinson, G. H. (2002) Ink technology past, present and future. *Surf. Coat. Int. B: Coat. Trans.*, **85**, 169–176.
12. Hatti-Kaul, R., Tornvall, U., Gustafsson, L. and Borjesson, P. (2007) Industrial biotechnology for the production of bio-based chemicals – a cradle-to-grave perspective. *Trends Biotechnol.*, **25**, 119–124.
13. Biermann, U., Friedt, W., Lang, S. *et al.* (2000) New syntheses with oils and fats as renewable raw materials for the chemical industry. *Angew. Chem. Int. Ed.*, **39**, 2206–2224.
14. Gunstone, F. D. (1997) Major sources of lipids, in *Lipid Technologies and Applications* (eds F. D. Gunstone and F. B. Padley), Marcel Dekker, Inc., New York.
15. Firestone, D. F. (ed.) (1999) *Physical and Chemical Characteristics of Oils, Fats and Waxes*, AOCS Press.
16. Smith, N. O., Maclean, I., Miller, F. A. and Carruthers, S. P. (1997) Crops for Industry and Energy in Europe, European Commission, DG XII E-2, Luxembourg.
17. Metzger, J. O. and Bornscheuer, U. (2006) Lipids as renewable resources: current state of chemical and biotechnological conversion and diversification. *Appl. Microbiol. Biotechnol.*, **71**, 13–22.
18. Spitzer, V. (1999) Screening analysis of unknown seed oils. *FETT-LIPID*, **101**, 2–19.
19. Dyer, J. M., Stymne, S., Green, A. G. and Carlsson, A. S. (2008) High-value oils from plants. *Plant J.*, **54**, 640–655.
20. Dyer, J. M. and Mullen, R. T. (2005) Development and potential of genetically engineered oilseeds. *Seed Sci. Res.*, **15**, 255–267.
21. Hill, K. (2000) Fats and oils as oleochemical raw materials. *Pure Appl. Chem.*, **72**, 1255–1264.
22. Zhang, P., Xue, Q., Du, Z. and Zhang, Z. (2000) The tribological behaviour of LB films of fatty acids and nanoparticles. *Wear*, **242**, 147–151.

23. Borchardt, J. K. (1997) The use of surfactants in de-inking paper for paper recycling. *Curr. Opin. Colloid Interface Sci.*, **2**, 402–408.
24. Quast, K. B. (2000) A review of hematite flotation using 12-carbon chain collectors. *Miner. Engng*, **13**, 1361–1376.
25. Theander, K. and Pugh, R. J. (2004) Surface chemicals concepts of flotation de-inking. *Colloids Surf. A: Physiochem. Engng Aspects*, **240**, 111–130.
26. Edidin, M. (2003) Lipids on the frontier: a century of cell-membrane bilayers. *Nat. Rev. Molec. Cell Biol.*, **4**, 414–418.
27. Akkerman, H. B., Blom, P. W. M., de Leeuw, D. M. and de Boer, B. (2006) Towards molecular electronics with large-area molecular junctions. *Nature*, **441**, 69–72.
28. Loste, E., Díaz-Martí, E., Zarbakhsh, A. and Meldrum, F. C. (2003) Study of calcium carbonate precipitation under a series of fatty acid Langmuir monolayers using Brewster angle microscopy. *Langmuir*, **19**, 2830–2837.
29. Teixeira, A. C. T., Fernandes, A. C., Garcia, A. R. *et al.* (2007) Microdomains in mixed monolayers of oleanolic and stearic acids: thermodynamic study and BAM observation at the air–water interface and AFM and FTIR analysis of LB monolayers. *Chem. Phys. Lipids*, **149**, 1–13.
30. Schlossman, M. L. and Tikhonov, A. M. (2008) Molecular ordering and phase behaviour of surfactants at water–oil interfaces as probed by X-ray surface scattering. *Annu. Rev. Phys. Chem.*, **59**, 153–177.
31. Duwez, A. S. (2004) Exploiting electron spectroscopies to probe the structure and organization of self-assembled monolayers – a review. *J. Electron Spectrosc. Relat. Phenom.*, **134**, 97–138.
32. Dutta, P. (2000) What X-rays tell us about the ordering of molecular and counter ion binding strength are found to decrease as the chain backbones in Langmuir monolayers. *Colloids Surf. A*, **171**, 59–63.
33. Ignes-Mullol, J., Claret, J., Riegada, R. and Sagues, F. (2007) Spread monolayers: structure, flows and dynamic self-organization phenomena, *Phys. Rep.*, **448**, 163–179.
34. Fujii, T., Yuasa, R. and Kawase, T. (1999) Biodetergent. Part 4. Monolayers of corynomycolic acids at the air–water interface. *Colloid Polym. Sci.*, **277**, 334–339.
35. Huda, M. S. (1997) Thermodynamical study of hydroxylated fatty acid monolayer films. *Colloid Surf. B*, **9**, 213–223.
36. Overs, M., Fix, M., Jacobi, S. *et al.* (2000) Assembly of new *vic*-dihydroxyoctadecanoic acid methyl esters at the air–water interface. *Langmuir*, **16**, 1141–1148.
37. Hac-Wydro, K. and Wydro, P. (2007) The influence of fatty acids on model cholesterol/phospholipid membranes. *Chem. Phys. Lipids*, **150**, 66–81.
38. Siegel, S., Vollhardt, D. and Cadenhead, D. A. (2005) Effect of the hydroxy group position on the monolayer characteristics of hydroxypalmitic acids. *Colloids Surf. A: Physicochem. Engng Aspects*, **256**, 9–15.
39. Alonso, C. and Zasadzinski, J. A. (2006) A brief review of the relationshops between monolayer viscosity, phase behaviour, surface pressure and temperature using a simple monolayer viscometer. *J. Phys. Chem. B*, **110**, 22185–22191.
40. Zwierzykowski, W. and Konopacka-Lyskawa, D. (1999) The interactions of saturated fatty acids at the dodecane/water interface and their sodium salts at the air/water interface. *Colloids Surf. A: Physicochem. Engng Aspects*, **160**, 183–188.
41. Bravo, B., Sánchez, J., Cáceres, A. *et al.* (2008) Partitioning of fatty acids in oil/water systems analysed by HPLC. *J. Surfactants Deterg.*, **11**, 13–19.
42. Varma, R. P. and Goel, H. (2000) Conductance behaviour of lithium soaps in aqueous methanol. *J. Surfactants Deterg.*, **3**, 527–532.
43. Neys, B. and Joos, P. (1998) Equilibrium surface tensions and surface potentials of some fatty acids. *Colloids Surf. A*, **143** (5), 467–447.

44. Yehia, A. (1997) Effect of hydrocarbon chain configuration on the surface activity of fatty acids–effect of solution pH. *Afinidad*, **54**, 315–320.
45. Small, D. M. (1986) *The Physical Chemistry of Lipids, Handbook of Lipid Research*, vol. **4**, Plenum Press, New York.
46. Johann, R., Vollhardt, D. and Möhwald, H. (1999) Texture features of long-chain fatty acid monolayers at high pH of the aqueous subphase. *Mater. Sci. Eng. C: Biomimetic Supramol. Syst.*, **8**, 35–42.
47. Miranda, P. B., Du, Q. and Shen, Y. R. (1998) Interaction of water with a fatty acid Langmuir film. *Chem. Phys. Lett.*, **286**, 1–8.
48. Wen, X. Y. and Lauterbach, J. F. (2000) Surface densities of adsorbed layers of aqueous sodium myristate inferred from surface tension and infrared reflection absorption spectroscopy. *Langmuir*, **16**, 6987–6994.
49. Kanicky, J. R. and Shah, D. O. (2002) Effect of degree, type, and position of unsaturation on the p*K*(a) of long-chain fatty acids. *J. Colloid Interface Sci.*, **256**, 201–207.
50. Kanicky, J. R. and Shah, D. O. (2003) Effect of premicellar aggregation on the p*K*a of fatty acid soap solutions. *Langmuir*, **19**, 2034–2038.
51. Kralchevsky, P. A., Danov, K. D., Pishmanova, C. I. *et al.* (2007) Effect of the precipitation of neutral-soap, acid-soap, and alkanoic acid crystallites on the bulk pH and surface tension of soap solutions source. *Langmuir*, **23**, 3538–3553.
52. Kanicky, J. R., Poniatowski, A. F., Mehta, N. R. and Shah, D. O. (2000) Cooperativity among molecules at interfaces in relation to various technological processes: effect of chain length on the p*K*(a) of fatty acid salt solutions. *Langmuir*, **16**, 172 .
53. Somasundaran, P., Ananthapadmanabhan, K. P. and Ivanov, I. B. (1984) Dimerization of oleate in aqueous solutions. *J. Colloid Interface Sci.*, **99**, 128–135.
54. Novales, B., Navailles, L., Axelos, M. *et al.* (2008) Self-assembly of fatty acids and hydroxyl derivative salts. *Langmuir*, **24**, 62–68.
55. Onitsuka, S., Kasai, Y. and Yoshimura, K. (1989) Quantitative structure–toxic activity relationshop of fatty acids and the sodium salts to aquatic organisms. *Chemosphere*, **18**, 1621–1631.
56. Prats, D., Rodriguez, M., Varo, P. *et al.* (1999) Biodegradation of soap in anaerobic digesters and on sludge amended soils. *Water Res.*, **33**, 105–108.
57. Saad, A. L. G., El-Kholy, S. A. and Barakat, Y. (1998) Relaxation behaviour and dipole moment of some nonionics. Part 1. CMC and thermo-complex commercial mixtures that are currently being used in dynamics of ethoxylated oleic acid in nonaqueous medium. *Z. Phys. Chem.*, **203**, 73–85.
58. O'Lenick Jr, A. J. and Parkinson, J. K. (1998) The effect of branching and unsaturation upon some properties of polyoxyethylene glycol diesters. *J. Surfactants Deterg.*, **1**, 529–532.
59. O'Lenick Jr, A. J. (2000) Evaluation of polyoxyethylene glycol esters of castor, high-erucic acid rapeseed, and soybean oils. *J. Surfactants Deterg.*, **3**, 201–206.
60. Cox, M. F. and Weerasooriya, U. (1998) Impact of molecular structure on the performance of methyl ester ethoxylates. *J. Surfactants Deterg.*, **1**, 11–22.
61. Cox, M. F. and Weerasooriya, U. (1997) Methyl ester ethoxylates. *J. Am. Oil Chem. Soc.*, **74**, 847–859.
62. Cox, M. F. and Weerasooriya, U. (2000) Partially saponified triglyceride ethoxylates. *J. Surfactants Deterg.*, **3**, 213–220.
63. Trathnigg, B., Hreczuch, W., Szymanowski, J. *et al.* (2004) Comparison of composition and properties of ethoxylated dodecanoic acid obtained in the presence of various catalysts. *Tenside Surfactant Deterg.*, **41**, 54–61.
64. Hreczuch, W., Trathnigg, B., Dziwinski, E. and Pyzalski, K. (2001) Direct ethoxylation of a longer-chain aliphatic ester. *J. Surfactants Deterg.*, **4**, 167–173.

65. Cox, M. F., Weerasooriya, U., Filler, P. and Mellors, W. (1998) Methyl ester propoxylates. *J. Surfactants Deterg.*, **1**, 167–175.
66. Alejski, K., Bortkowiak, P. and Szymonowski, J. (2004) Modelling of fatty acid methyl ester oxyethylation in a semi-batch reactor with successive dosing of ethylene oxide. *Tenside Surfactant Deterg.*, **41**, 130–134.
67. Alejski, K., Bialowas, E., Hreczuch, W. *et al.* (2003) Oxyethylation of fatty acid methyl esters. Molar ratio and temperature effects. Pressure drop modelling. *Ind. Engng Chem. Res.*, **42**, 2924–2933.
68. Bialowas, E. and Szymanowski, J. (2004) Catalysts for oxyethylation of alcohols and fatty acid methyl esters. *Ind. Engng Chem. Res.*, **43**, 6267–6280.
69. Doll, K. M., Moser, B. R., Sharma, B. K. and Erhan, S. Z. (2006) Current uses of vegetable oil in the surfactant, fuel, and lubrication industries. *Chim. Oggi-Chem. Today*, **24**, 41–44.
70. Renkin, M., Fleurackers, S., Szwach, I. and Hreczuch, W. (2005) Rapeseed methyl ester ethoxylates: a new class of surfactants of environmental and commercial interest. *Tenside Surfactants Deterg.*, **42**, 280–287.
71. Gauvrit, C., Muller, T., Milius, A. and Trouve, G. (2007) Ethoxylated rapeseed oil derivatives as non-ionic adjuvants for glyphosate. *Pest Manage. Sci.*, **63**, 707–713.
72. Makowska, D., Hreczuch, W., Zimoch, J. *et al.* (2001) Surface activity of mixtures of oxyethylated methyl dodecanoate and sodium dodecylbenzenesulfonate. *J. Surfactants Deterg.*, **4**, 121–126.
73. Littau, C. and Miller, D. (1998) Optimization of surfactant systems containing methyl ester ethoxylates. *SÖFW – Seifen, Öle, Fette, Wachse*, **124**, 690–697.
74. Hama, I., Sakaki, M. and Sasamoto, H. (1997) Effects of ethoxylate structure on surfactant properties of ethoxylated fatty methyl esters. *J. Am. Oil Chem. Soc.*, **74**, 823–827.
75. Hama, I., Sasamoto, H., Tamura, T. *et al.* (1998) Skin compatibility and ecotoxicity of ethoxylated fatty methyl ester nonionics. *J. Surfactants Deterg.*, **1**, 93–97.
76. Ishitobi, M. and Kunieda, H. (1998) Effect of chain length distribution on esterification reactions. *J. Surfactants Deterg.*, **1**, 343–344.
77. Duerr-Auster, N., Kohlbrecher, J., Zuercher, T. *et al.* (2007) Microstructure and stability of a lamellar liquid crystalline and gel phase formed by a polyglycerol ester mixture in dilute aqueous solution. *Langmuir*, **23**, 12827–12834.
78. Kawakatsu, T., Tragardh, G., Tragardh, C. *et al.* (2001) The effect of the hydrophobicity of microchannels and components in water and oil phases on droplet formation in microchannel water-in-oil emulsification. *Colloids Surf. A: Physicochem. Engng Aspects*, **179**, 29–37.
79. Izquierdo, P., Acharya, D. P., Hirayama, K. *et al.* (2006) Phase behaviour of pentaglycerol monostearic and monooleic acid esters in water. *J. Dispersion Sci. Technol.*, **27**, 99–103.
80. Ai, S. and Ishitobi, M. (2006) Effects of the number of fatty acid residues on the phase behaviours of decaglycerol fatty acid esters. *J. Colloid Interface Sci.*, **296**, 685–689.
81. Kato, T., Nakamura, T., Yamashita, M. *et al.* (2003) Surfactant properties of purified polyglycerol monolaurates. *J. Surfactants Deterg.*, **6**, 331–337.
82. Kawaguchi, M., Yamamoto, M., Nakamura, T. *et al.* (2001) Surface properties of mono-, di-, and triglycerol monostearate monolayers spread at the air–water interface. *Langmuir*, **17**, 4677–4680.
83. Shrestha, L. K., Shrestha, R. G., Iwanaga, T. and Aramaki, K. (2007) Aqueous phase behavior of diglycerol fatty acid esters. *J. Dispersion Sci. Technol.*, **28**, 883–891.
84. Shrestha, L. K., Shrestha, R. G., Solans, C. and Aramaki, K. (2007) Effect of water on foaming properties of diglycerol fatty acid ester–oil systems. *Langmuir*, **23**, 6918–6926.
85. Shrestha, L. K. and Aramaki, K. (2007) Phase behaviour of diglycerol monomyristate in different nonpolar organic solvent systems. *J. Dispersion Sci. Technol.*, **28**, 1236–1241.

86. Shrestha, L. K., Acharya, D. P., Sharma, S. C. *et al.* (2006) Aqueous foam stabilized by dispersed surfactant solid and lamellar liquid crystalline phase. *J. Colloid Interface Sci.*, **301**, 274–281.
87. Krog, N. (1997) Food emulsifiers, in *Lipid Technologies and Applications* (eds F. D. Gunstone and F. B. Padley), Marcel Dekker, Inc., New York.

2
Nitrogen Derivatives of Natural Fats and Oils

Ralph Franklin
AkzoNobel Surfactants, Brewster, USA

2.1 Introduction

In this chapter the manufacture, use, benefits and challenges facing nitrogen-based surfactants derived from renewable resources are discussed. While there are many natural fats and oils in existence only a few are used in significant quantities for the manufacture of oleo-based surfactants; thus, the potential for innovation through the development of the less explored oils is high. Historically, tallow has been the raw material of choice for surfactant manufacturers due to its ready availability, low cost and versatility. A by-product from the meat processing industry, it has been the primary feedstock for fatty amine derivatives since the late 1940s.

Fatty acids are obtained by saponification of the corresponding fat or oil, yielding a mixture of chain lengths, predominantly even numbered and containing varying degrees of unsaturation [1]. Hydrogenation of the natural blend followed by fractionation affords saturated acids of a single chain length. Reaction of the acid with ammonia at high temperature over a metal oxide catalyst leads to the fatty nitrile, which provides a springboard to a wide range of amines and amine derivatives.

Tallow affords derivatives with a predominantly C16–C18 chain length; for the shorter C10–C14 chain length derivatives coconut oil is the preferred material. As other oils have become available in quantity and competitive in pricing they too have been adopted by surfactant manufacturers. Today, tallow and coconut are joined by palm, rapeseed and soya bean oils as fatty acid feedstocks for nitrogen surfactants using the nitrile process [1, 2]. New oils on the horizon such as meadowfoam, crambe and lesquerella have yet to make any significant impact, while other oils remain unexplored [3].

Surfactants from Renewable Resources Edited by Mikael Kjellin and Ingegärd Johansson

Nitrogen derivatives of fatty acids in the context of this article refer for the most part to surfactants derived via conversion of the acid to an amine, or via condensation of the fatty acid with an amine such as an alkanolamine. The manufacture of such derivatives has been reviewed in detail by the author and co-workers [4] and will be discussed in less detail here.

As the chemical industry has responded to meet society's demands for better and environmentally friendlier products so has the surfactant industry. A clear effort has been made to demonstrate the biodegradability of its products and to develop new products with a 'greener' profile. For the personal care market products have been developed based solely on vegetable oils to ease the concern over tallow after outbreaks of bovine spongiform encephalopathy (BSE). However, work funded by the oleochemical industry has shown that the infective agent is destroyed during fatty acid production [5]. A recent study indicates that hydrogenation of tallow under industrial conditions is also effective at destroying the infectivity [6].

While BSE is certainly a concern for the oleochemical industry, another significant threat to surfactants based on renewable fats and oils is their use as alternative energy sources. The escalation in crude oil prices had an impact on natural fats and oils in that they are now viewed as a viable alternative energy source and a way to reduce greenhouse gas emissions. Fats and oils may be used for biodiesel, co-refining with crude oil or direct combustion. The calorific value of tallow is about 90% that of fuel oil and it requires minimal changes in burner design to combust it, while the emissions in the form of sulfur oxides and particulates is lower than for fossil fuels. Subsidies provided by governments to energy companies developing renewable fuels from fats and oils have increased demand, causing prices to rise and follow the swings in petroleum pricing. Tallow is particularly affected in that it has a finite supply, being a by-product from the rendering industry and production cannot be expanded in the way an agricultural-based product can. Tallow is the primary raw material for the US oleochemical industry and concerns over the impact of fuel subsidies on the industry were expressed by the Surfactant and Detergent Association to the US House of Representatives Ways and Means Committee, 19 April 2007.

2.2 Manufacture of Fatty Nitrogen Derivatives

2.2.1 Fatty Amines

There are two common methods for the manufacture of fatty amines: from fatty acid via the nitrile route or from fatty alcohols. For the production of primary and secondary amines the nitrile route is favoured while for tertiary amines both nitrile and alcohol routes may be used. As already mentioned the manufacture of fatty amine has been previously reviewed in detail by the author [4].

2.2.1.1 The Nitrile Route

The conversion of fatty acid to fatty nitrile involves reaction of the acid with ammonia at high temperatures over a metal oxide catalyst. Suitable metal oxides include alumina and zinc oxide although other catalysts are cited in the literature. The process can be done in batch mode or via continuous processing, the acid being contacted with ammonia and catalyst at temperatures in excess of 250 °C. In both cases post-treatment of the nitrile is usually required to remove impurities such as soaps and amides. Distillation is one option for purification whereas others use clay treatment to remove the impurities. The nitrilation process has little impact on the unsaturation of the fatty acid as measured by its Iodine Value. The resulting nitriles find limited use as products in themselves and are thus further processed almost entirely into amines via catalytic hydrogenation.

Hydrogenation of fatty nitrile to a primary amine is done by either a batch or continuous process. Batch processing uses a nickel catalyst in the presence of ammonia to suppress formation of dialkyl and trialkylamines. Figure 2.1 illustrates the reaction pathways for the hydrogenation of fatty nitriles. Both Raney and supported nickel catalysts are used in commercial operations; precious metal catalysts are not typically used. The nickel may be incorporated in a matrix of a primary fatty amine or nitrile to make the catalyst easier to handle [7]. Under these conditions some loss of unsaturation can be expected, but catalysts with varying degrees of selectivity are available to control this; recent examples include modification of the catalyst surface by deposition of carbonaceous residue [8] and the use of aromatic amide modifiers [9]. If retention of unsaturation (Iodine Value) is required then other catalysts may be used, although this requires higher processing temperatures. Continuous hydrogenation of nitrile to primary amines is practised commercially and continuous production of secondary and tertiary amines has been described [4].

To manufacture dialkyl and trialkyl amines ammonia has to be removed from the reactor to shift the equilibrium in favour of the desired product. Another option is to react the dialkyamine with an alcohol to introduce other chain lengths or functionality. Methylation of both primary and secondary amines can be done via reductive amination with formaldehyde to give alkyldimethylamines and dialkylmethylamines. Reduction of primary amine in the presence of aldehydes such as 2-ethylhexaldehyde or benzaldehyde leads to unsymmetrical secondary amines, while reaction of primary amine with

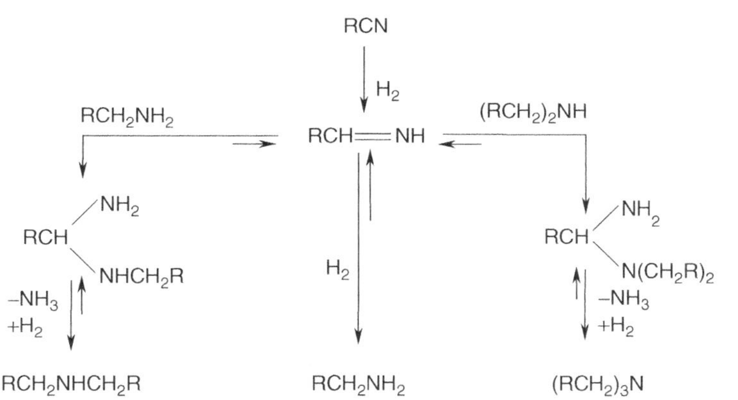

Figure 2.1 Hydrogenation of fatty nitrile.

formaldehyde leads to a hexahydrotriazine, which reduces to an alkylmethylamine under catalytic hydrogenation conditions.

2.2.1.2 *The Alcohol Route*

2.2.1.2.1 *Amines*

Direct conversion of fatty alcohols to primary amines by reaction with ammonia is not commercially practised but new catalyst developments show improved yields with this technology [10]. However, production of dialkylamines from alcohols is practised particularly with C8–C10 alcohols due to the limited availability of the corresponding acids. Alcohols are commonly used in the manufacture of tertiary alkyldimethylamines either through reductive amination with dimethylamine or by conversion to the alkylhalide followed by reaction with dimethylamine. Both primary and secondary amines can be reacted with alcohols to produce tertiary trialkyl amines [11]. The chain branching seen with some synthetic alcohols means that the derived amines are not identical to those from natural sources.

The reaction of alcohols with acrylonitrile followed by hydrogenation affords etheramines, which can then be further derivatized. The distinct characteristic of etheramines compared to alkylamines is their liquidity and oil solubility.

2.2.2 Amidoamines and Esteramines

Nitrogen derivatives can be made directly from fatty acids and esters by reaction with polyamines and alcoholamines. Reaction of fatty acids with polyamines such as diethylenetriamine or aminoethylethanolamine leads to the formation of amidoamines, which can be cyclized through further condensation to give alkyl imidazolines [12–15]. Condensation with dimethylaminopropylamine affords alkylamidodimethylaminopropylamines, which are popular precursors to betaines and amine oxides (Figure 2.2). While imidazolines and amidoamine derivatives have lost ground to esterquats in fabric

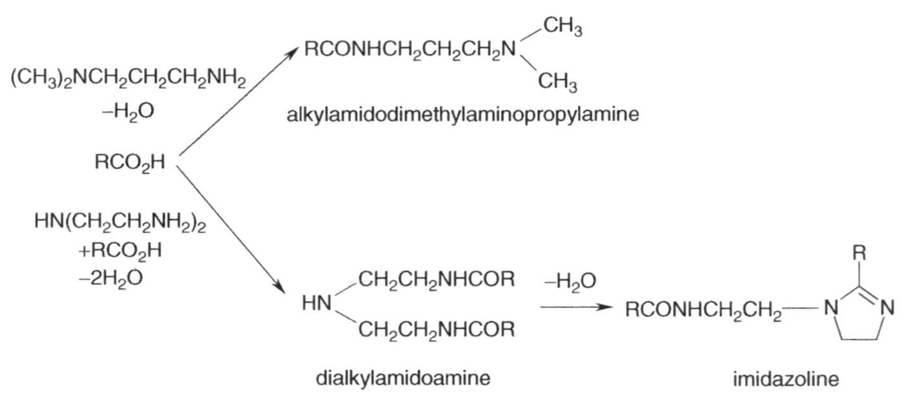

Figure 2.2 *Amidoamines.*

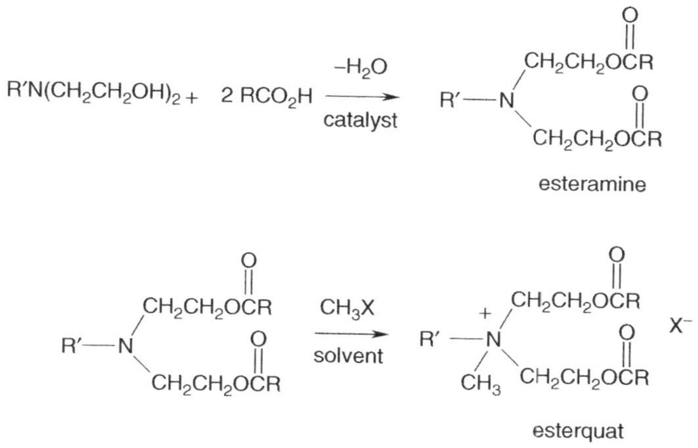

Figure 2.3 Ester quat synthesis.

softening applications, they continue to hold strong positions in industrial applications such as corrosion control, mineral flotation and asphalt applications – particularly tall oil fatty acid condensates.

Alcohol amines such as methyldiethanolamine and triethanolamine react with fatty acids to form esteramines. The esteramines have found limited utility and are mostly quaternized to yield the corresponding quats (Figure 2.3). Esterquats are used predominantly in fabric softening, paper softening and hair conditioning applications due to their substantivity and biodegradability [16].

2.2.3 Polyamines

Michael addition of a primary amine to acrylonitrile followed by hydrogenation affords an alkyl 1,3-propanediamine. Addition of a second mole of acrylonitrile and hydrogenation yields either a linear or a branched triamine, depending on the reaction conditions. One can continue with this acrylonitrile/hydrogenation cycle to make tetramines and so on, but with each addition the side reactions become more prevalent and complex mixtures of amines and polyamines are produced (Figure 2.4). These polyamines have an increased cationic character due to multiple nitrogens and their salts and quaternary ammonium derivatives have enhanced water solubility. The polyamines and derivatives are useful as asphalt emulsifiers, corrosion inhibitors, chain lubricants, pigment dispersants and mineral flotation aids. The C12–C14 diamines also have strong biocidal properties.

2.2.4 Fatty Amides

Fatty amides can be made directly from either fatty acids or fatty esters by ammonolysis and dehydration. The reaction is done at high temperature over a boric acid or metal oxide catalyst, and the crude product can be distilled and stabilized with antioxidants and chelants to reduce colour and odour. Substituted amides can be made by replacing

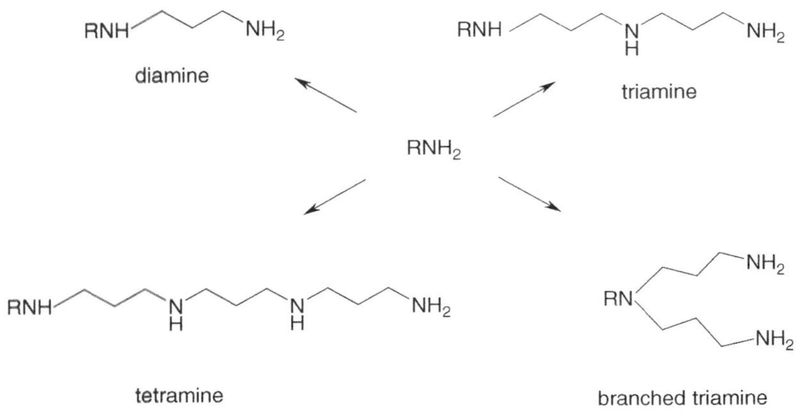

Figure 2.4 *Fatty polyamines.*

ammonia with amines, while ethoxylated amides can be made from primary amides by base-catalysed addition of ethylene oxide. Many amides such as oleamide and stearamide are considered to be readily biodegradable, further enhancing their attractiveness as surfactants from renewable resources. The chemistry and properties of fatty acid amides has been comprehensively reviewed by Johansson and should be referred to for additional references [17]. Simple fatty amides such as stearamide and erucamide are valued for their lubricating properties and thus find use in applications such as friction modifiers for engine oils and slip additives and mould release agents for polymers. Ethylene bis-stearamide is widely used as an internal lubricant and flow modifier for polymers such as polyvinyl chloride (PVC).

2.2.5 Alkanolamides

Alcoholamines such as ethanolamine and diethanolamine condense with acids to form mono- and diethanolamides [18, 19]. Due to the competing reactions between the amine group and the alcohol group these materials contain a distribution of products from these competing reactions and are thus strongly impacted by stoichiometry and the reaction conditions used. Reaction of fatty acid with diethanolamine in a 1 : 2 molar ratio at 150–180 °C yields the Kritchevsky product, which comprises about 60% of the alkyl diethanolamide with 20% of the diethanolmine remaining unreacted. The Kritchevsky alkanolamides are recognized for their detergent properties and have been widely used in detergents, cleaners and shampoos. In a 1 : 1 molar ratio of acid to amine, a water-insoluble product containing significant esteramide is obtained. By using fatty acid methyl ester in a 1 : 1.1 molar ratio with diethanolamine at 105 °C and a base catalyst, the diethanolamide is produced in about 90% yield; this is known as the 'Superamide' process.

The current health concern over by-products in diethanolamides has encouraged the personal care industry to look at monoethanolamides as alternatives to diethanolamides as formulation foam boosters and viscosity builders (Figure 2.5). Mono alkanolamides

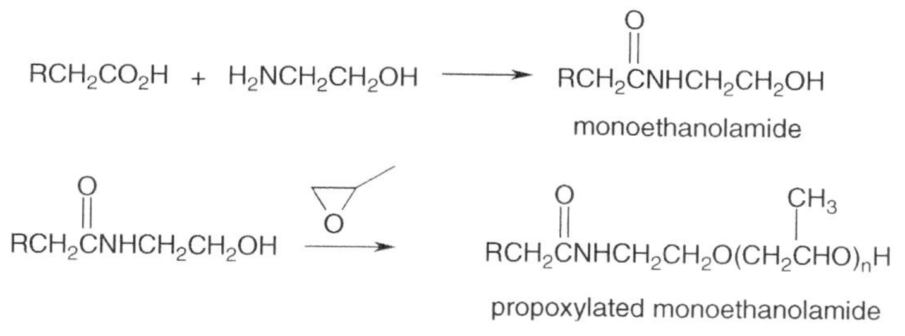

Figure 2.5 Monoethanolamides.

are produced commercially by reaction of monoethanolamine or monoisopropanolamine with fatty acid, ester or triglyceride under conventional condensation conditions. However, a solvent free enzymatic process has been described in a recent paper [20]. Monoethanolamides are solids with low water solubility but better hydrolytic stability than the diethanolamides; ethoxylation of monoethanoloamides further improves their hydrolytic stability. The solid nature of monoethanolamides is a problem for some formulators and liquid forms are preferable; this can be achieved through propoxylation of the monoethanolamides [21], by the formation of emulsion concentrates [22] or by the addition of crystal modifiers [23]. Alkanolamides when formulated with alkylarylsufonates and alkylsulfates boost the viscosity of formulations and enhance the foaming; thus they have been widely used in formulations requiring high levels of stable foam.

2.2.6 Alkoxylated Amines

Alkoxylation of primary fatty amines is widely practised and the alkoxylates find widespread use as emulsifers, lubricants, corrosion inhibitors and thickening agents in many industries including textiles, metal working, agrochemicals and oilfield applications [24]. Ethoxylated amines are also converted to a variety of quaternized derivatives for use as viscosifying agents, hydrotropes, corrosion inhibitors and conditioning aids (Figure 2.6). Ethylene oxide (EO) is the main alkoxylating agent used, although some propoxylates and mixed alkoxylates are also produced. The addition of propylene oxide affords a less hydrophilic amine but one with reduced foaming and better wetting characteristics than ethoxylates. The addition of the first two molecules of ethylene oxide is uncatalysed and affords the bis(2-hydroxyethyl)alkylamine with high selectivity. Higher ethoxylates are typically achieved by base-catalysed addition of EO to the bis(2-hydroxyethyl) adduct at 150–180 °C at pressures below 50 psi. The resulting ethoxylated amine shows a broad distribution of polyoxyethylene chains typical of base-catalysed ethoxylations. While sodium and potassium hydroxides are the catalysts commonly used for ethoxylation of amines other bases have been employed in an attempt to control the colour formation associated with higher ethoxylates [25]. Narrow-range ethoxylation has focused on metal oxide and hydrotalcite-type catalysts [26–28], but more recently the development of narrow-range ethoxylated amines has been reported based on a BF_3 catalyst [29].

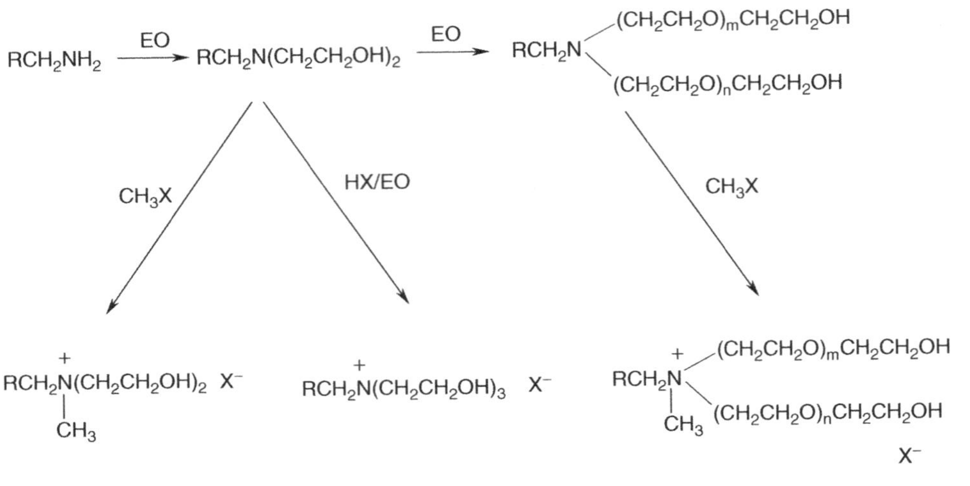

Figure 2.6 *Ethoxylated fatty amines and derivatives.*

2.2.7 Amine Oxides

Amine oxides are manufactured by the hydrogen peroxide oxidation of tertiary amines. The hydrogen peroxide is typically supplied in 35–70% aqueous solution and the reaction is conducted in a water miscible solvent such as isopropanol, ethanol or propylene glycol. Due to the low solubility of the higher alkylamines in water, amine oxide solutions are often quite low in actives content, 25–40%. Amine oxides are thermally unstable, degrading via the Cope reaction at temperatures above 120 °C and by reversion to amine at slightly lower temperatures. The reactions are therefore often catalysed to lower the reaction temperature and time. Carbon dioxide is the most common catalyst but carbon dioxide liberating salts such as ammonium bicarbonate or sodium bicarbonate are also used [30–34].

2.2.8 Quaternary Ammonium Salts

Quaternization of fatty tertiary amines is typically done in solvent using either methyl chloride, benzyl chloride or dimethyl sulfate as the quaternizing agent. In the case of higher ethoxylated amines the need for a solvent is eliminated and the reaction can be done neat. Common solvents for methyl chloride quaternizations are water, ethanol, isopropanol and propylene glycol. The reactions are run at 60–100 °C and pressures of about 90 psi. Dimethyl sulfate reactions are run at atmospheric pressure and relatively mild temperatures depending on the reactivity of the amine; water is not usually used as a solvent here due to the high reactivity of dimethyl sulfate, but may be post-added as a diluent. Dimethyl sulfate is a far more reactive alkylating agent than methyl chloride and is favoured for use with less reactive amines such as the esteramines based on triethanolamine. Methosulfate quats are not considered corrosive to carbon steel and have better heat stability than chlorides; thus they find favour in applications such as

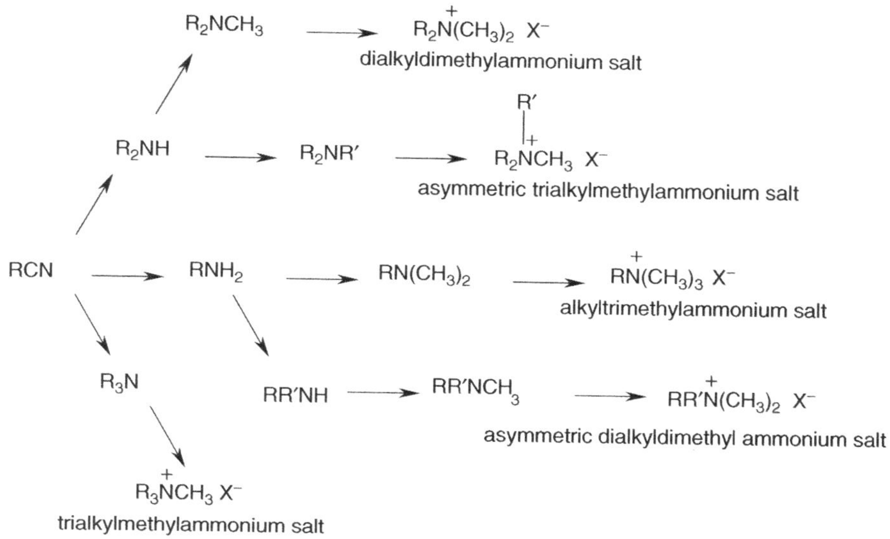

Figure 2.7 *Quaternary ammonium compounds from fatty nitrile.*

drier sheet softeners and paper treatments. Figure 2.7 illustrates typical pathways to alkyl quaternary ammonium salts starting from fatty nitrile.

Dimethylcarbonate (DMC) has found some use as an alkylating agent to produce either methocarbonate quats or bicarbonate quats (after hydrolysis). DMC is a fairly weak alkylating agent and reactions are often run at elevated temperatures with an excess of DMC and methanol as a solvent/catalyst. Other carbonates can be used to make bicarbonate and carbonate quats by an in situ transesterification–quaternization process [35]. The attraction of dimethyl carbonate is that it has a good environmental profile and the quats produced do not have the corrosivity shown by chloride quats, making them favoured for wood preservation where corrosion of nails, screws and staples is a concern.

There has been much written about gemini surfactants but these materials have yet to make any significant impact in commercial applications. A review by Shukla and Tyagi summarizes the current situation with regards to cationic gemini surfactants [36]. Gemini quats are recognized for their significant difference in performance compared to monomeric quats. In general gemini surfactants have low critical micelle concentration (CMC), produce low surface tension values, are often strong biocides, are highly substantive and can be highly foaming. While geminis can be produced in a number of ways the common methods include: reaction of a fatty tertiary amine with an alkyl dihalide, reaction of a tetramethylpolyalkylene diamine with a fatty alkyl halide, and reaction of an alkyldimethylamine with epichlorohydrin. The performance of the gemini quats is related to both the nature of the head group and the space chain between the nitrogen centres. Gemini quats incorporating C10–12 alkyl groups, for example, are superior to their monomeric equivalents in lowering surface tension, foaming and biocidal activity. The foam height and foam stability of some C12–C14 gemini quats are better than that of sodium dodecyl sulfate (SDS). The biocidal activity of gemini quats falls off as the number of atoms between the cationic groups (spacer unit) is increased [37]. Gemini quats can display poor biodegradability and the development of ester-based gemini quats has been an attempt to address this; however, ester gemini quats are not always readily biodegradable even though they are hydrolytically unstable [38].

2.3 Production Data

In 2004 it was estimated that about 80 000 metric tonnes of natural fatty acids were consumed for manufacture of quaternary ammonium salts, in North America, Europe and Japan, using the nitrile process. Another 85 000 tonnes were converted to amines and other derivatives (Table 2.1). In 2006, North America consumed 35 000 tonnes of quats made via the nitrile process and another 44 000 produced from fatty alcohols or olefins [39]. There has been growing competition between oleochemical- and petrochemical-based detergent alcohols for surfactant use, driven by the high price of crude oil and the increased supply of natural oils, particularly palm.

About 60% of the fatty alcohols produced in 2006 were oleo based and this is expected to grow to about 65% by 2011. Oleochemical alcohols are particularly competitive in Asia due to the abundance of palm and coconut oils [39]. The consumption of cationic surfactants made by the various processes is shown in Table 2.2, and the impact of ester quats for fabric softening is clearly reflected in the numbers for cationics made via the fatty acid route.

2.4 Ecological Aspects

The reasons for using renewable materials to make surfactants are both social and economic. Today's society is increasingly demanding that products be not only safe and

Table 2.1 *Fatty acid consumption in nitrile-based amine derivatives – 2004 thousands of metric tonnes*

	Quaternary ammonium salts	Amines and others
North America	45	46
Europe	31	35
Japan	4	4

Data from SRI consulting: Reference [39].

Table 2.2 *Cationic surfactants by process route – 2006*

Thousands of metric tonnes on a 100% actives basis			
	Fatty acid route	Nitrile route	Alcohol or olefin
North America	183	115	143
Western Europe	151	49	63

Data from SRI Consulting: Reference [39].

easy to use but also to have a low environmental impact. These demands have to be balanced with the need of the supplier to provide products that are high performing and cost effective. Natural fats and oils have proven to be a valuable raw material for the surfactant manufacturer on a cost, performance and environmental basis. However, in many cases the resulting surfactant is only realized through reaction of the natural component with petroleum-based chemicals such as ethylene oxide or methyl chloride (although it is possible to manufacture ethylene oxide from ethanol, which can be derived from grain or sugar). Some key surfactants in use today can only be obtained through petrochemical sources, for example linear alkyl benzene sulfonates, and these provide a reminder that a transition to all natural surfactants is unlikely and may be impractical. Pittinger and co-workers suggest that major shifts to oleochemical feedstock sourcing would have a minimal impact on fossil fuel consumption but a major impact on the environment in terms of land use [40].

In the manufacture of surfactants, and in particular amines and amine derivatives, the most important resources are tallow, palm, soya bean, rapeseed and coconut. All of these materials are considered carbon neutral in that their carbon dioxide emissions equal that consumed during their growth. While the supply of vegetable oils can be adjusted to meet demand by further plantings the supply of tallow is constrained by the consumption of beef and has remained fairly consistent from year to year. Tallow is a by-product from the beef rendering industry and until recently has been an inexpensive feedstock; this is now changing with recent developments in alternative fuels technology, causing tallow prices to increase although the benefits of biofuels, for example CO_2 emission reduction, is already achieved by the use of these renewable resources to manufacture surfactants.

In looking at the life cycle analysis of surfactants based on renewable materials versus petrochemical feedstocks there are pros and cons to each side. Petrochemical processes use more energy but generate less waste whereas agricultural processes generate more waste and gaseous emissions. However, the opportunities for efficiency improvements are greater in the oleochemical and allied industries than in the petrochemical industry [40].

An eco-efficiency study on the manufacture of dimethyloctadecylamine from tallow or alpha olefin concluded that the process based on tallow used less energy and had lower global warming potential [41]. It was also noted that the largest potential for improving the eco-profile of the tallow process is in the processing of the tallow itself at the renderers, supporting the earlier observations of Pittinger. Providing a subsidy on tallow for biodiesel production places the surfactant manufacturer at a clear cost disadvantage and reduces the overall eco-efficiency of the total process.

2.5 Biodegradation

A key concern for formulators is the biodegradability of the surfactants they use. This has been driven by a growing public awareness of potential environmental and health issues related to chemicals in general. Certainly this has been the case in the consumer products market for some time, where people are concerned that chemicals ending up in the sewage system pose a threat to their local rivers and water supply. Today, the concern has spread beyond the consumer products market and is inherent in all industries using surfactants.

The most common standard applied in determining a surfactant's suitability is that of ready biodegradability. The term 'readily biodegradable' means that the surfactant has passed a 28 day screening test indicating that under aerobic conditions it degrades rapidly [42]. There are several approved test methods for determining ready biodegradation such as the closed bottle test and the Sturm test. However, these screening tests are not without limitation and materials that fail the test may still degrade under environmental conditions; these materials are considered inherently biodegradable. The test methods for inherent biodegradability use an activated sludge to simulate wastewater treatment plant conditions [42].

Biodegradation tests do not in themselves demonstrate complete mineralization or ultimate biodegradation of the surfactant, that is complete breakdown into carbon dioxide, water and mineral salts. For this metabolic studies have to be conducted. Metabolic studies also eliminate the possibility of toxic metabolites being formed as a result of the degradation pathway.

Biodegradation tests and metabolic studies have shown that many cationic surfactants undergo complete degradation [43–47]. Primary, secondary and tertiary amines are degraded, as are monoalkyl and dialkyl quaternary ammonium salts, monoalkylbenzyl quaternary ammounium salts, ethoxylated quaternary ammonium salts and ester quats. The failure of long-chain dialkyl quaternary ammonium salts to meet the readily biodegradable requirements has been shown to be the results of low bioavailability and not resistance to biodegradation. These materials are poorly soluble in water and strongly absorbed on surfaces (including the glass bottles used for standard testing), which limits their bioavailability to bacterium. By extending the term of biodegradation testing significant degradation is observed [44].

Amine oxides are generally considered readily biodegradable. In a recent study it was found that while alkyldimethylamine oxides and an amidoamine oxide all showed ready biodegradation under aerobic conditions it was only the amidoamine oxide that degraded under anaerobic conditions. The amidoamine oxide was also found to have a lower aquatic toxicity than the alkyldimethylamine oxides [45]. In a similar study with amphoteric surfactants it was found that they were all readily biodegradable under aerobic conditions and with the exception of alkyl betaines were biodegradable under anaerobic conditions; aquatic toxicity was found to increase with the alkyl chain length [46].

Many cationics are believed to degrade via scission of the alkyl–N bond [44]. By breaking the R–N bond the molecule loses its surfactancy and becomes less toxic to the microorganisms. Ester quats are different in that they undergo hydrolysis first. Interestingly, it has been reported in a recent study of monomeric and gemini ester quats that the gemini quats were found to be not readily biodegradable, whereas the monomeric quats degraded readily, and that in fact there was no correlation between the rate of chemical hydrolysis and the rate of biodegradation; both carboxylate and betaine-type ester quats were studied and behaved in the same way [38].

Polyethoxylated amines are not readily biodegradable. Degradation occurs by initial attack on the alkyl chain and is independent of chain length. In general, the biodegradation of ethoxylated amines worsens with increasing EO content. The converse is true for ethoxylated quats. Asymmetric polyethoxylated amines exhibit better biodegradability.

2.6 Properties of Nitrogen-Based Surfactants

2.6.1 Hydrophobe Effects

A surfactant comprises a hydrophobic tail and a hydrophilic head. The balance of hydrophobic and hydrophilic character is what defines the properties and performance of the surfactant. Hydrophobes from fatty acids consist of linear alkyl chains from C8–C22. Coconut oil provides a valuable source of the shorter C8–C14 acids while tallow, palm oil and soya bean oil are rich in C16–C18 acids. The longer C20–C22 chains are obtained from rapeseed. Natural fats and oils are triglycerides comprised of a distribution of different fatty acids and therefore, unless fractionated, the surfactants produced from them also comprise a mixture of chain lengths. While the chain length distribution of an oil or fat varies on a lot-to-lot basis, the differences in composition do not significantly impact the performance of the surfactant in most applications.

Within the natural distribution of chain lengths there are also differences in the degree of unsaturation and isomer distribution. Of these the most important is the degree and type of unsaturation. For example, soya bean oil is a highly unsaturated acid with an Iodine Value of about 140. In this case, most of this unsaturation is provided by the C18 acids: oleic acid, linoleic acid and linolenic acid, containing one double bond, two double bonds and three double bonds, respectively. Unsaturation in the hydrophobe affects liquidity, oxidative stability and solubility of the surfactant [47]. The oxidative stability of the hydrophobe is affected not only by unsaturation but also by the type of unsaturation; linolenic acid is about 10 times less oxidatively stable than linoleic acid which in turn is less stable than oleic acid [48]. This means that surfactants containing polyunsaturated chains will be sensitive to oxidation and will typically have poor colour and odour stability. Conversely, linoleic acid is more readily hydrogenated than linolenic which is more readily hydrogenated than oleic. Thus, the oxidative stability of a surfactant can be improved by selective hydrogenation of the polyunsaturated C18s to oleyl without too much loss of liquidity. Unsaturation also provides improved solubility in hydrocarbons [47], and oleyl derivatives are popular choices for applications in petroleum and allied industries.

The hydrophobe may contain a variety of isomers both positional and geometric but the most discussed isomer effect is centred on the *cis*:*trans* configuration of the oleyl chains. In nature the predominant configuration is *cis*, but exposure to a hydrogenation catalyst will result in isomerization and the formation of some *trans* isomers. By careful selection of catalyst and process conditions, it is possible to exercise a degree of control over the *cis*:*trans* isomerization during the hydrogenation procedure. In some applications, such as concentrated fabric softener formulations, the resulting ratio of *cis*:*trans* has been found to be critical [49].

Natural fats and oils provide predominantly linear hydrophobes although low levels of branched C18 have been detected in tallow [48]. Hydrophobes based on petrochemical sources contain a higher degree of branching and are fully saturated. Branching in the hydrophobe can impact the physical form of the derivative; for example, isostearic acid is a liquid whereas stearic acid is a solid at room temperature. The level of branching in synthetic alcohols varies and is dependent on the manufacturing procedure. Isostearic acid is a by-product from dimer acid production and has a very high level of methyl branching. The dimer acids themselves can also be converted to nitriles and amines in a similar way to that already discussed [50]. Branching of the acid chain can be done catalytically but this is not practised on a commercial basis [51, 52]. The variety of chain lengths, the degree of unsaturation and the ability to chemically modify the chain through hydrogenation, oxidation and other chemistries, makes natural fatty acids a highly versatile source of surfactant hydrophobes.

2.6.2 Hydrophile Effects

The hydrophilic character of the surfactant is provided via amino and amido groups. The cationic character is incorporated by adding a cationic charge through either salt formation or quaternization. Alkoxylation provides a way of adjusting hydrophilic properties of amines and amine derivatives. While ethoxylation adds considerably to hydrophilicity, propoxylation adds little but provides other attributes such as increased liquidity and reduced foam characteristics. An estimate of how changes in the head group and chain length impact the surfactant can be estimated through hydrophilic lipophilic balance (HLB) calculations. The higher the HLB the more water loving the surfactant will be. For cationic surfactants the HLB on the Davies scale is conveniently calculated using published group contributions and can be used as a guide in selecting emulsifiers [53]. Table 2.3 illustrates the effect on the HLB of varying the head group by alkoxylation and Table 2.4 shows the effect on the HLB of varying the chain length in quaternary ammonium salts. James and co-workers examined head group effects on

Table 2.3 *HLB and pour point of alkoxylated fatty amines*

Product	HLB – Davies scale	Pour point (°C)
Cocoamine + 2EO	12.2	7
Cocoamine + 15EO	16.8	−4
Tallowamine + 2EO	10.1	32
Tallowamine + 15EO	14.7	−1
Tallowamine + 20EO	17.0	7
Soyaamine + 15EO	14.7	−7
Cocoamine + 2PO	11.0	−15
Tallowamine + 2PO	9.2	7
Tallow 1,3-propanediamine + 3EO	19.0	18–21

Source: AkzoNobel Surface Chemistry.

Table 2.4 *HLB of quaternary ammonium salts*

Product	HLB-Davies Scale
Cocotrimethyl ammonium chloride	22.9
Octadecyltrimethyl ammonium chloride	20.5
Dicocodimethyl ammonium chloride	17.3
Dihydrogenated tallow dimethyl ammonium chloride	13.2
Cocomethyl[ethoxylated (2)] ammonium chloride	25.8
Cocomethyl[ethoxylated (15)] ammonium chloride	30.4

Source: AkzoNobel Surface Chemistry.

the hydrophilicity of nitrogen-based surfactants by measuring the partition coefficients between heptane and water [54]. Any change to the head group will also affect its size and charge density, and therefore how it will pack in a micelle and react at an interface. This in turn impacts surfactant properties such as solubility, foaming, gelling, dispersion, emulsification and substantivity. As already noted for fatty amine derivatives, a change in the head group can also impact biodegradability by restricting attack at the C–N bond.

2.6.3 Surfactant-surfactant Interactions

Mixed surfactant systems are used in many applications. By combining surfactants, unique performance attributes not achievable with a single surfactant can be realized. Common combinations often involve nonionic surfactants with cationic or anionic surfactants or with other nonionics. Mixing cationic surfactants with anionic surfactants is more problematical in that, being oppositely charged, they are attracted to one another, forming a highly hydrophobic complex that often results in precipitation. However, if the concentrations are kept within the solubility product of the complexes, then clear, stable solutions with unique properties can be obtained. To reduce the electrostatic interaction steric hindrance can be introduced, for example by using ethoxylated quaternary ammonium salts in place of alkyltrimethyl quaternary ammonium salts or by using long-chain trialkyl quaternary ammonium salts in place of mono- or dialkyl quats. This has been a common practice for formulating 2-in-1 shampoos and detergents. The synergism obtained by mixing surfactants is strongest when the charge difference between the surfactants is at its greatest [55]. Thus, the highest synergies are seen with cationic/anionic combinations. Such combinations have been shown to be more surface active than the individual components, for example having lower critical micelle concentration (CMC) and surface tension.

A study of the effect of various quaternary ammonium salts in combination with sodium lauryl sulfate (SLS) and sodium lauryl ether sulfate (SLES-3) (where SLES-3 indicates that there are three EO groups in the molecular structure) found that in general the ether sulfate provided a broader mixing range than SLS and that alkylamido quats were more compatible with anionics than alkyl quats [56]. Betaines were also examined and found to provide good synergies for foam and viscosity control [57].

2.7 Applications

Cationic surfactants are noted for their affinity for natural surfaces and their strong biocidal activity. Thus, they find use in applications such as fabric softening, mineral flotation, corrosion inhibition, cleaning and disinfecting. A few examples illustrating applications and benefits afforded by natural fat and oil-based cationic surfactants are given here.

2.7.1 Fabric Softeners

Fabric softeners represent the single largest application for cationic surfactants, about 250 000 metric tonnes worldwide [58]. This market was for many years dominated by formulations based on di(hydrogenated tallow)dimethyl ammonium chloride, but over the last 15 years the desire for more biodegradable products has resulted in the conversion of much of the industry to the more readily biodegradable ester quats. In the Americas and Europe the major commercial products are tallow esters of triethanolamine or methyldiethanolamine, tallow being preferred due to its availability and pricing, but other sources such as palm oil, soya bean and rapeseed are used in other markets.

While ester quats provide benefits such as biodegradability and wash off the fabric better than conventional dialkyl quats they are not considered to be as effective at softening as the dialkyl quats. In a study of ester and amidoester quats it was shown that for good biodegradability two ester groups are needed and that unsaturation in the alkyl chain also provides a slight benefit. However, the same article concluded that the more biodegradable the product the worse the softening performance [59]. Thus, other technologies such as silicones and cationic starches are employed to boost softener performance, improve colour control, provide antiwrinkle benefits and generally increase the value of the formulation to the consumer [60, 61].

Diquaternary ammonium salts based on esters of ethoxylated hexamethylenediamine were shown to produce softener actives that offered good softening and better compatibility with detergent carryover in the rinse cycle than other quats. Unfortunately, while these diquats were classified as readily biodegradable under standard test conditions they were not ultimately biodegradable, stopping at about 90% degradation [62]. Clearly, there remain opportunities for the development of softener actives that offer gains in both softening and environmental profile.

2.7.2 Biocides

There are three common categories of biocidal quaternary ammonium salts: alkytrimethyl ammonium, alkylbenzyldimethyl ammonium and dialkyl dimethyl ammonium. The North American, European and Japanese markets consume about 55 000 metric tonnes (2004), with the largest piece being the alkyldimethylbenzyl quats; although in the USA didecyldimethyl ammonium carbonate also commands a significant volume due to the wood preservative business [63]. The biocidal efficacy of quats is greatest at the C10–C14 carbon chain length and quats with these chain lengths are commonly used in cleaning, disinfecting and preservative formulations. A comparison of chain length effects on benzalkonium chlorides found C14 to be optimal for biocidal activity, whereas for dialkyl quats C10 is the optimal length. The benzyl quats and dialkyl quats offer enhanced activity over simple alkyl quats [64]. At typical use levels quats offer advantages over other biocides in that they have relatively low toxicity, low corrosivity, are nonstaining, have reasonable hard water tolerance and are considered biodegradable. Hard water tolerance of benzalkonium chlorides has been shown to be improved by narrowing the chain length distribution over a natural coconut distribution. They are effective against both Gram-positive and Gram-negative bacteria. Their mode of action is believed to be via adsorption and penetration of the cell wall followed by interaction with the proteins and subsequent disruption of the cell. Care has to be taken when formulating cationic biocides with anionic surfactants in that complexation will reduce the biocidal activity. However, certain ethoxylated anionics show good compatibility with cationic biocides.

Amphoterics offer an alternative to quaternaries for some applications [65]. They are considered less irritating and also have better alkali tolerance, detergent and wetting characteristics; making them attractive for use in cleaning in place (CIP) applications in food processing industries such as breweries and dairies. Their activity increases with increasing nitrogen content but is reduced by increased carboxymethylation. Betaines have very little biocidal activity, but some mixtures of ampholytes and betaines show a strong biocidal synergy. The increase in regulatory activity in the biocides market will lead to additional research in synergistic blends of biocides and other performance enhancers rather than in the development of new molecules.

2.7.3 Wood Preservation

The voluntary withdrawal of chromated copper arsenate (CCA) as a wood preservative for domestic applications was driven by public concerns over its health and environmental profile. In its place have risen two competing systems, the amine copper quat (ACQ) system and the copper azole system. Both systems avoid using heavy metals such as chromium and arsenic and rely on the co-biocidal effects of copper and organic biocides. The elimination of CCA has created many opportunities for oleochemicals as preservative companies try to develop formulations that are not only environmentally friendly but can also match the preservative performance of CCA [66]. The ACQ system is based on didecyldimethyl ammonium bicarbonate and has produced a significant demand for C10-based amine and quat. The azole systems use biocides such as tebuconazole and propiconazole in combination with copper ethanolamine complexes. Ethoxylated amines [67] and amine oxides [68] have been described as providing improved performance in azole-based systems. Other copper systems have employed ethoxylated diamines [69] and amine oxides [70] to enhance performance.

Hydrophobic quats such as di(hydrogenated tallow)dimethyl ammonium chloride have been claimed to provide waterproofing [71] and improvements in copper leach resistance in pressure-treated lumber [72], as have amine oxides [73]. Alkoxylated amines and diamines can also be used to improved water repellency in pressure-treated lumber [74].

A trend towards metal-free systems for pressure treating lumber will provide additional opportunities for nitrogen derivatives as both biocides and synergists.

2.7.4 Asphalt Application

Cationic surfactants are widely used in the manufacture of asphalt emulsions. Today, over 8 million tonnes of asphalt emulsions are produced worldwide and over 3 million tonnes in the USA. A recent overview of asphalt emulsion technology by James clearly describes the state of the technology and the varied uses of asphalt emulsion in road construction and maintenance [75].

Typical cationic emulsifier formulations are based on tallow amine and tallow diamine. The type of emulsifier used depends on the specific asphalt, the aggregate and the intended application; most asphalt emulsions are used in road maintenance applications. The advantages of emulsions over hot asphalt are quite significant. The oil-in-water emulsions have a much lower viscosity than asphalt and can be applied at lower temperatures than hot asphalt, thus making for easier and safer handling. They require less energy to make and apply. The lower working temperatures required by emulsions also means less fumes are produced, thus providing additional health, safety and environmental benefits. In an eco-efficiency study on the use of asphalt emulsions for road maintenance it was concluded that emulsions had a lower environmental impact than a thin hot-mix overlay [76].

The use of solvent cut-back asphalts has largely been eliminated from road construction and maintenance operations but they are still favoured for prime coat operations. However, for both economic and environmental reasons solvent-free emulsions are now being developed with promising results. Another environmental initiative is in recycled asphalt; here emulsion technology is applied in the form of cold-mix recycling, where old road surfaces are ground up, mixed with asphalt emulsion, relayed and compacted.

Fatty amine derivatives act not only as emulsifiers but can also act as adhesion promoters in hot-mix applications. As an adhesion promoter the surfactant is attracted to the surface of the aggregate to provide a more hydrophobic surface for the asphalt to bond to. In some cases it may also displace moisture from the surface of the aggregate. The improved adhesion of the asphalt to the aggregate leads to a stronger and more durable road surface.

2.7.5 Agrochemicals

Nitrogen-containing surfactants have been used in pesticide formulations for many years. In particular, fatty amine alkoxylates are well known for their adjuvancy in glyphosate herbicide formulations. They can be used in both tank-mix and in-can formulations. More recently, the fatty amine alkoxylates have found use in fungicide formulations [77]. The growth in adjuvants is linked to both the pesticide and the crop, and as new markets have developed, for example the introduction of transgenic (glyphosate ready) crops, so has the need for more or new adjuvants. Another factor impacting the growth in adjuvants is the increased use of generic pesticides as many of the major pesticide chemistries came off patent in the USA in 2002 [78].

Amine alkoxylates are not the only nitrogen derivatives to find utility in agrochemical applications; quaternary ammonium salts [79], amine oxides [80], alkylamidoamines [81] and betaines [82] are also reported to have similar bioefficacy-enhancing properties to fatty amine alkoxylates. There has also been development of adjuvants with improved biodegradability and certain fatty esteramine and amidoamine deriviatives, for example, have been developed that exhibit enhanced biodegradability [81].

2.7.6 Petroleum Applications

Production and refinery operations both employ surfactants for a variety of functions such as corrosion inhibitors, acid thickeners, drilling fluid additives, emulsifiers, cleaners, foamers and so on [83]. The oil industry is aware of its environmental responsibilities and the need for greener solutions for some of its chemistries, water-based drilling fluids have replaced solvent-based ones for some operations and strict regulations exist for discharges in off-shore operations. For fat-based chemicals, one area of great interest has been viscoelastic fluids for fracturing and other well stimulation techniques. Long-chain fatty quaternary ammonium salts and amine oxides have found utility in this application, replacing polymeric rheology modifiers that can result in permanent damage to the formation [84, 85]. The desire to use this technology in deeper wells where the temperature and pressures are higher resulted in the development of systems based on amphoteric and zwitterionic surfactants. These surfactants are considered to have improved stability at high temperatures ~200 °F [86–89]. Most recently, amide carboxylates based on C18–C22 amines have been described that provide viscoelastic fracturing fluids, where the average-use temperature is in excess of 200 °F [90]. The use of viscoelastic fluids extends well beyond oilfield applications and these surfactant systems can be applied to thickeners for industrial cleaners, cleansing gels, cements and grouts, drag reducers for district heating and cooling, and many more [91].

2.7.7 Ionic Liquids

Ionic liquids are considered a class of novel solvents that offer a huge potential for industrial applications. They have strong solvating power, a broad range of liquidity, negligible vapour pressure and their properties can be readily tailored to meet specific requirements by modification of either the cation or anion and their substituents. Ionic liquids are viewed as green solvents for the future which can replace high volatile organic compound (VOC) solvents. Many fatty quaternary ammonium salts can be classified as ionic liquids and these are now beginning to attract attention, not least because of their performance and lower toxicity compared to other commercial ionic liquids. Ethoxylated cocoamine quats are being studied as ionic liquid solvents for the enzymatic production of monoacyl glycerols [92, 93]. As an electrolyte for chrome electroplating, ethoxylated quats overcome many of the health and safety concerns associated with current chrome plating methods using chromic acid, which contains highly carcinogenic chromium VI compounds [94].

2.8 Conclusions

The challenges presented by government regulations and various consumer and environmental groups to make products and processes environmentally safer have driven innovation in surfactant technology for many years. As a result, many new surfactants with increased biodegradability and better ecotox profiles have been developed. Today, sustainability is the new driver for the surfactant industry, and while surfactants made from renewable materials have been available for many years, it is only recently with the increased public attention on declining fossil fuel supplies and heightened ecological awareness that the focus on them has sharpened. As discussed here, natural fats and oils are both renewable and versatile materials that provide access to a broad range of amine derivatives used in industry today, for example viscoelastic surfactants for oilfield fracturing, quaternary ammonium salts for fabric softening and amine-based adjuvants for agrochemicals. The trend away from solvent-based systems to water-based ones is expected to afford new growth opportunities for surfactants, many of which by design will be based on renewable materials. While some markets such as Personal Care favour vegetable-based surfactants the requirements of many others continue to be met with tallow derivatives. Thus, natural fats and oils will remain an attractive source of renewable hydrophobes for surfactants in the near future. One concern, however, is the consumption of these valuable resources as fuel, and this is something that must be addressed by the oleochemical industry if natural oil and fat-based amines and surfactants are to remain competitive.

References

1. Gunstone, F. D. (2001) Basic oleochemical products and new industrial oils, in *Oleochemical Manufacture and Applications* (eds F. D. Gunstone and R. J. Hamilton), Sheffield Academic Press.
2. Biswas, A., Sharma, B. K., Willet, J. L. *et al.* (2008) Soybean oil as a renewable feedstock for nitrogen containing derivatives. *Energy Environ. Sci.*, **1**, 639–644, Advance Articles, DOI: 10.1039/b809215j.
3. Pryde, E. H., Princen, L. H. and Mukherjee, K. D. (eds) (1981) *New Sources of Fats and Oils*, American Oil Chemist Society.
4. Franklin, R., Hoey, M. D. and Zachwieja, J. (2001) Cationic and amine-based surfactants, in *Oleochemical Manufacture and Applications* (eds F. D. Gunstone and R. J. Hamilton), Sheffield Academic Press.
5. Muller, H., Stitz, L. and Riesner, D. (2006) Risk assessment for fat derivatives in case of contamination with BSE. *Eur. J. Lipid Sci. Technol.*, **108**, 812–826.
6. Muller, H., Stitz, L. and Riesner, D. (2008) Prion decontamination during the oleochemical process of fat hydrogenation. *Eur. J. Lipid Sci. Technol.*, **110**, 392–399.
7. Ottens, T. J., Van Den Berg, J. and Van Poecke, P. (2007) US 7208448, AkzoNobel N.V.
8. Ostgard, D. J., Olindo, R., Berweiler, M. *et al.* (2007) The chemoselective hydrogenation of tallow nitriles to unsaturated primary amines with carbon modified nickel catalysts. *Catal. Today*, **121**, 106–114.
9. Makimura, M. and Izumoto, H. (2008) US 20080167497, Lion Akzo Co., Ltd.
10. Fukushima, T., Jono, M., Suzuki, Y. and Terasaka, M. (2007) PCT Int. Appl. WO 2007 77903, Kao Corporation.
11. Hasegawa, S., Nishimura, T. and Hirota, A. (2007) PCT Int. Appl. WO 2007 74922, Kao Corporation.
12. Friedli, F. (1990) Amidoamine surfactants, *Cationics*, Surfactant Science Series, vol. 34, Marcel Dekker, New York, pp. 51–99.
13. Earl, G. (1990) Imidazoline surfactants, *Cationics*, Surfactant Science Series, vol. 34, Marcel Dekker, New York, pp. 101–144.
14. Bajpai, D. and Tyagi, V. K. (2006) Fatty imidazolines: chemistry, synthesis, properties, and their industrial applications. *J. Oleo Sci.*, **55**, 319–329.
15. Bajpai, D. and Tyagi, V. K. (2008) Surface-active and performance of cationic imidazolinium surfactants based on different fatty acids. *Surf. Rev. Lett.*, **15** (4), 361–367.
16. Kruger, G., Boltersdorf, D. and Overkempe, K. (1998) Esterquats, in *Novel Surfactants*, Surfactant Science Series, vol. 74 (ed. K. Holmberg), Marcel Dekker, Inc., New York, pp. 115–138.

17. Johansson, I. (2004) Amides, fatty acid, *Kirk-Othmer Encyclopedia of Chemical Technology*, 5th edn, vol. 2, John Wiley & Sons, Ltd.
18. Shelmerdine, B., Garner, Y. and Nelson, P. (1990) The preparation and application of alkanolamides and their derivatives, in *Industrial Applications of Surfactants II* (ed. D. R. Karsa), The Royal Society of Chemistry, pp. 132–149.
19. Lif, A. and Hellsten, M. (1998) Nonionic surfactants containing an amide group in *Nonionic Surfactants: Organic Chemistry*, Surfactant Science Series, vol. 72 (ed. N. M. Van Os), Marcel Dekker, Inc., New York (Shell International Chemicals B.V.).
20. Tufvesson, P., Annerling, A., Hatti-Kaul, R. and Adlercreutz, D. (2007) Solvent-free enzymatic synthesis of fatty alkanolamides. *Biotechnol. Bioengng*, **97** (3), 447–453.
21. Perella, J. E., Komor, J. A., Frost, D. L. and Katstra, R. D. (2003) US 6531443, Mona Industries.
22. Gray, J. and D'Aversa, E. (2007) US 7282524, McIntyre Group, Ltd.
23. Griswold, K. E. (2001) US 6310174, Huntsman Petrochemical Corporation.
24. Gadberry, J. F. and Hoey, M. (1998) Polyoxyethylene alkylamines, in *Nonionic Surfactants*, Surfactant Science Series (ed. N. M. Van Os), Marcel Dekker, Inc., New York, pp. 163–175.
25. Yamashita, S. (2007) Jpn Kokai Tokkyo, Koho JP 2007 262249, Sanyo Chemical Industries.
26. Wolf, G., Burkhardt, B., Lauth, G. *et al.* (1998) US 5741947, BASF Aktiengesellschaft.
27. Hama, I., Okamoto, T., Sasamoto, H. and Nakamura, H. (1998) US 5750796, Lion Corporation.
28. Breuer, W. and Raths, H. C. (1994) US 5326891, Henkel Komm.
29. Zhu, S., Nguyen, G., Nguyen, K. *et al.* (1994) WO 07109791, AkzoNobel N.V., MonsantoTechnology LLC.
30. Murata, A., Tsuchiya, S., Konno, A. *et al.* (1994) US 4247480, Nissan Chemical Industries Ltd.
31. Borland, J., Impastato, F. J. and Smith, K. R. (1993) US 5208374, Ethyl Corporation.
32. Thomas, D., Krzystowczyk, D. H. and Six, L. G. (2000) US 6037497, Albermarle Corporation.
33. Laurenzo, K. and Bauer, D. (1981) US 4889954, Ethyl Corporation.
34. Smith, K. and Sauer, J. (1990) US 4960934, Ethyl Corporation.
35. Walker, L. E. (2006) In-situ process for preparing quaternary ammonium bicarbonates and quaternary ammonium carbonates, US 6989459, Lonza Inc., January 24.
36. Shukla, D. and Tyagi, V. K. (2006) Cationic gemini surfactants: a review. *J. Oleo Sci.*, **55**, 381–390.
37. Kim, T. S., Kida, T., Nakatsuji, Y. *et al.* (1996) Surface active properties of novel cationic surfactants with two alkyl chains and two ammonium groups. *J. Am. Oil Chem. Soc.*, **73**, 907–912.
38. Tehrani-Bagha, A. R., Oskarsson, H., van Ginkel, G. C. and Holmberg, K. (2007) Cationic ester containing gemini surfactant: chemical hydrolysis and biodegradation. *J. Colloid Interface Sci.*, **312**, 444–452.
39. Malveda, M. P., Blagoev, M., Gubler, R. and Yagi, K. (2006) Natural fatty acids, *CEH Marketing Research Report*, SRI Consulting.
40. Pittinger, C. A., Sellers, J. S., Janzen, D. C. *et al.* (1993) Environmental life-cycle inventory of detergent-grade surfactant sourcing and production. *J. Am. Oil Chem. Soc.*, **70** (1), 1–15.
41. AkzoNobel Surface Chemistry, Internal Report.
42. Painter, H. A. (1995) Biodegradability testing, in *Biodegradability of Surfactants* (eds D. R. Karsa and M. R. Porter), Blackie Academic and Professional, pp. 65–118.
43. van Ginkel, C. G. (1995) Biodegradability of cationic surfactants, in *Biodegradability of Surfactants* (eds D. R. Karsa and M. R. Porter), Blackie Academic and Professional, pp. 183–204.
44. van Ginkel, C. G. (2004) Biodegradation of cationic surfactants: an environmental perspective, in *Handbook of Detergents: Part B Environmental Impact*, Surfactant Science Series, vol. 121, Marcel Dekker, Inc., New York, pp. 523–549.

45. Garcia, M. T., Campos, E. and Ribosa, I. (2007) Biodegradability and ecotoxicity of amine oxide based surfactants. *Chemosphere*, **69**, 1574–1578.
46. Garcia, M. T., Campos, E. and Ribosa, I. (2007) Ecological properties of amphoteric surfactants. *Comunicaciones Presentadas a las Journadas del Comite Espanol de la Detergencia*, **37**, 369–382.
47. Forno, M. W. (1979) Physical properties of fats and fatty acids, in *Baileys Industrial Oil and Fat Products*, 4th edn, vol. 1 (ed. D. Swern), Wiley.
48. Sonntag, N. O. V. (1979) Composition and characterization of individual fats and oils, in *Baileys Industrial Oil and Fat Products*, 4th edn, vol. 1 (ed. D. Swern), Wiley.
49. Wahl, E., Bacon, D. R., Baker, E. S., *et al.* (1994) WO 94/20597, Procter & Gamble.
50. Leonard, E. C. (1975) *The Dimer Acids*, Humko Sheffield Chemical, Connecticut, Memphis.
51. Zhang, S. and Zhang, Z. (2004) US 6723862, AkzoNobel N.V.
52. Zhang, S., Zhang, Z. and Steichen, D. (2004) US 6831184, AkzoNobel N.V.
53. HLB and Emulsification, Publication SC 08-06, AkzoNobel Surface Chemistry.
54. James, A. D., Wates, J. M. and Wyn-Jones, E. (1993) Determination of the hydrophilicities of nitrogen based surfactants by measurement of partition coefficients between heptane and water. *J. Colloid Interface Sci.*, **160**, 158–165.
55. Kume, G., Gallotti, M. and Nunes, G. (2008) Review on anionic/cationic surfactant mixtures. *J. Surfactants Deterg.*, **11**, 1–11.
56. O'Lenick Jr, A. J. (2005) Anionic/cationic complexes. *Cosmet. Toiletries*, **120** (11), 63–64.
57. O'Lenick, T. and Anderson, L. (2006) Amphoteric anionic interactions. *Cosmet. Toiletries*, **121** (3), 67–74.
58. Janshekar, H., Chang, R. J., Yokose, K. and Ma, X. (2007) *Surfactants, SCUP Report*, SRI Consulting.
59. Friedli, F. E., Keys, R., Joe Toney, C. *et al.* (2001) Novel new ester quaternaries for improved performance benefits as rinse cycle softeners. *J. Surfactants Deterg.*, **4** (4), 401–405.
60. Friedli, F. E., Koehle, H. J., Fender, M. *et al.* (2002) Upgrading triethanolamine esterquat performance to new levels. *J. Surfactants Deterg.*, **5** (3), 211–216.
61. Corona, A., Ward, A. M., Gizaw, Y. *et al.* (2006) US 7135451, Procter & Gamble.
62. Keys, R. (2007) The search for softer fabric softeners. *J. Surfactants Deterg.*, **10** (1), A11–A13.
63. Modler, R. F., Janshekar, H. and Yagi, K. (2005) *Biocides, SCUP Report*, SRI Consulting.
64. Stephenson, R. A. (1990) Optimization of the performance of quaternary ammonium compounds, in *Industrial Applications of Surfactants II* (ed. D. Karsa), The Royal Society of Chemistry, pp. 236–275.
65. Davis, B. and Jordan, P. (1990) An overview of the biocidal activity of cationics and ampholytes, in *Industrial Applications of Surfactants II* (ed. D. Karsa), The Royal Society of Chemistry, pp. 195–210.
66. Freeman, M., Shupe, T. F., Vlosky, R. P. and Barnes, H. M. (2003) Past, present and future of the wood preservation industry. *Forest Prod. J.*, **53** (10), 8–15.
67. Van der Flaas, M. A. J., Tiedink, J. G. and Paris, E. P. T. (2006) EP 1473996, Janssen Pharmaceutica N.V., April 26.
68. Chuen-ing, T. and Walker, L. E. (2003) US 6527981, Lonza Inc.
69. Bell, J. P. (1995) US 5426121 AkzoNobel N.V.
70. Walker, L. (2000) WO 2000 071311, Lonza Inc., November 30.
71. Walker, L. (2005) EP 1114704, Lonza Inc., November 23.
72. Jin, L., Preston, A. F., Archer, K. J. *et al.* (2006) US 2006 0251915, Chemical Specialities, November 9.
73. Walker, L. E. (2003) US 6572788, Lonza Inc.
74. Van der Flaas, M. A. J. (2007) WO 2007 026008, Janssen Pharmaceutica N.V., March 8.

75. James, A. D. (2006) Overview of Asphalt Emulsions, Transportation Research Circular No. E-C102, August.
76. Takamura, K., Lok, K. P., and Wittinger, R. (2001) Microsurfacing for preventive maintenance: eco-efficiency strategy, in Proceedings of the 39th ISSA Annual Meeting, March 2001, Maui, Hawaii.
77. Hulsof, W. T. (1993) US 5226943, AkzoNobel N.V.
78. Underwood, A. K. (2000) Adjuvant trends for the new millenium. *Weed Technol.*, **14**, 765–772.
79. Berk, H. C. and Kassebaum, J. W. (1994) WO 94/02021, Monsanto Company.
80. Gioia, P., Chuah, P. H. and Sclapari, T. (2007) WO 2007054540A2, Rhodia Recherches et Technologies.
81. Gustavsson, B. and Weuste, B. (2004) US 6747164, AkzoNobel N.V., June 8.
82. Bramati, V., Coret, J., Edwards, S. *et al.* (2005) US 20050170965A1 Rhodia Inc.
83. Schramm, L. L. (ed.) (2000) *Surfactants: Fundamentals and Applications in the Petroleum Industry*, Cambridge University Press, Cambridge.
84. Brown, E. J., Card, R. J. and Nelson, E. B. (1999) US 5964295, Schlumberger Technology Corporation, Dowell Division.
85. Franklin, R., Hoey, M. and Premachandran, R. (2002) The use of surfactants to generate viscoelastic fluids, in *Chemistry in the Oil Industry VII* (eds T. Balson, H. A. Craddock, J. Dunlop *et al.*), The Royal Society of Chemistry, Cambridge, pp. 96–106.
86. Dahanayake, M. S., Yang, J., Niu, J. H. Y. *et al.* (2007) US 7238648, Schlumberger Technology Corporation, July 3.
87. Dahanayake, M. S., Yang, J., Niu, J. H. Y. *et al.* (2001) US 6258859, Rhodia Inc.
88. Fang, L., Dahanayake, M. and Colaco, A. (2005) WO 2005 118743, Rhodia Inc.
89. George, L., Morris, L. and Daniel, S. (2002) Function and application of oilfield chemistry in open hole sand control completions, in *Chemistry in the Oil Industry VII* (eds T. Balson, H. A. Craddock, J. Dunlop *et al.*), The Royal Society of Chemistry, Cambridge, pp. 107–122.
90. Hughes, T., Jones, T. G. J., Tustin, G. J. and Zhou, J. (2008) US 7410934, Schlumberger Technology Corporation.
91. Yang, J. (2002) Viscoelastic wormlike micelles and their applications. *Curr. Opin. Colloid Interface Sci.*, **7** (5–6), 276–281.
92. Guo, Z. and Xu, X. (2006) Lipase-catalyzed glycerolysis of fats and oils in ionic liquids: a further study on the reaction system. *Green Chem.*, **8**, 54–62.
93. Guo, Z. and Xu, X. (2005) New opportunities for enzymatic modification of fats and oils with industrial potential. *Org. Biomol. Chem.*, **3**, 2615–2619.
94. Kuzmanovic, B., Van Strien, C. J. G., Bartel, C. E. *et al.* (2007) WO 07 093574, AkzoNobel N.V.

3

Surface-Active Compounds as Forest-Industry By-Products

Bjarne Holmbom, Anna Sundberg and Anders Strand
Process Chemistry Centre, Åbo Akademi University, Finland

3.1 Introduction

Forests are a major source of renewable organic material on earth. One third of the world's forests are primarily used for production of wood and non-wood products [1]. Global wood removals were in 2005 estimated to be 3.1 billion cubic metres. Europe and North and Central America accounted for about half of the wood removals. Most of the wood in Europe and North America is used as industrial roundwood, while most of the wood in Africa, Asia and South America is still used as fuel for cooking and heating.

In the industrialized countries in Europe and North America most of the harvested wood goes to production of pulp and paper. Along with the slow growth in pulp and paper markets, the industry is presently developing forest biorefineries, which will have production at the same site not only of pulp and paper but also of biofuels, as well as high-value specialty biomaterials and biochemicals. Forests are a good base for biorefineries, since wood can be harvested all year around in a steady flow with the harvesting and raw material logistics well established. Furthermore, forest biorefineries do not compete with food use as agriculture-based biorefineries do.

Wood is composed of cellulose, hemicelluloses (also called 'non-cellulosic polysaccharides') and lignin. There is also a wide variety of minor constituents, commonly called 'extractives', typically amounting to 1–5% of the wood. Concentrations as high as 10–30% can, however, be found in heartwood, knots and bark of many tree species [2–4].

Surfactants from Renewable Resources Edited by Mikael Kjellin and Ingegärd Johansson

The pulp and paper industry in Europe and North America has been producing chemicals from the extractives for almost a century, primarily as tall oil and turpentine products.

Surface-active compounds can be produced from all wood constituents. For instance, water-soluble cellulose derivatives and lignin, in the form of lignosulfonates, can be used as dispersing agents in many applications. This chapter focuses on the production of resin and fatty acids as well as sterols and sterol ethoxylates as forest-industry by-products, and their use as surface-active components. Moreover, the potential production of hemicelluloses and their possible utilization as steric stabilizers of oil-in-water emulsions is also outlined.

3.2 Resin and Fatty Acids

Resin acids are diterpenoid acids that constitute 70–80% of the oleoresin in coniferous (softwood) tree species [2, 3]. The main eight resin acids all have the same three-ring hydrocarbon skeleton with the carboxyl group in the same position, but they differ in the number and positions of double bonds (Figure 3.1). Furthermore, abietic-type acids have an isopropyl side group, while pimaric-type acids have a vinyl and a methyl group in that position. Abietic acid is the dominating resin acid in most species [3].

All tree species, both softwoods and hardwoods, contain fatty acids, typically 0.5–1% of the wood, mainly in the form of triglycerides or steryl esters. The predominant fatty acids are the oleic, linoleic and pinolenic acids with 18 C-atoms and one, two and three double bonds, respectively (Figure 3.1).

3.2.1 Resin and Fatty Acids in Different Wood Species

Pines contain especially large amounts of resin acids, from about 1–2 wt% of the wood up to 30% in the heartwood of old trees. Pine knots, that is the branch roots inside the tree stem, also contain much resin acids, typically 15–25% [4].

All trees contain fatty acids, mainly as constituents in triglycerides and steryl esters [2, 3]. Certain hardwood species, including birch and aspen species, are so-called fat-storing species and typically contain 1–2% of fatty acids, while most other hardwood species contain only 0.2–0.5% of fatty acids. Pines and spruces typically contain 0.5–1% of fatty acids. Oleic acid and linoleic acid are the predominant fatty acids in most tree species (Figure 3.1). Pines, spruces, firs and larches contain the rare pinolenic acid, which is not found in nature elsewhere than in the *Pinaceae* family. In addition to these major fatty acids, trees contain a large number of other minor fatty acids. For example, in spruce, more than 30 fatty acids, with chain lengths from C-12 to C-24 and with up to four double bonds in different positions, have been identified [3]. Certain hardwoods contain considerable amounts of long-chain fatty acids, that is saturated acids with 26–30 carbon atoms. *Acacia mangium*, an important plantation species used as pulpwood in South-East Asia, contains over 1% of such long-chain acids, commonly called 'policosanoic acids' [5]. The C-26 and C-28 fatty acids occur in the wood as surface-active monoglycerides. Aspen species contain monoglycerides, with C-24 and C-26 fatty acids [6].

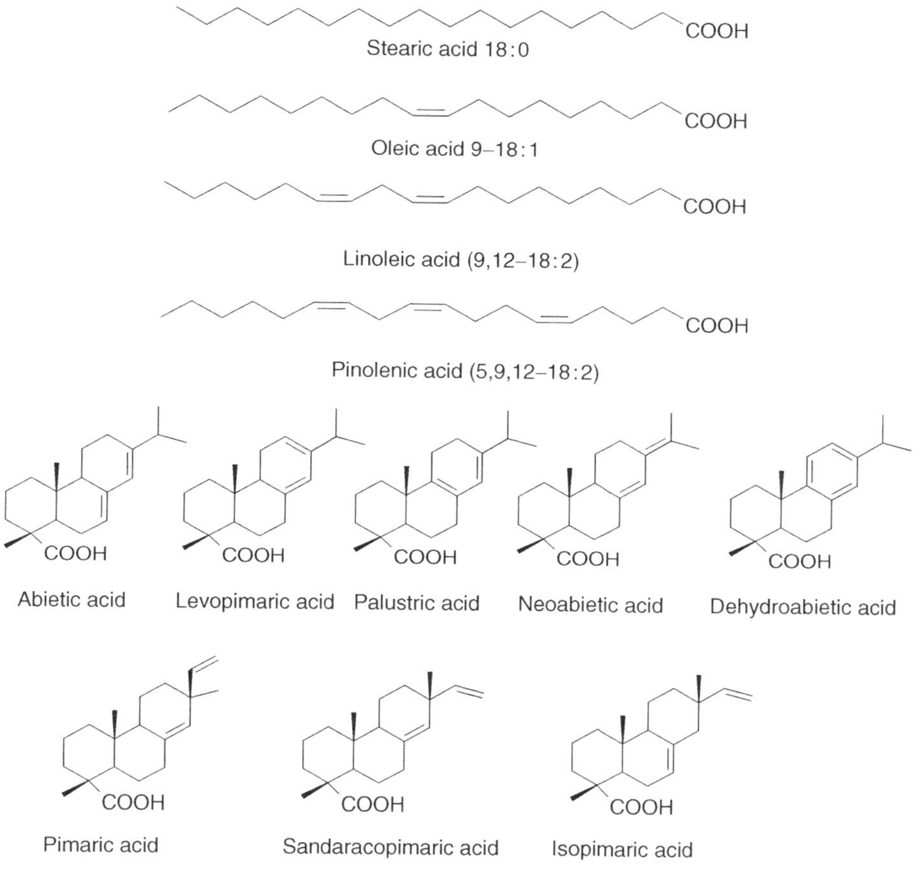

Figure 3.1 *Major fatty and resin acids in tree species used by the pulp and paper industry.*

3.2.2 Industrial Production of Resin and Fatty Acids

When coniferous wood is pulped in the strongly alkaline Kraft pulping process, also called the 'sulfate pulping process', the predominant chemical pulping process today, the resin and fatty acids are converted into sodium soaps, and are after cooking recovered as so-called 'sulfate soap' from the surface of the pulping liquor, the black liquor. The sulfate soap is a lamellar crystalline phase where the resin and fatty acids form the lamellae, and neutral components, such as sterols, are included between the lamellae [7].

The sulfate soap is processed to crude tall oil (CTO) by acidification with sulfuric acid at the pulp mills. CTO is then delivered by the pulp mills to tall oil distillation plants, where CTO is fractionated by vacuum distillation to yield the main products: tall oil rosin (TOR), tall oil fatty acid (TOFA) and distilled tall oil (DTO) (Figure 3.2).

The global annual production of CTO is about 1.4 million tonnes, out of which about 750 000 tonnes is produced in the USA [8]. In the Nordic countries, the CTO production

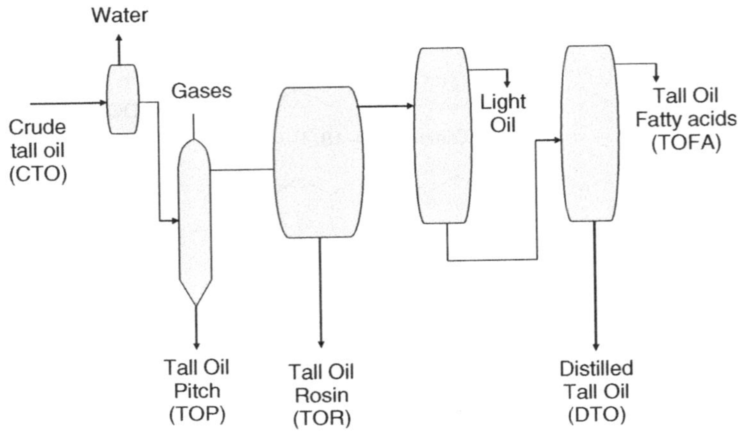

Figure 3.2 *Principal scheme and main products of crude tall oil distillation.*

is approximately 400 000 tonnes annually. Most of the CTO is further processed by distillation, but there is an increasing competition with energy use, including production of biofuels [8]. During the current economical crisis, pulp production has been cut both in the USA and the Nordic countries, accordingly also decreasing CTO production.

In 2005, the production of tall oil distillation products was approximately as follows: TOR 400 000 tonnes, TOFA 450 000 tonnes and DTO 60 000 tonnes [9]. Gum rosin, a product similar to TOR, originating from tapping oleoresin from pine trees, is produced on a large industrial scale, especially in China. The Chinese production was in 2005 close to 700 000 tonnes [9].

The composition of resin and fatty acids in CTO, and in the distilled tall oil products, is dependent mainly on the tree species used for pulping. However, storage of pulpwood can lead to degradation of fatty and resin acids and thus decreases the CTO yield considerably and also alters the resin and fatty acid composition [10, 11]. The resin and fatty acids are also to some extent isomerized in pulping and soap acidification [10, 11]. Abietic acid is partly isomerized to other abietic-type acids with conjugated double bonds, and polyunsaturated fatty acids are isomerized to conjugated fatty acids in the Kraft cooking process. Furthermore, in CTO distillation at temperatures of 250–270 °C certain resin and fatty acids undergo thermal conversions [10]. Typical compositions of distilled tall oil products are compiled in Table 3.1.

American CTOs contain smaller amounts of polyunsaturated fatty acids than Nordic CTOs. This difference is also reflected in the tall oil distillates, with lower amounts of linoleic and pinolenic acids than in corresponding Nordic products. This difference is related to the warmer climate in the southeastern US, but this trend can also be noticed even within Finland and Sweden.

3.2.3 Resin and Fatty Acids as Surfactants

Both resin and fatty acid soaps have been used as surfactants since ancient times. Today, about 10% of the total TOFA and DTO production is consumed in soap products [12].

Table 3.1 Typical compositions of Nordic and American tall oil distillation products

Acid	Tall oil rosin (TOR)		Distilled tall oil (DTO)		Tall oil fatty acids (TOFA)	
	Nordic	America	Nordic	America	Nordic	America
Stearic (18 : 0)			1	1	1.5	2
Oleic (9–18 : 1)			15–17	14–16	25–28	45–55
Linoleic (9,12–18 : 2)			24–28	20–22	40–45	35–40
Pinolenic (5,9,12–18 : 3)			3–4	3–4	7–10	2–3
Sciandonic (5,11,14–20 : 3)			6–8	4–5	1	
Minor fatty acids, sum			8–13	20–25	10–15	10–15
Total fatty acids	1–2	1–2	60–65	65–70	96–97	96–97
Pimaric	4–5	2–4	5–7	3–5		
Sandaracopimaric	1–2		1–2			
Isopimaric	3–5	8–12	3–5	4–7		
Palustric	10–12	8–12	4–6	2–3		
Dehydroabietic	21–22	20–25	3–5	4–5		
Abietic	42–43	35–40	4–6	5–7		
Neoabietic	4–5	3–5	1			
Minor resin acids, sum	8–12	10–15	4–8	11–15		
Total resin acids	94–96	92–96	25–30	25–30	1–2	1–2
Total resin and fatty acids	95–97	95–97	94–97	95–98	97–99	97–99
Neutral components	3–5	3–5	3–6	2–5	1–2	1–2

Minor components (<1%) are not tabulated. Compilation of data is from various sources [10–12] and analyses in companies.

TOFA and DTO soaps compete with soaps from other vegetable oils, especially from coconut and tallow. TOFA and DTO have the advantage of being ready in free acid form and soaps can be prepared in rather simple equipment, at low cost. TOFA and DTO soaps are also used as emulsifiers in cutting oils. Fatty acid soaps have the disadvantage that they do not function well in hard waters, which is because insoluble calcium soaps are formed. Therefore, cationic and nonionic derivatives such as amidoamines and ethoxylates are also produced. Fatty acid esters with glycerol (mono- and diglycerides) and with sorbitol are also manufactured by several companies.

Soaps of resin and fatty acid mixtures have very different surfactant properties compared to fatty acids and resin acids alone, especially in waters containing salts. In an early study, tall oil resin acids were found to be much less effective in lowering surface tension than TOFAs [13]. This was studied in more detail later, with sodium oleate and sodium abietate as model compounds [14]. Both sodium oleate (NaOl) and sodium abietate (NaAb) form small micelles at low concentrations with a clearly defined critical micelle concentration (CMC). As expected, the CMC decreases with increasing salt concentration (Table 3.2).

The solubility profiles with increasing NaCl concentration were very different for the three models (Figure 3.3).

The solubility decreases for sodium abietate as salt concentration is increased, while it first increases and then decreases for sodium oleate. The 1 : 1 mixture has a good solubility up to a high salt concentration. It was concluded that sodium oleate and sodium abietate together form very stable mixed micelles. The good solubility of these mixed micelles in salt solutions also enables solubilization of insoluble fatty components at

Table 3.2 *Critical micelle concentrations for sodium oleate (NaOl) and sodium abietate (NaAb) determined at 60 °C*

NaCl M	CMC		
	NaOl mM	NaAb mM	NaOl : Na Ab (1 : 1) mM
0	0.48	10	1.0
0.25	0.035	2.9	0.058
0.75	Insoluble		0.028

Reproduced with permission from H. Palonen, P. Stenius and G. Ström, Surfactant behaviour of wood resin components. The solubility of rosin and fatty acid soaps in water and in salt solutions. *Svensk Papperstidning*, **85**, R93–R99 (1982).

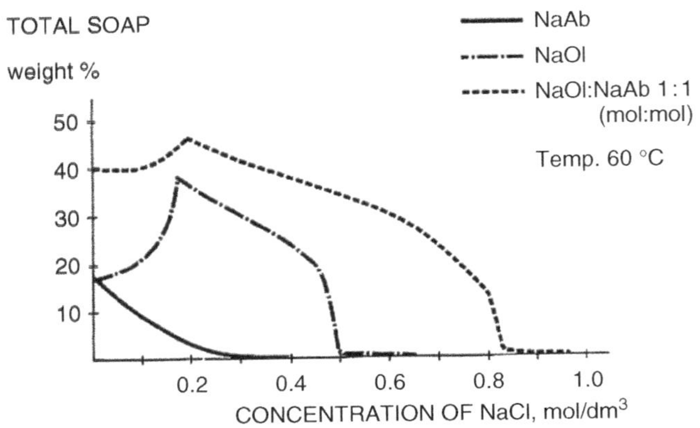

Figure 3.3 *Solubility of sodium oleate (NaOl) and sodium abietate (NaAb) and their 1 : 1 mixture at 60 °C and different NaCl concentrations.*

Reproduced with permission from H. Palonen, P. Stenius and G. Ström, Surfactant behaviour of wood resin components. The solubility of rosin and fatty acid soaps in water and in salt solutions. *Svensk Papperstidning*, **85**, R93–R99 (1982).

fairly high salt concentrations. Another study concluded that the removal of lipophilic extractives in pulp washing is efficient when the fatty acid to resin acid ratio is between 1 : 1 and 1 : 2 by weight at a 0.1–0.3 M NaCl concentration [15, 16]. A high temperature and high pH is favourable. In hardwood pulping, resin acids are commonly added in the form of softwood-derived tall oil soap or TOR to promote the removal of neutral extractives, such as sterols and triterpenyl alcohols, which are troublesome in papermaking. By tradition in Finland, it is known that soap made from DTO, which in addition to fatty acids contains 20–30% resin acids, is a better agent than mere fatty acid soaps for washing of cloth and carpets in brackish waters.

To function as surfactants the resin and fatty acids must be dissolved as soaps in the water phase. The dissolution is strongly dependent on the pH.

In studies of films of fatty acids it was found that a pH exists where various interfacial properties show a distinct peak [17]. This optimum was near the pK_a of the fatty acid salts as determined by titration with NaOH. The increase in pK_a for saturated fatty acids with increasing chain length was explained by the possibility of closer packing and stronger binding forces of longer fatty acid chains. Unsaturated fatty acids had lower pK_a

values compared to the saturated acid of the same chain length [18]. This association of hydrocarbon chains is reflected also in the much higher melting points for saturated fatty acids compared to unsaturated acids. Unexpectedly high 'apparent' pK_a-values for fatty acids were found also for fatty acids in dilute aqueous solutions, at concentrations clearly below the CMC [19]. The high pK_a-values were explained to be due to association between ionized and unionized carboxyl groups into premicellar aggregates, where the ionized group is associated to the unionized carboxyl group. In such an aggregate it is more difficult to remove the acid proton, leading to an increase in the pK_a.

More recently, McLean *et al.* [20] determined the pK_a of individual fatty and resin acids by titration using a similar titration method to that of Kanicky and Shah [19]. The values, which they termed 'colloidal pK_a-values', were slightly lower than those obtained by Kanicky and Shah, especially for the unsaturated C-18 acids. The differences were, however, not discussed. Resin acids gave pK_a-values of 6.2–7.3, values that are in the range of earlier reported values [21–23].

In a recent study, the authors studied a model system relevant to lipophilic wood pitch in mechanical pulp and further manufacturing of wood-containing paper [24]. Wood pitch, which is colloidally dispersed as a dilute emulsion of pitch droplets, was analysed for the distribution of resin and fatty acids between the lipophilic phase, mainly composed of triglycerides and steryl esters, and the water phase in the pH range 3–11. The study aimed at obtaining better understanding of the behaviour of pitch in pulp and paper mills. These phenomena are important with regard to pitch removal, but have implications also on the formation of sticky pitch deposits on paper machines, and even on the properties of the produced paper.

It was found that individual resin and fatty acids are dissolved in the water phase, that is they migrate out from the colloidal pitch droplets at very different pH levels (Figure 3.4).

The pH where 50% of an acid is retained in the lipohilic colloidal droplets (l) and the other half is dissolved in the water phase (w), was termed pK_{lw}. Dehydroabietic acid having an aromatic ring had the lowest pK_{lw}, of 5.3. The other resin acids were rather similar, with pK_{lw} values of 6.8–7.1. The pinolenic acid with three double bonds behaved similarly to the resin acids, and linoleic acid similarly to palmitic acid. However, oleic acid dissolved at a clearly higher pH than the polyunsaturated fatty acids. The saturated stearic acid was preferentially in the colloidal phase, even at pH 10. The long-chain fatty acids with 20–24 C-atoms were dissolved only to a small extent, even at pH 11.

By addition of NaCl the dissolution into the water phase was moved towards a higher pH, as presented for abietic acid and oleic acid in Figure 3.5.

Abietic acid was still dissolved completely at pH 9–10, while oleic acid was only partly dissolved, even at pH 11. Corresponding studies with addition of calcium chloride showed that calcium ions lower the solubility even more dramatically, especially for saturated acids and oleic acid [25]. Resin acids were less affected by the addition of calcium, and dehydroabietic acid was only a little influenced, even by addition of 10 mM of calcium ions. This implies that resin acids can also function well as surfactants in hard waters.

From these studies it is obvious that the ratio of resin acid and fatty acid soaps dissolved in the water phase is very dependent on the pH. The resin acids are dissolved at a lower

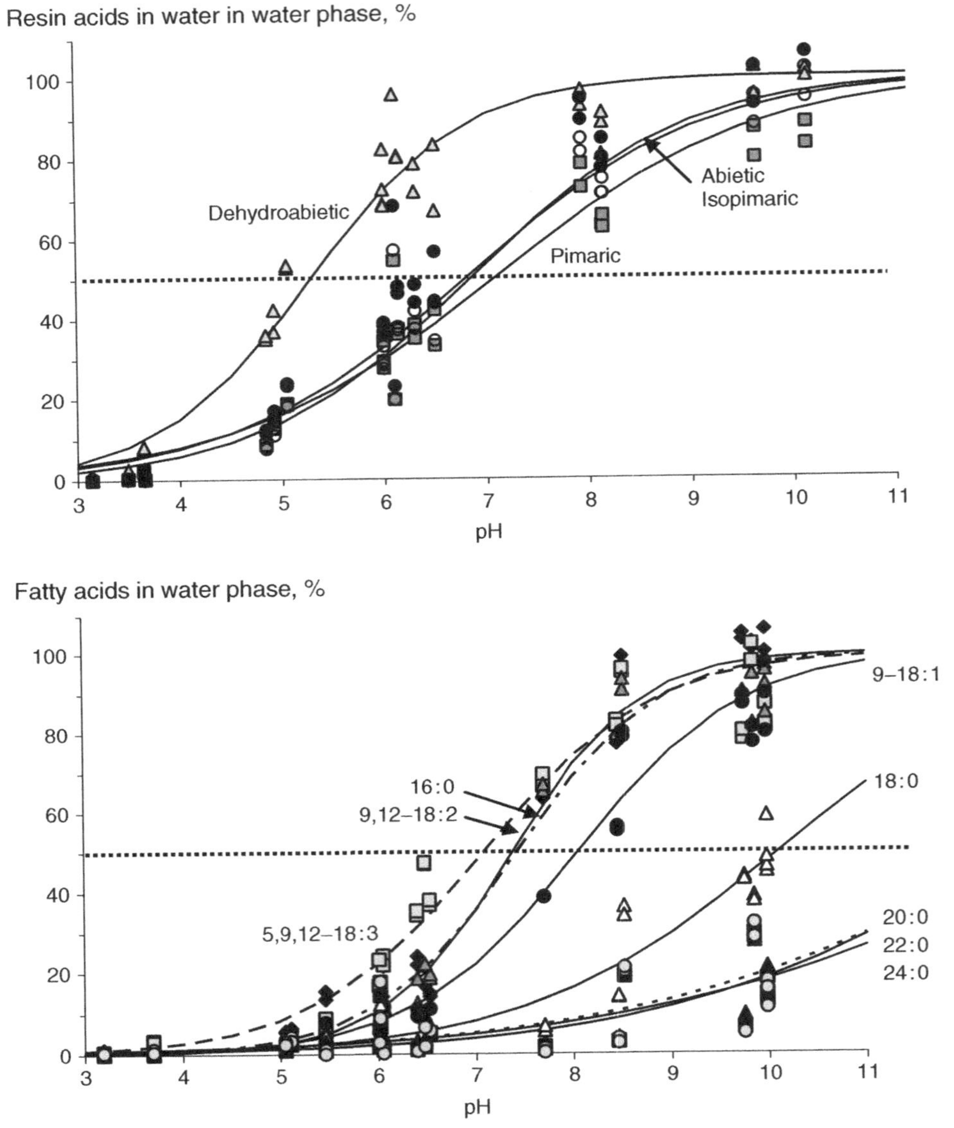

Figure 3.4 *Distribution of resin acids and fatty acids between lipophilic phase and water phase at 50 °C and low NaCl concentration in wood pitch emulsion.*

A. Sundberg, A. Strand, L. Vähäsalo and B. Holmbom, Phase distribution of resin and fatty acids in colloidal pitch emulsions at different pH-levels, *Journal of Dispersion Science and Technology*, **30**, 912–919, 2009. Reprinted by permission of the Taylor & Francis Group, http://www.informaworld.com.

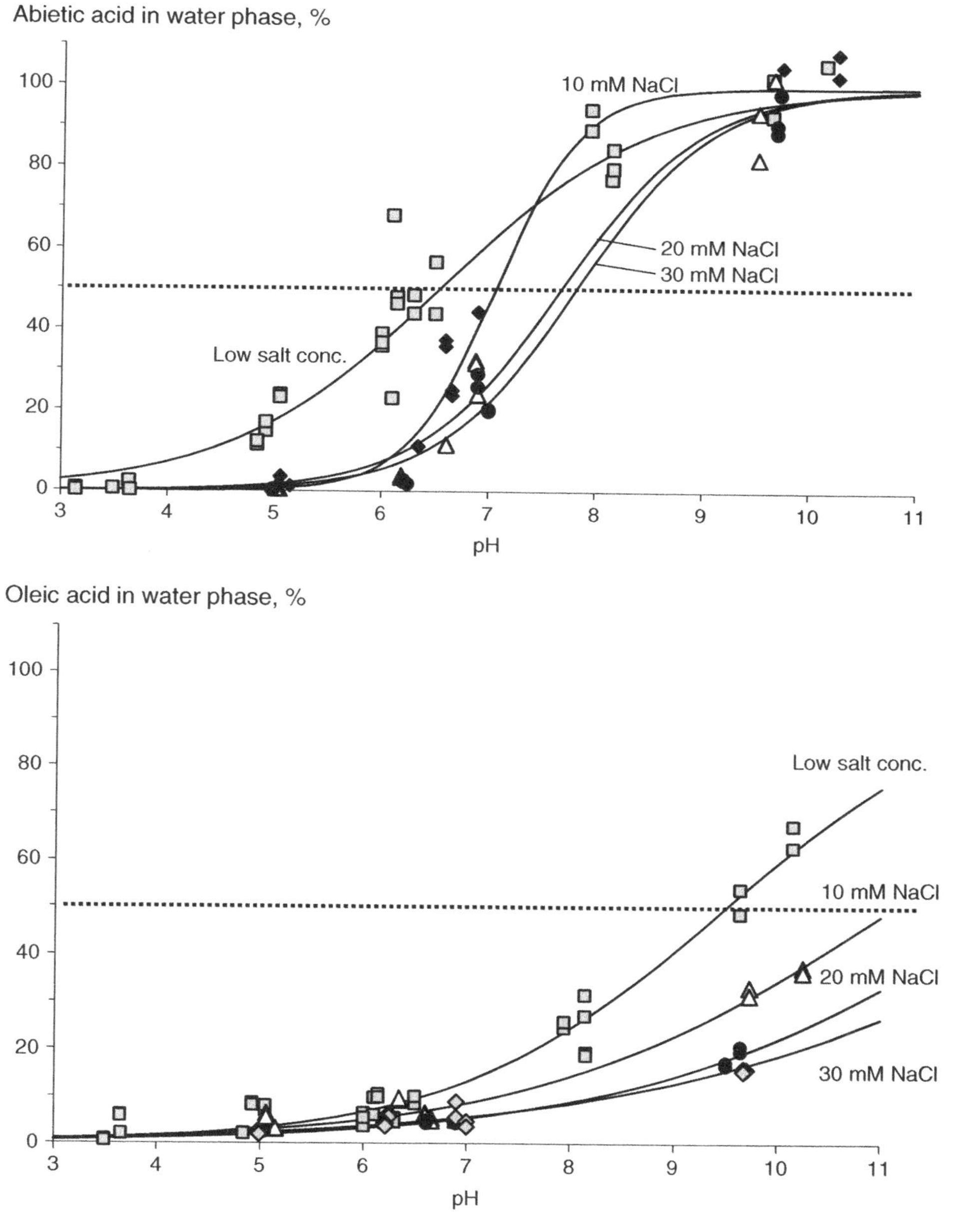

Figure 3.5 Distribution of abietic acid and oleic acid between lipophilic phase and water phase in a wood pitch emulsion at different NaCl concentrations. The temperature was 50 °C.

A. Sundberg, A. Strand, L. Vähäsalo and B. Holmbom, Phase distribution of resin and fatty acids in colloidal pitch emulsions at different pH-levels, *Journal of Dispersion Science and Technology*, **30**, 912–919, 2009. Reprinted by permission of the Taylor & Francis Group, http://www.informaworld.com.

pH than the fatty acids. Among the fatty acids, the saturated stearic acid clearly has a lower solubility than the unsaturated C-18 acids. Long-chain saturated fatty acids with 20 or more C-atoms do not dissolve in the water phase, even at very high pH.

TOR and gum rosin, derived from tapping oleoresin, as well as DTO and other TOR–TOFA mixtures, are also good emulsifiers, and are used widely in emulsion polymerization of synthetic rubber, especially styrene-butadiene rubber (SBR).

3.2.4 Toxicity and Biodegradation of Resin and Fatty Acids

Resin acids are toxic to fish at concentrations of 0.5–1 mg/l, but are not toxic to humans and other mammals. The toxicity of the resin acids is related to their uptake from water and accumulation in the fish [26]. The resin acids are then metabolized in the liver by formation of their glucuronic acid esters, which are excreted through the bile. The glucuronyltransferase enzyme system in the liver can be overloaded and seriously affect the function of the liver [27]. However, resin acids, as well as fatty acids, are to a large extent removed in modern water treatment plants with mechanical and biological stages [28].

Another concern with resin acids is that they may cause contact allergy [29]. Oxidation products of abietic and dehydroabietic acids are contact allergens. The prevalence of contact allergy is, however, rather low.

3.3 Sterols and Sterol Ethoxylates

3.3.1 Production and Uses of Phytosterols

Trees contain considerable amounts of sterols [2, 3], present in the parenchyma cells in the form of fatty acid esters. Sterols are found both in softwoods and hardwoods. As well as sterols, many hardwoods also contain nonsteroidal triterpenyl alcohols. Sitosterol (Figure 3.6) is the main sterol in temperate-zone tree species [2, 30]. It is accompanied by varying amounts of its saturated analogues, sitostanol and campesterol, as well as various triterpenyl alcohols.

In the Kraft pulping process, the steryl esters are largely, although not completely, hydrolysed to free sterols and fatty acids [31]. The sterols are incorporated in the sulfate soap and consequently in the CTO. CTO contains 3–5% of sterols [30], which is a higher concentration than in vegetable oils from annual plants. In distillation of CTO,

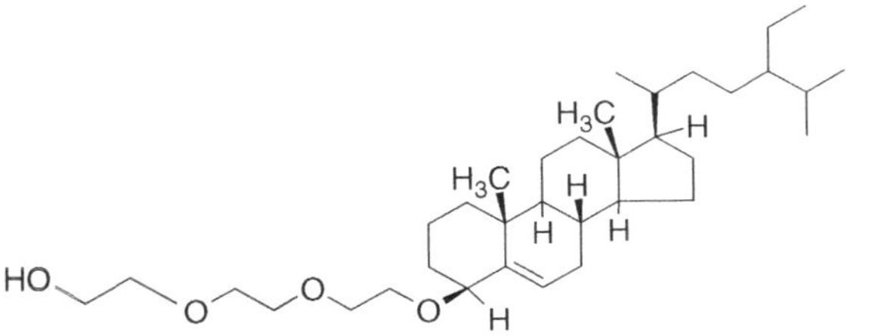

Figure 3.6 *Sitosterol ethoxylate with three ethoxylate groups.*

the sterols are concentrated in the tall oil pitch residue, occurring partly in free form and partly as steryl esters.

Sterols were produced from tall oil pitch on a small scale in a plant in France in the early 1970s. A new process for extraction of sterols and other neutral components from sulfate soap was developed in Finland in the 1970s [32]. An industrial plant was built and purified sterols were produced starting in 1980. The sterols were used as emulsifiers in cosmetics in the form of ethoxylates, and as a marker in butter to prove that the butter was produced in the EU [33, 34]. In 1995, sitostanol esters were launched as a new cholesterol-lowering food additive by the Finnish company Raisio, under the trade mark Benecol. Today, plant sterols and stanols, also referred to as 'phytosterols', are included in a large range of functional food products all over the world [35]. The sterols are extracted from tall oil, especially from tall oil pitch, as well as from certain vegetable oils [36]. Phytosterol production is presently at a level of 10 000 tonnes per year and is growing by about 10% per year.

3.3.2 Sterol Ethoxylates

Phytosterol ethoxylates (Figure 3.6) obtained by reaction of sterols with ethylene oxide in the presence of catalysts were produced in the early 1980s, for use mainly as emulsifiers in cosmetics [34, 35, 37]. Plant sterol ethoxylates are produced with an average degree of ethoxylation from 5 to 30 [38], but with a very broad homologue distribution.

More recently, the surfactant properties of ethoxylates prepared from soya bean phytosterol (a mixture of sitosterol, stimasterol and campesterol in a 2 : 1 : 1 ratio) have been studied [39, 40]. The degree of ethoxylation largely determines the surfactant properties (Figure 3.7).

The CMC values of sterol ethoxylates at room temperature are very low, in the range of 3–10 μM, which is much lower than those of corresponding ethoxylated nonylphenols.

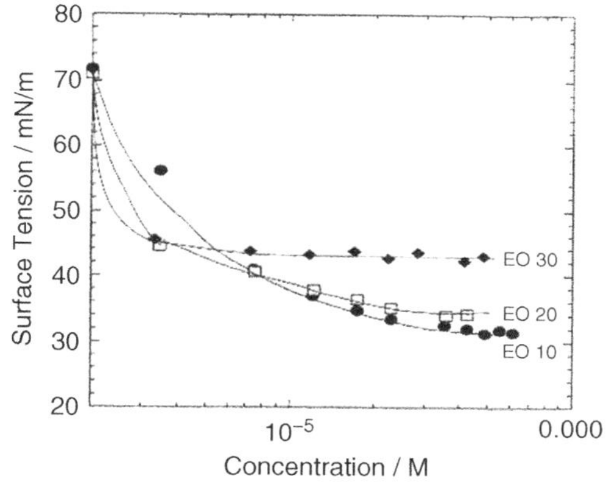

Figure 3.7 *Surface tension of plant sterol ethoxylates with different ethoxylation degrees, 150 minutes after addition of the surfactant.*

Reprinted from *J. Colloid Interface Sci.*, **213**, B. M. Folmer, M. Svensson, K. Holmberg and W. Brown, The physicochemical behavior of phytosterol ethoxylates, 112–120. Copyright 1999, with permission from Elsevier.

CMC decreases with increasing oxyethylene chain length, contrary to the normal trend for other ethoxylated surfactants.

There has been a growing interest in using phytosterol ethoxylates in the cosmetic industry [38], and they are used in many cosmetic formulations. The lower ethoxylates can serve as typical water-in-oil co-emulsifiers, while the higher ethoxylates are typical oil-in-water co-emulsifiers.

3.4 Hemicelluloses

Hemicellulose is a collective name for the non-cellulosic polysaccharides, which are a large constituent in the structure of the wooden cell wall. Hemicelluloses are generally considered to cover the elementary cellulose fibrils and function as an amorphous glue binding the cellulose fibrils together. Hemicelluloses comprise typically 25–35% of the wood, and are thus the second most abundant component group in wood. Hemicelluloses have only a very weak surface activity but are, nevertheless, of scientific and technical interest because of their ability to function as emulsion stabilizers through a steric stabilization mechanism.

3.4.1 Hemicelluloses in Wood

The predominant hemicellulose type in softwoods is galactoglucomannan (GGM), having a linear backbone of mannose and glucose units with single galactose units as side groups [41, 42] (Figure 3.8).

About every second mannose unit carries an acetyl group at the C-2 or C-3 hydroxyl group. The acetyl groups make the GGMs amorphous, and also make them soluble in water. In hardwoods, the predominant hemicellulose is 4-*O*-methyl glucuronoxylan, having a linear backbone of xylose units with single side groups of 4-*O*-methyl glucuronic acid (Figure 3.8). Acetyl groups are found at the C-2 and C-3 hydroxyl groups of the xylose units. Also xylans are water-soluble when the acetyl groups are preserved in the structure. Xylans are considered to have an analogous function in hardwoods: to bind cellulose fibrils together, like GGMs do in softwoods. GGMs are nonionic, but xylans are anionic, with a pK_a value for the glucuronic acid carboxyl group of approximately 3.7. The molar mass distribution of GGMs and xylans in the wood cannot be determined accurately, because dissolution of the hemicelluloses also involves chain cleavage. Based on studies on extracted hemicelluloses it is generally considered that they have a degree of polymerization in the range of 100–300 [2], corresponding to a molar mass range of 16–48 kDa for GGMs and 13–39 kDa for xylans.

3.4.2 Extraction and Purification of Hemicelluloses

In the manufacture of mechanical pulps, either by grinding of logs or by refining of chips, part of the hemicelluloses will be dissolved in the process waters. Spruce mechanical

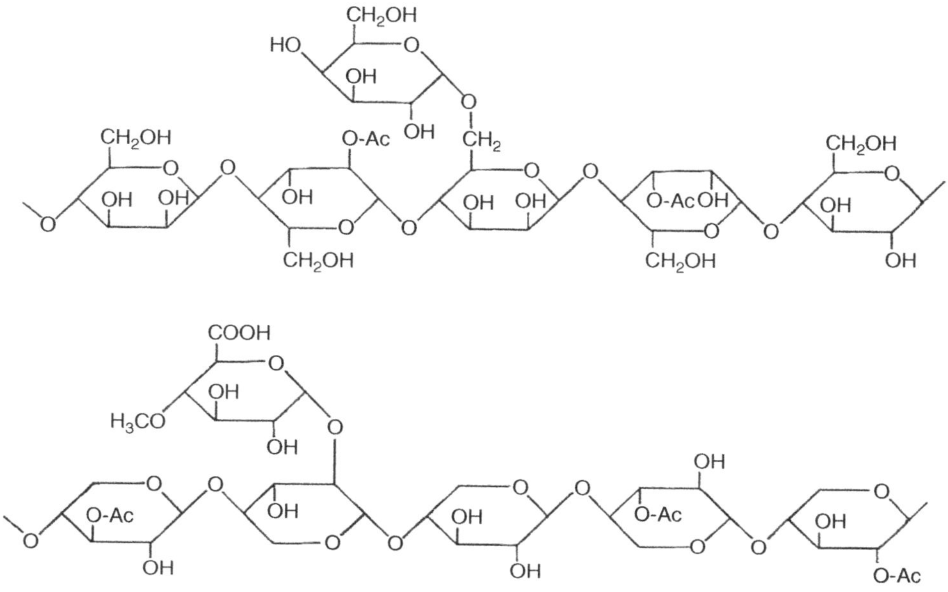

Figure 3.8 Principal structure of the main softwood hemicellulose type, galactoglucomannan, and of the main hardwood hemicellulose type, 4-O-methyl glucuronoxylan.

pulp waters contain GGMs, still in acetylated form, as the major component [41]. Pitch is dispersed as colloidal droplets in the waters and is stabilized sterically by the GGMs [42].

GGMs can be concentrated and recovered by ultrafiltration of the process waters [44–46]. GGMs can also be extracted directly from softwoods using plain water at temperatures of 160–170 °C [47], at conditions where the acetyl groups are preserved as much as possible. However, hydrolytic chain cleavage cannot be avoided and the dissolved GGMs exist partly as monomers and oligomers.

Xylans can be extracted from hardwoods, also using plain water [48]. Moreover, xylans can be extracted from hardwood black liquors, especially from liquor taken out in the early stage of the cook [49].

Today, there is still no production of GGMs, nor of xylans, but production would be technically possible if a valuable use could be found for the hemicelluloses [43].

3.4.3 Sterical Stabilization of Oil-in-Water Emulsions

GGMs in mechanical pulping waters were documented to play a key role for the stability of colloidal pitch droplets against aggregation with simple electrolytes [43]. The mechanism is evidently steric stabilization. It has been stated that polysaccharides such as galactomannans (locust bean gum (LBG) or guar gum (GG)) can be adsorbed to semi-solid interfacial layers, a phenomenon induced by a salting-out effect, and can, therefore, also be considered as emulsifiers [50]. In a later study, GGMs extracted from TMP

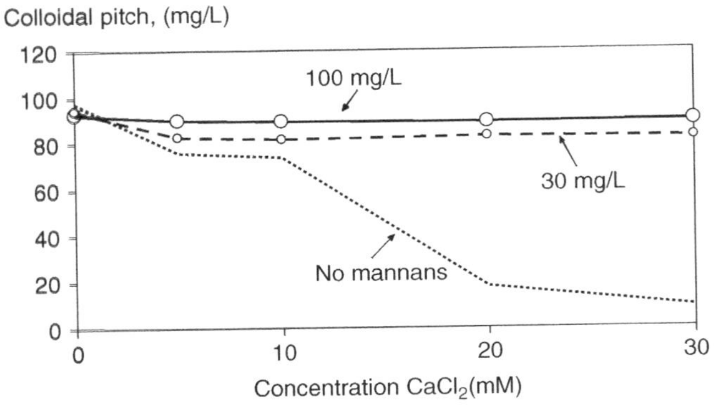

Figure 3.9 *Stabilization by native acetylated spruce galactoglucomannan, with an average molar mass of 20 kDa, of a colloidal spruce pitch emulsion against aggregation with calcium ions.*

Reproduced with permission from T. Hannuksela and B. Holmbom, Stabilization of wood resin emulsions by dissolved galactoglucomannans and galactomannans, *J. Pulp Pap. Sci.*, **30**, 159–164 (2004).

waters were compared to guar gum galactomannans as wood pitch stabilizers [51]. Both acetylated GGMs and guar gum galactomannans (unacetylated) were able to stabilize wood pitch droplets (Figure 3.9).

The higher the molar mass of the mannan, the better was its stabilizing effect. Deacetylated GGMs, in turn, were poor stabilizers. A birch xylan was also an excellent stabilizer, while neutral and cationic starches were not very effective. However, in another similar study birch xylan was found to have only a minor stabilizing effect [52].

Spruce GGMs isolated from TMP waters by ultrafiltration and precipitation with ethanol were also studied in a model beverage system with orange oil as 0.16% oil-in-water emulsion, together with sodium benzoate, citric acid and sucrose as additives [53]. Comparison was made to GG and LBG galactomannans, konjac glucomannan (KGM) and corn arabinoxylan (cAX). The polysaccharides (PS) were added in a PS-to-oil ratio of 0.05 (w/w). Initial emulsion turbidity and its stability were used as measures of emulsifying ability and emulsion stability, respectively. GGMs enhanced the emulsion formation and increased the emulsion stability during 14 days at room temperature (Figure 3.10).

Ethanol-precipitated GGM was more effective than the other studied mannans, and was almost as effective as corn arabinoxylan, which is considered to be a good stabilizer. The study gave the first indication of spruce GGM as a potential stabilizer of food emulsions and was much in line with the previous stabilizing studies of pitch emulsions [51].

Acknowledgements

Financial support to the Process Chemistry Centre within the Centre of Excellence programme of the Academy of Finland and support from the Finnish Funding Agency for Technology and Innovation (Tekes) are acknowledged.

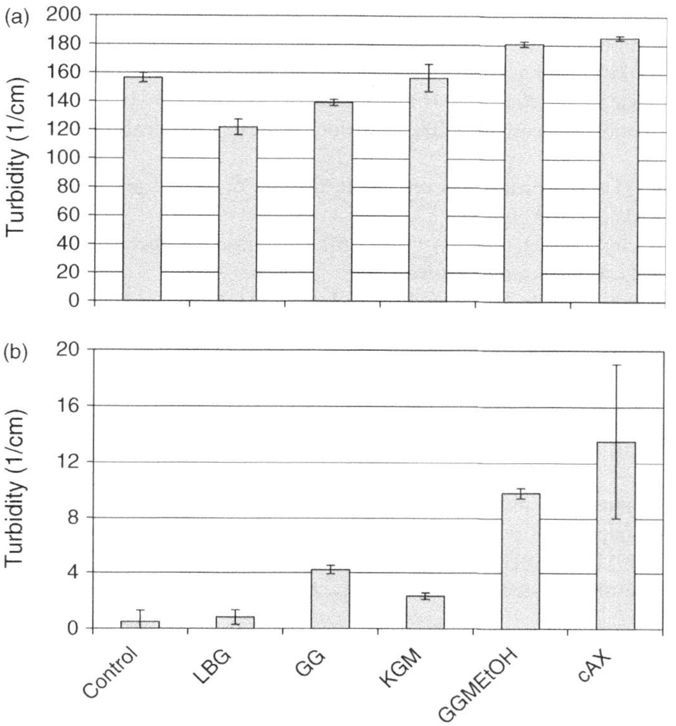

Figure 3.10 *Turbidities of emulsions containing different mannans and corn xylan at a PS-to-oil ratio of 0.05 (a) immediately after preparation and (b) after 14 days at room temperature. LBG: locust bean gum galactomannan, GG: guar gum galactomannan, KGM: konjac glucomannan, GGMEtOH: galactoglucomannan precipitated with ethanol, and cAX: corn arabinoxylan. The averages are based on three absorbance measurements at 650 nm wavelength and the error bars indicate standard deviations.*

Reproduced from K. S. Mikkonen, M. Tenkanen, P. Cooke, C. Xu, H. Rita, S. Willför, B. Holmbom, K. B. Hicks and M. P. Yadav, Mannans as stabilizers of oil-in-water beverage emulsions, *LWT - Food Science and Technology*, **42**, 849–855 (2009).

References

1. FAO (2005) Global forest resources assessment 2005, www.fao.org/forestry/fra2005.
2. Fengel, D. and Wegener, G. (1989) *Wood – Chemistry, Ultrastructure, Reactions*, de Gruyter, Berlin, New York, 182–226.
3. Ekman, R. and Holmbom, B. (2000) The chemistry of wood resin, in *Pitch Control, Wood Resin and Deresination* (eds E. Back and L. H. Allen), TAPPI Press, Atlanta, Georgia, pp. 37–76.
4. Willför, S., Hemming, J., Reunanen, M. and Holmbom, B. (2003) Phenolic and lipophilic extractives in Scots pine knots and stemwood. *Holzforschung*, **57**, 359–372.
5. Pietarinen, S., Willför, S. and Holmbom, B. (2004) Wood resin in *Acacia mangium* and *Acacia crassicarpa* wood and knots. *Appita J.*, **57** (2), 146–150.
6. Pietarinen, S., Hemming, J., Willför, S. *et al.* (2005) Wood resin in bigtooth and quaking aspen wood and knots. *J. Wood Chem. Technol.*, **25**, 27–39.

7. Roberts, K., Österlund, R. and Axberg, C. (1976) Liquid crystals in systems of rosin and fatty acids: implications for tall oil recovery. *Tappi J.*, **59** (6), 156–159.
8. Wong, A. (2007) Tall oil as feedstock for second-generation biodiesel production in southeastern USA. *Forest Chem. Rev.*, **117** (6), November–December, 5–10.
9. (2007) International Yearbook 2005 Edition, Forest Chemicals Review, New Orleans, Louisiana.
10. Holmbom, B. (1978) Constituents of tall oil, a study of tall oil processes and products, PhD Thesis, Åbo Akademi, Åbo.
11. Zinkel, D. F. and Russel, J. (eds) (1989) *Naval Stores; Production, Chemistry, Utilization*, Pulp Chemicals Association, New York.
12. McSweeney, E. E., Arlt Jr, H. G. and Russel, J. (1987) *Tall Oil and Its Uses–II*, Pulp Chemicals Association, New York.
13. Snell, F. D. and Reich, I. (1950) Physical–chemical properties of tall-oil soap solutions. *J. Am. Oil Chem. Soc.*, **27**, 73–74.
14. Palonen, H., Stenius, P. and Ström, G. (1982) Surfactant behaviour of wood resin components. The solubility of rosin and fatty acid soaps in water and in salt solutions. *Sven. Papperstidn.*, **85**, R93–R99.
15. Ström, G., Stenius, P., Lindström, M. and Ödberg, L. (1990) Surface chemical aspects of the behaviour of soaps in pulp washing. *Nord. Pulp Pap. Res. J.*, **5**, 44–51.
16. Ström, G. (2000) Physico-chemical properties and surfactant behaviour, in *Pitch Control, Wood Resin and Deresination* (eds E. Back and L. H. Allen), TAPPI Press, Atlanta, Georgia, pp. 139–149.
17. Kanicky, J. R., Poniatowsky, A. F., Mehta, N. R. and Shah, D. O. (2000) Cooperativity among molecules at interfaces in relation to various technological processes: effect of chain length on the pK_a of fatty acid solutions. *Langmuir*, **16**, 172–177.
18. Kanicky, J. R. and Shah, D. O. (2002) Effect of degree, type and position of unsaturation on the pK_a of long-chain fatty acids. *J. Colloid Interface Sci.*, **256**, 201–207.
19. Kanicky, J. R. and Shah, D. O. (2003) Effect of premicellar aggregation on the pK_a of fatty acid solutions. *Langmuir*, **19**, 2034–2038.
20. McLean, D. S., Vercoe, D., Stack, K. R. and Richardson, D. (2005) The colloidal pK_a of lipohilic extractives commonly found in *Pinus radiata*. *Appita J.*, **58**, 362–366.
21. Back, E. and Steenberg, B. (1950) Simultaneous determination of ionization constant, solubility product and solubility for slightly soluble acids and bases. Electrolytic constants for abietic acid. *Acta Chem. Scand.*, **4**, 810–815.
22. Nyrén, V. and Back, E. (1958) The ionization constant, solubility product and solubility of abietic and dehydroabietic acid. *Acta Chem. Scand.*, **12**, 1516–1520.
23. Bruun, H. (1952) An interpretention of the effect of ionization upon monolayer properties of rosin acids. *Acta Chem. Scand.*, **6**, 955–957.
24. Sundberg, A., Strand, A., Vähäsalo, L. and Holmbom, B. (2009) Phase distribution of resin and fatty acids in colloidal pitch emulsions at different pH-levels. *J. Dispersion Sci. Technol.*, **30**, 912–919.
25. MacNeil, D., Sundberg, A., Vähäsalo, L. and Holmbom, B. (2009) Effect of calcium on the phase distribution of resin and fatty acids in pitch emulsions (submitted).
26. Oikari, A., Ånäs, E., Kruzynski, G. and Holmbom, B. (1984) Free and conjugated resin acids in the bile of rainbow trout, *Salmo gairdneri*. *Bull. Environ. Contam. Toxicol.*, **33**, 233–240.
27. Oikari, A. and Holmbom, B. (1986) Assessment of water contamination by chlorophenolics and resin acids with the aid of fish bile metabolites, in *Aquatic Toxicology and Environmental Fate*, vol. 9, ASTM Special Technical Publication 921, American Society for Testing and Materials, Philadelphia, Pennsylvania, pp. 252–267.

28. Kostamo, A., Holmbom, B. and Kukkonen, J. (2004) Fate of wood extractives in wastewater treatment plants at kraft pulp mills and mechanical pulp mills. *Water Res.*, **38**, 972–982.
29. Karlberg, A. T. and Gäfvert, E. (1996) Extractives in mechanical pulp can cause allergic contact dermatitis. *Nord. Pulp Pap. Res. J.*, **11**, 70–73.
30. Vikstroem, F., Holmbom, B. and Hamunen, A. (2005) Sterols and triterpenyl alcohols in common pulpwoods and black liquor soaps. *Holz. Roh. Werkst.*, **63**, 303–308.
31. Paasonen, P. (1966) On the nonvolatile ethyl ether extractives of birch wood and the changes in the composition effected by aging and sulfate pulping, PhD Thesis, Helsinki University, Helsinki.
32. Holmbom, B. and Avela, E. (1976) Method for refining of soaps using solvent extraction, US Patent no 3,965,085.
33. Lövberg, C.-E. (1983) Wood-derived β-sitosterol – a raw material for the cosmetic industry. *Seifen, Oele, Fette, Wachse*, **109**, 444–445.
34. Hamunen, A. and Hirschfeldt, A. (1986) Sterols from unsaponifiable constituents of tall oil – a wood-based raw material for the cosmetic, pharmaceutical, and food industries. *Seifen, Oele, Fette, Wachse*, **112**, 261–262.
35. Cantrill, R. (2008) Phytosterols, phytostanols and their esters: chemical and technical assessment, http://www.fao.org/ag/agn/agns/jecfa/cta/69/Phytosterols_CTA_69.pdf.
36. Fernandes, P. and Cabral, J. M. S. (2007) Phytosterols: applications and recovery methods. *Bioresour. Technol.*, **98**, 2335–2350.
37. Schrader, K. (1983) Use of phytosterols in cosmetic products. *Drug Cosmet. Ind.*, **133** (3), 34–35, 100–101.
38. Folmer, B. M. (2003) Sterol surfactants: from synthesis to applications. *Adv. Colloid Interface Sci.*, **103**, 99–119.
39. Folmer, B. M., Svensson, M., Holmberg, K. and Brown, W. (1999) The physicochemical behaviour of phytosterol ethoxylates. *J. Colloid Interface Sci.*, **213**, 112–120.
40. Folmer, B. M. and Nyden, M. (2008) Structure and dynamics of micelles and cubic phase structures with ethoxylated phytosterol surfactant in water. *Langmuir*, **24**, 6441–6446.
41. Thornton, J., Ekman, R., Holmbom, B. and Örså, F. (1994) Polysaccharides dissolved from Norway spruce in thermomechanical pulping and peroxide bleaching. *J. Wood Chem. Technol.*, **14**, 159–175.
42. Sundberg, K., Thornton, J., Holmbom, B. and Ekman, R. (1996) Effects of wood polysaccharides on the stability of colloidal wood resin. *J. Pulp Pap. Sci.*, **22**, J226–J230.
43. Willför, S., Sundberg, K., Tenkanen, M. and Holmbom, B. (2008) Spruce-derived mannans – a potential raw material for hydrocolloids and novel advanced natural materials. *Carbohydr. Polym.*, **72**, 197–210.
44. Willför, S., Rehn, P., Sundberg, A. *et al.* (2003) Recovery of water-soluble acetylgalactoglucomannans from mechanical pulp of spruce. *Tappi J.*, **2** (11), 27–32.
45. Andersson, A., Persson, T., Zacchi, G. *et al.* (2007) Comparison of diafiltration and size-exclusion chromatography to recover hemicelluloses from process water from thermomechanical pulping of spruce. *Appl. Biochem. Biotechnol.*, **136–140**, 971–983.
46. Persson, T., Nordin, A.-K., Zacchi, G. and Jönsson, A.-S. (2007) Economic evaluation of isolation of hemicelluloses from process streams from thermomechanical pulping of spruce. *Appl. Biochem. Biotechnol.*, **136–140**, 741–752.
47. Song, T., Pranovich, A. and Holmbom, B. (2008) Extraction of galactoglucomannan from spruce wood with pressurised hot water. *Holzforschung*, **62**, 659–666.
48. Liu, S. and Amidon, T. E. (2007) Essential components of a wood-based biorefinery. *O. Papel.*, **68**, 54–75.
49. Dahlman, O., Tomani, P., Axegard, P. *et al.* (2007) Method for separating polymeric pentose from a liquid/slurry. PCT Int. Appl., WO 2007120091 A1.

50. Garti, N. and Leser, M. E. (2001) Emulsification properties of hydrocolloids. *Polym. Adv. Technol.*, **12**, 123–135.
51. Hannuksela, T. and Holmbom, B. (2005) Stabilization of wood resin emulsions by dissolved galactoglucomannans and galactomannans. *J. Pulp Pap. Sci.*, **30**, 159–164.
52. Johnsen, I. A., Lenes, M. and Magnusson, L. (2004) Stabilization of colloidal wood resin by dissolved material from TMP and DIP. *Nord. Pulp Pap. Res. J.*, **19** (1), 22–28.
53. Mikkonen, K. S., Tenkanen, M., Cooke, P. *et al.* (2009) Mannans as stabilizers of oil-in-water beverage emulsions. *LWT – Food Sci. Technol.*, **42**, 849–855.

Part 2

Renewable Hydrophiles

4

Surfactants Based on Carbohydrates and Proteins for Consumer Products and Technical Applications*

Karlheinz Hill
Cognis GmbH, Monheim, Germany

4.1 Introduction

Carbohydrate-based surfactants are the final result of a product concept that is based on the greatest possible use of renewable resources. While the derivatisation of fats and oils to produce a variety of different surfactants for a broad range of applications has a long tradition and is well established [1, 2], the production of surfactants based on fats and oils and carbohydrates on a bigger industrial scale is relatively new. Today the most important carbohydrate-based surfactants are alkyl polyglycosides, sorbitan esters and sucrose esters [3].

4.2 Raw Materials

Natural fats and oils, carbohydrates and proteins are key raw materials for the chemical industry using renewable resources. Although in general biomass is available in large amounts (e.g. cellulose), the annual production volumes of selected bio-based

*Chapter based on K. Hill, O. Rhode, Sugar-based surfactants for consumer products and technical applications, *Fett/Lipid* 1999, **101**, 25–33. Copyright Wiley-VCH Verlag GmbH & Co. KGaA. Reproduced with permission.

Surfactants from Renewable Resources Edited by Mikael Kjellin and Ingegärd Johansson

Table 4.1 Annual production of commodities worldwide (2004, million tonnes)[a]

Wheat	Rice	Starch	Sugar[b]	Fats and oils[c]	Crude oil	Coal[d]
610	610	40	145	131	3600	3800

[a] Sources: OilWorld, USDA, Industrieverband Agrar, Wikipedia.
[b] From beet and cane.
[c] Vegetable and animal based, with 13 million tonnes for industrial use (including soap production).
[d] As SKE (1 kg SKE = 0.984 kg bituminous coal = 29.307.6 kJ).

commodities are still small compared to coal or crude oil (Table 4.1). Until now, availability and use was quite balanced and the quantities could be adjusted according to different demands. For example, in the case of natural oils and fats, the production volume was steadily increased from 30 million tonnes in 1960 to 131 million tonnes in 2004. Most of it was used for food (81% in 2004), with a minor amount for animal feed (6% in 2004) and chemistry (10% in 2004). However, what we have observed for some time is a shift towards an increasing use of renewable raw materials for bioenergy and biofuels. In the case of fats and oils, the expected share for energy is estimated to grow to 15% of the total annual capacity in 2012 compared to 3% in 2004. This is one consequence of political measures such as the European Biofuel Directive 2003/30/EC. Biodiesel production volumes were expanded significantly in the recent past and this trend is expected to continue in Europe and other regions such as South East Asia and South America with a further increase in production capacities forecasted at least for the next 5–10 years. When the so-called second generation products, such as sundiesel or biomass-to-liquid fuels, are ready to be launched on the market, the demand on fats and oils for biofuels will not increase further or might decrease again. These new technologies are definitely needed assuming that, even with increasing production volumes for fats and oils, the future bioenergy and biofuel demand cannot be satisfied by this source alone.

In the meantime, the high demands for biodiesel, still further stimulated by subsidies, will create strong competition with the established uses for vegetable oils for nutrition and also for the chemical industry (oleochemistry). A very similar situation is being observed in the case of bioethanol from carbohydrates and/or sugar. The competition between the use of agricultural products for nutrition and energy is one of the reasons why market prices of such agricultural commodities are often subject of extremely high volatility. Other reasons are the increasing demand for food in various regions of the earth, crop yield and financial speculations by investment funds.

The use of renewable resources is only one important part of the future 'green' strategy in industry. What must also be considered are sustainability practices across the entire value chain. This strategy is already applied to palm oil. It is the first time an expert group (The Round Table of Sustainable Palm Oil, RSPO) involving all participants in the industrial agricultural commodity value chain has defined what sustainable agriculture really should mean. The challenging goal to develop, implement and verify credible global standards for sustainable palm oil products has finally been achieved. The principles and criteria for sustainable palm oil are implemented and certificated for products, which are produced according to the standards are available. Cognis was the first chemical supplier in membership and is until today one of the few. Its expertise in natural raw materials enables Cognis to develop concepts with its customers on how to make renewable raw materials sustainable, as almost all of its raw materials from the palm tree are being sourced from RSPO members [4–6].

4.3 Products and Applications

Considering the amphiphilic structure of a typical surfactant with a hydrophilic head group and a hydrophobic tail, it has always been a challenge to attach a carbohydrate molecule as an alternative to polyethylene glycol to a fat and oil derivative, such as a fatty acid or a fatty alcohol [7–11]. Although science has reported numerous ways of making such linkages and has also described a large number of different carbohydrates used in such reactions, it is clear from an industrial perspective that only a few carbohydrates fulfil the criteria of price, quality and availability to be an interesting raw material source. Those are mainly sucrose from sugar beet or sugar cane, glucose derived from starches and sorbitol as the hydrogenated glucose derivative (Table 4.2) [11]. More recently lactose, xylose and the carbohydrate-based residues from straw and hemicellulose processing have been used as well as starting materials for the respective derivatives.

4.3.1 Sorbitan Esters

Sorbitan esters have been known for a couple of decades when the first industrial chemical processes were established for their manufacture. One differentiates between a two-step and a one-step process (Scheme 4.1). In the first process, dehydration of sorbitol occurs in the presence of acid (e.g. NaH_2PO_3) to form 1,4-sorbitan as the main isomer, which is subsequently esterified with fatty acids in a second reaction step with an alkaline catalyst (e.g. K_2CO_3) at 200–250 °C. In the one-step process both reactions are carried out simultaneously [12–14]. Both methods have been developed for industrial-scale production. Depending on the type and amount of fatty acid used, different product compositions, consisting of mixtures of mono-, di- or trisorbitan esters (e.g. laurates, oleates or stearates) are produced with hydrophilic/lipophilic balance (HLB) values in a typical range of 1–8. To modify these relatively hydrophobic materials it is common technology to further derivatize the sorbitan esters by reaction with ethylene oxide to produce sorbitan ester ethoxylates – or polysorbates – with HLB values of 10–17, depending on the number of ethylene oxide units attached (Figure 4.1) [12].

The main manufacturers of sorbitan ester products today are listed in Table 4.3. The total market size for sorbitan esters (including the ethoxylated products) is estimated to

Table 4.2 *Availability of carbohydrate raw materials*[a]

	Production volume (t/a)	Average price (€/kg)[b]
Sucrose	150 000 000	0.25
Glucose	30 000 000	0.30
Sorbitol	650 000	1.80

[a]Source: From Lichtenthaler [11].
[b]Due to various reasons market prices of agricultural commodities are of high volatility – therefore the figures shown can be only rough indications.

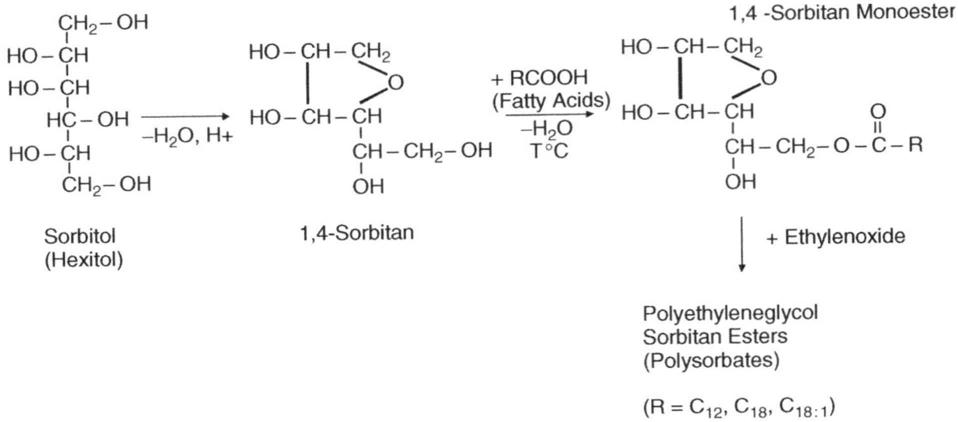

Scheme 4.1 Synthesis of sorbitan esters by intramolecular dehydration of sorbitol and subsequent base-catalysed esterification with fatty acids.

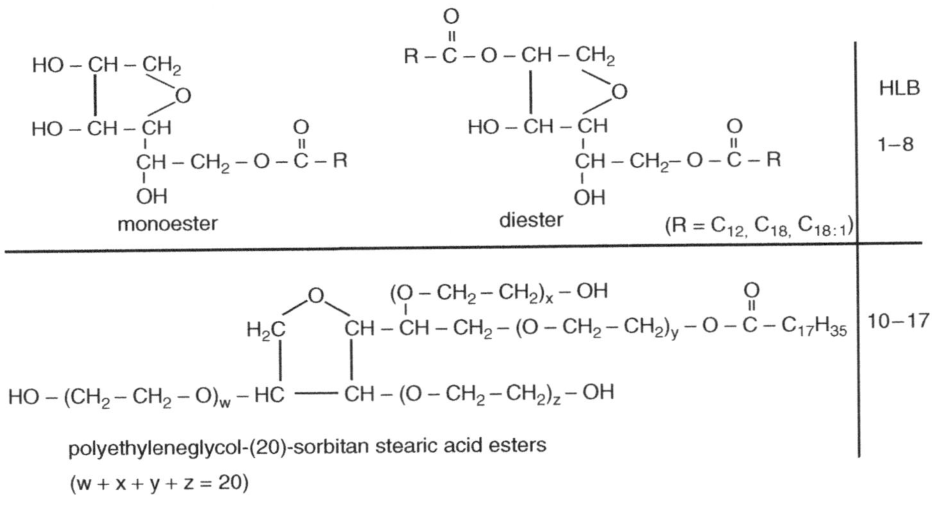

Figure 4.1 Hydrophilicity of sorbitan esters.

be approximately 20 000 tonnes/year. Mainly used as emulsifiers in pharmaceuticals, foods, cosmetic products, pesticides, for emulsion polymerization and explosives and other technical applications, the market size seems to be relatively stable and there is obviously no attempt and no need for further development of this mature technology.

4.3.2 Isosorbide Derivatives

Just recently activities have been started to investigate isosorbide, which is produced from sorbitol by twofold intramolecular ringforming during removal of two moles of water

Table 4.3 Selected suppliers and main fields of application for carbohydrate-based surfactants

	Selected suppliers	Fields of applications	Production capacity, world (t/a)[a]
Sorbitan esters	AkzoNobel, Cognis, Dai-ichi Kogyo Seiyaku, Kao, Riken Vitamin, SEPPIC	Pharmaceuticals, personal care, food, fibre, agrochemicals, coatings, explosives	20 000
Sucrose esters	Cognis, Croda, Dai-ichi Kogyo Seiyaku, Evonik/Goldschmidt, Mitsubishi-Kagaku, Sisterna, Jiangsu Weixi	Food, personal care, pharmaceuticals	<10 000
Alkyl polyglycosides	AkzoNobel, BASF, China Research Institute of Daily Chemical Industry, Cognis, Dai-Ichi Kogyo Seiyaku, Kao, LG, SEPPIC	Personal care, detergents, agrochemicals, I + I	85 000
Others			
Methylglucoside esters	Lubrizol/Noveon	Personal care, pharmaceuticals	<10 000
Anionic alkyl polyglycoside derivatives	Cognis, Cesalpina	Personal care	

[a] Estimated figures based on private communications and literature data, references given in the text.

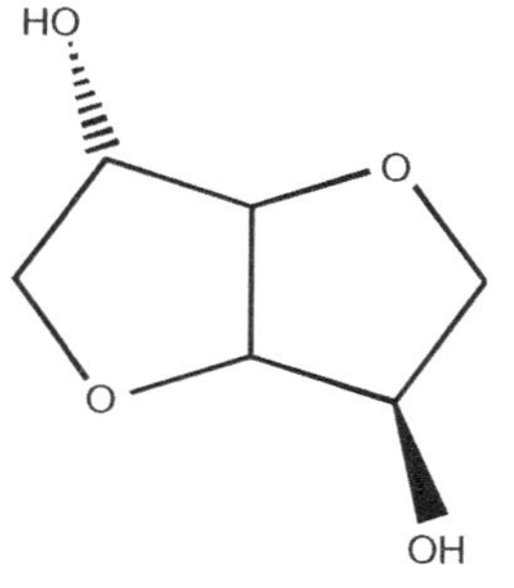

Figure 4.2 *Structure of isosorbide.*

(Figure 4.2). The product is already available on an industrial scale and some derivatives are well known and have already been used for some time, such as the pharmaceutical isosorbide dinitrate [11]. The present research programme, which is named Biohub®, is focusing on the extended use of isosorbide as a starch-based raw material to produce a variety of derivatives for different areas of industrial applications – within an integrated bio-refinery. The programme is funded by the French Agency for Innovation (OSEO) and was approved by the European Commission in 2006. The consortium of the Biohub® programme, led by Roquette, includes major European companies and also small- and medium-sized enterprises, as well as public research laboratories [15].

4.3.3 Sucrose Esters

Sucrose esters are described as very mild with regard to their dermatological properties and are approved as food additives in many countries. As a consequence these products seem to be perfect raw materials for food and cosmetic formulations and their use in those applications as a specialty emulsifier has a long tradition [16–18]. In Asia one can find sucrose esters with a low degree of esterification (or degree of substitution, DS) as well as in special detergent products. The octaesters have been developed by Procter & Gamble and are used as noncaloric fat substitutes in food technology.

Sucrose is a nonreducing disaccharide containing multifunctionalities: three primary alcohols, five secondary alcohols and two anomeric carbons. The manufacture of sucrose esters is quite challenging because the sucrose molecule is very temperature sensitive and, due to its high functionality with eight hydroxyl groups, selectivity in the esterification reaction is difficult to achieve (Scheme 4.2). General chemical pathways are either direct esterification with fatty acid or transesterification of sucrose with fatty acid methyl ester. Typically complex product mixtures consisting of mono-, di-, tri-, tetra- and pentaesters are formed. These products are very hydrophobic and of limited application potential, except for specific emulsification needs. Historically, the studies carried out by the Sugar Research Foundation gave a real boost to the applied research of sugar esters, leading to relatively more hydrophilic products (= higher content of mono- and diester in the product mixture). Those were first commercialized by Dai-Nippon Manufacturing (now Mitsubishi-Kagaku Food Corporation, Ryoto® esters) in the late 1960s for use in the food industry. Manufacturing is based on a transesterification process in solvents, such as dimethylformamide (DMF) or dimethylsulfoxide (DMSO), to solubilize the sucrose, and in the presence of an alkaline catalyst like K_2CO_3 [19, 20]. The purification in order to achieve the permissible levels of residual DMF/DMSO is a critical point and requires a relatively complex process. Therefore several methods have been developed to achieve a higher selectivity in the reaction or provide economical purification procedures using liquid–liquid extraction and crystallization technologies.

Several improvements of the process, such as the reaction with acyl chlorides or the application of two-phase reaction systems with propylene glycol and an emulsifier in order to build a microemulsion, have been described in the literature and patents [21]. Another approach for the synthesis of sugar esters is the use of enzymes. Enzymatic catalysis in the field of carbohydrate chemistry has been actively explored over years in laboratory

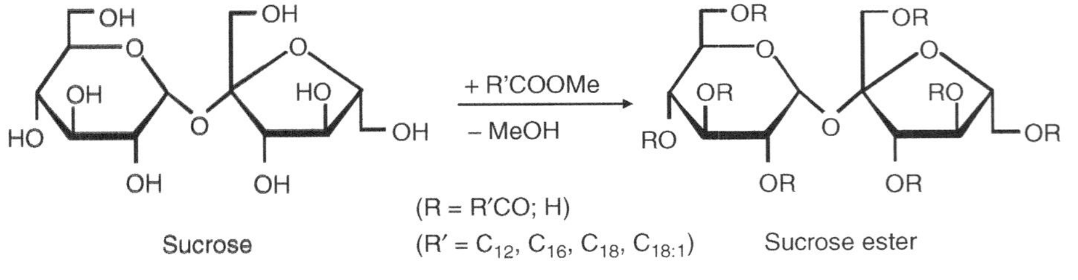

Scheme 4.2 Synthesis of sucrose esters by base-catalysed trans-esterification with fatty acid methyl esters (R'COOMe), usually carried out in solvents (e.g. dimethyl formamide), microemulsions or solvent free.

scale, resulting in a higher degree of selectivity. Whether optimized enzymes, e.g. those that can operate in a low water environment, can revolutionize this type of synthesis to make it industrially feasible is still not known [22–26]. Therefore, solvent-free processes based on transesterification of the molten sucrose and fatty acid ester at temperatures of about 130 °C and in the presence of an alkaline catalyst and an emulsifier (fatty acid soap and/or the sucrose esters themselves) are still preferred [22]. However, the purification step in order to eliminate the residual soap content remains as a difficult process step here. An optimized solvent-free and water-free process was described recently by Cognis [24]. The synthesis is carried out in an organized medium under heterogeneous catalysis (K_2CO_3) with sucrose ester as emulsifier [22]. In a first step, so the theory proposes, the continuous phase is composed of methyl esters. These methyl esters are adsorbed on the catalyst surface (Figure 4.3) and react with the sucrose esters already present to form a sucrose polyester as an intermediate [22, 24]. In a second step, the sucrose polyesters react with the molten sucrose in an acyl transfer reaction to obtain the final sucrose ester composition with an optimized DS of about 2.5 (Figure 4.4). The final product is mixed with a cosmetic emollient and refined. Applied as a cosmetic emulsifier, it improves

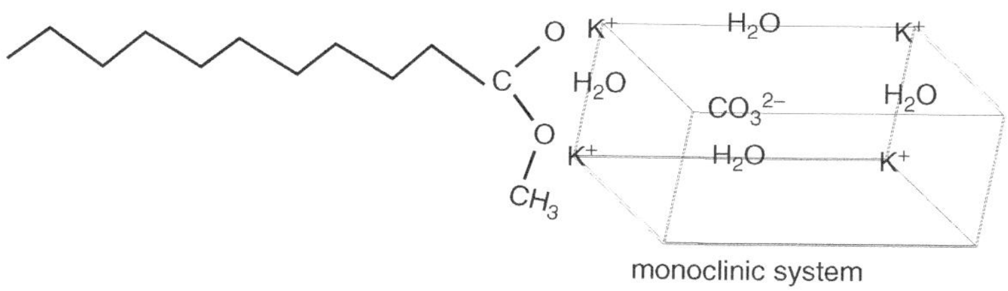

Figure 4.3 Surface reaction with preadsorption of fatty acid methylesters on K_2CO_3, according to Mouloungui, 1987 25.

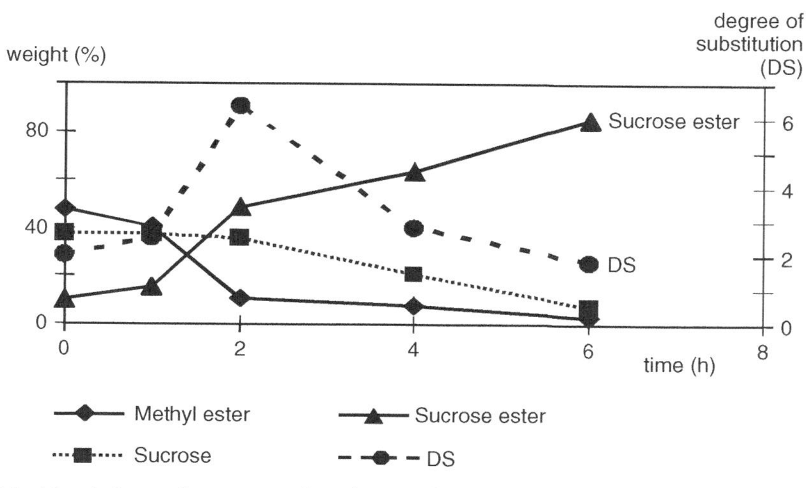

Figure 4.4 Mass balance of reagents and products in the esterification of sucrose with fatty acid methyl ester under solvent- and water-free conditions (125 °C, K_2CO_3, sucrose ester as emulsifier).

the stability of the final product formulation and the sensorial performance by forming lamellar liquid crystal structures [23].

Today, the major producers for sucrose esters are Dai-Ichi Kogyo Seiyaku and Mitsubishi in Japan, Croda and Procter & Gamble in the US, Sisterna in the Netherlands and Cognis in Germany (Table 4.3). It seems that the existing production capacities are much higher than the actual market potential, which is estimated to be less than 10 000 tonnes/year. However, demand and market volume could increase substantially, if reaction processes – especially for the synthesis of high mono- and diester products – can be further optimized.

4.3.4 Glucose-Derived Surfactants

The first step in overcoming the problem of nonselective derivatisation of carbohydrates was achieved when Emil Fischer discovered the reaction of glucose with alcohol to form alkyl glucosides [27]. The glucosidation reaction is highly selective due to the hemiacetal function in the glucose molecule and the resulting high reactivity of the hydroxyl-group at C1. The same is true for the synthesis of fatty acid glucamides. Here the glucose molecule reacts initially with methylamine, which, after hydrogenation, gives selectively the glucamine as an intermediate [28, 29]. Further derivatisation with fatty acid methyl ester leads to the desired product.

4.3.4.1 Alkyl Polyglycosides

The first syntheses of alkyl polyglycosides were carried out more than 100 years ago [27]. In the course of further developments the reaction of glucose with alcohols was applied to longer-chain alcohols with alkyl chains from C8 to C16. The result of the reaction is a complex mixture of alkyl mono-, di-, tri- and oligoglycosides as a mixture of α- and β-anomers (Scheme 4.3). Therefore, the industrial products are called alkyl polyglycosides. The products are characterized by the length of the alkyl chain and the average number of glucose units linked to it – the degree of polymerization (DP) [30–35].

The crucial point with regard to the development of an industrial process was to establish reaction conditions, which allowed the manufacturing of high-quality products under safe and economically acceptable conditions. This was achieved by optimizing the reaction parameters of temperature, pressure, reaction time and ratio of glucose to fatty alcohol. Of equal importance was the design of a special distillation technology to remove the excess fatty alcohol as smoothly as possible, as well as an appropriate bleaching and stabilization in the final treatment step (Scheme 4.4). This so-called direct synthesis of alkyl polyglycosides nowadays is the preferred manufacturing mode.

The breakthrough in the production of medium-chain ($C_{12/14}$) alkyl polyglycoside occurred in 1992 with the inauguration of a large-scale production plant for APG® surfactants by Henkel Corporation in the USA (now Cognis Corp.) and in 1995 with the opening of a second plant of equal capacity by Henkel in Germany (now Cognis GmbH). Cognis was the first company to offer medium-chain ($C_{12/14}$) alkyl polyglycosides on an industrial scale at the required quality. Currently Cognis is the supplier with the largest

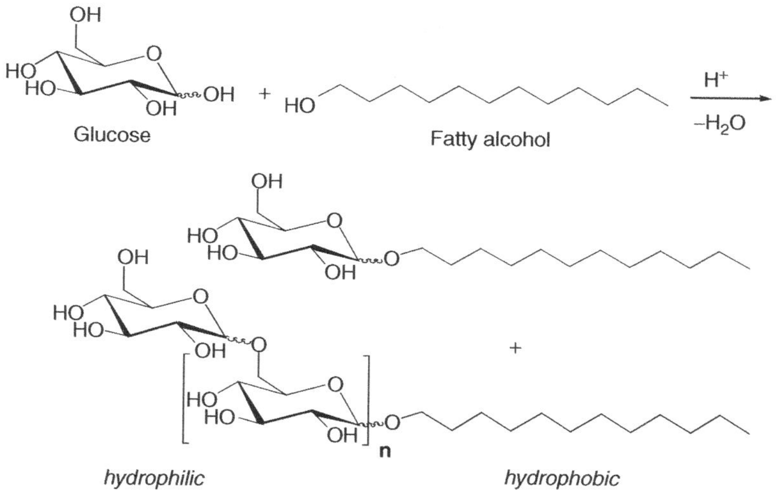

Scheme 4.3 Synthesis of alkyl polyglycosides by acid-catalysed acetalization of glucose in molar excess of fatty alcohol and removal of water under vacuum at 100–120 °C.

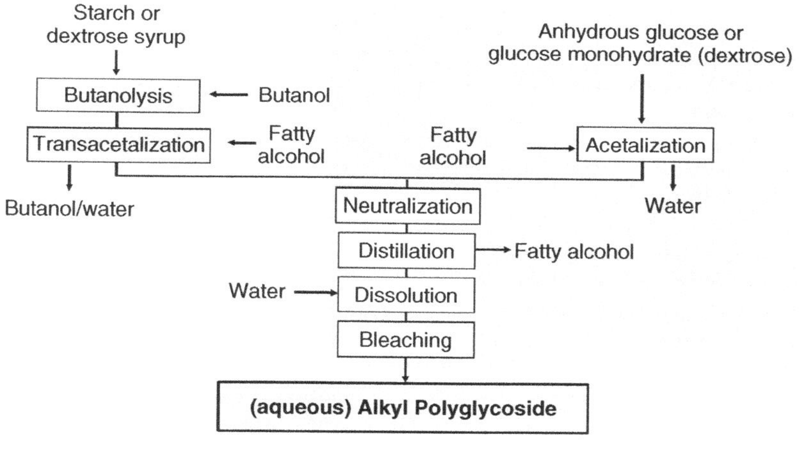

Scheme 4.4 Manufacturing process for alkyl polyglycosides.

capacity worldwide. Other manufacturers include AkzoNobel, BASF, China Research Institute of Daily Chemical Industry, Dai-Ichi Kogyo Seiyaku, Kao, LG and SEPPIC. The overall capacity available worldwide is estimated to be approximately 85 000 tonnes/year (Table 4.3).

4.3.4.2 Properties of Alkyl Polyglycosides

By combining fatty alcohol from vegetable oil (e.g. from palm kernel or coconut) and glucose from starch (e.g. from potato, wheat or corn) as raw materials it has for the first time become possible to offer commercially significant amounts of nonionic surfactants that are completely based on renewable resources without making any compromises in performance. With regard to their ecological, toxicological and dermatological properties, alkyl polyglycosides can be considered as surfactants with extraordinary product safety. This has been proven in a series of detailed investigations. The results are published in several papers by industry and research institutes [30–36]. The best practice manufacturing includes:

- solvent-free reaction;
- continuous removal and recycling of excess fatty alcohol;
- high yield and maximum atom economy;
- very low emissions.

The main characteristics of process and product regarding environmental impact are summarized in Table 4.4, applying the '12 Principles of Green Chemistry' according

Table 4.4 *The 12 principles of green chemistry – applied to alkyl polyglycosides (Glucopon®, Plantacare® and Plantaren® types)*[a]

1. Prevention – prevent waste instead of clean-up	√	Fully optimized manufacturing process with the re-use of the excess fatty alcohol
2. Atom economy – synthetic methods to maximize the incorporation of all materials used	√	Maximum utilization: reaction 100% – water => >90%
3. Less hazardous chemical syntheses – reagents and products with low human harm	√	The process is safe including auxiliaries (no solvent used, see principle 5: proven favourable tox and ecotox data (cf. References [45–47])
4. Designing safer chemicals – guarantee desired function while minimizing toxicity	√	By introducing glucose, ethylene oxide could be avoided
5. Safer solvents and auxiliaries – avoid wherever possible	√	No solvent used in the process; only water for the dilution of the final product
6. Design for energy efficiency – preferably ambient pressure and temperature if possible	√	Reaction under ambient pressure and continuous process for the distillation of excess fatty alcohol minimize the energy consumption
7. Use of renewable feedstocks	√	Raw materials used are 100% renewable (glucose and vegetable fatty alcohol)
8. Reduce derivatives – avoid blocking/protection groups, if possible	√	No protection groups are used
9. Catalysis	√	Acid is used in catalytic quantities
10. Design for degradation – favourable biodegradation properties after use	√	Aerobic and anaerobic degradation proven (cf. Reference [39])
11. Real-time analysis for pollution prevention – process monitoring	√	Process control via process information (PI) system
12. Inherently safer chemistry for accident prevention	√	Process is inherently safe, no runaway possible due to raw materials

[a] Cognis, unpublished results.

to Anastas and Warner for Glucopon®, Plantaren® and Plantacare® types [36]. As shown, these alkyl polyglycosides fulfil all the criteria. In fact, within the framework of international regulations concerning eco-friendly products, alkyl polyglycosides meet the requirements for highly accepted green labels such as Ecocert, the EU Eco-Flower, Green Seal and many others. In addition to its ecological footprint [37–39], alkyl polyglycosides are not toxic or harmful to human health and show a lower skin irritation than other surfactants [40–42]. A comparative study [42] of different surfactants showed that alkyl polyglycosides possess superior mildness to other surfactants found in the market, confirming the common association of 'greenness' with mildness. In two of the tests applied (red blood cell test (RBC) and epicutaneous patch testing (ECT)) the mucous membrane/ocular irritation potential of the selected surfactants and the primary skin irritation in humans, respectively, was assessed.

The main applications for the $C_{12/14}$ alkyl polyglycosides are liquid dishwashing agents and detergents and personal care products and for the $C_{8/10}$ (or branched C_8) alkyl polyglycosides it is hard surface cleaners, agrochemicals and products for industrial and institutional cleaning. In general, alkyl polyglycosides have been shown to be very efficient components in laundry detergents, manual dishwashing detergents and cleaners with distinct synergies with other surfactants and mildness [43, 44]. Table 4.5 summarizes the various applications for short-chain ($C_{8/10}$) and medium-chain ($C_{12/14}$) alkyl polyglycosides in the detergents field. In personal care products, alkyl polyglycosides represent a new concept in compatibility and care. They may be combined with conventional components and can even replace them in new types of formulations, leading to a rich

Table 4.5 *Potential application areas of alkyl polyglycosides in home care products*

Application area	Types of application	Properties[a]
Laundry	Powder detergent Laundry tabs Liquid detergent Special detergent	Liquid laundry Cleaning (dirt/stain removal) Good detergency for special stains (lip stick, oils) Fast penetration into oils and fibres Excellent hydrotrope performance Care – mild to the skin Convenience – rapid dissolution
Dishwash	Manual dishwashing detergents	Dishwash and cleaners Excellent emulsification properties for different types of oily soils to avoid re-soiling Safe to skin
Hard surface cleaners	All purpose cleaners Bathroom cleaners Kitchen cleaners, power cleaners Glass cleaners, floor cleaners Wipes	Streak-free cleaning Good foam profile
Industrial and institutional	Alkaline industrial cleaners	

[a] Selected properties relevant for the application in home care products.

Table 4.6 Manifold applications of alkyl polyglycosides in personal care products

Types of uses and claims	Properties[a]
Main or co-surfactant for all kind of cleansing preparation: shampoo, shower bath, foam bath, facial wash, shampoo concentrates	Extremely mild Readily biodegradable Good foamer Viscosity modifier
Emulsions: hair rinses, hair colours	Emulsifying
Setting lotions, Styling gels	Setting properties Curl retention
Permanent wave formulations	No hydrolysis at alkaline pH value
Volume and body formulations Shampoos for fine or damaged hair	Improves tensile strength of hair Modifies interactions of dry hair
Pore and wrinkle deep cleansing	Cleansing power

[a]Selected properties relevant for the application in personal care products.

spectrum of supplementary effects [45–47]. With regard to foam, alkyl polyglycosides are comparable to betaines and sulfosuccinates and can stabilize the foam of anionic surfactants in hard water and in the presence of sebum. The broad spectrum of applications in the personal care field is summarized in Table 4.6.

To demonstrate the large performance spectrum of alkyl polyglycosides, one more application should be mentioned as well. Alkyl polyglycosides ($C_{8/10}$ and $C_{12/14}$) have been shown to be substitutes for alkyl phenol ethoxylates in agrochemical formulations. They lead to higher salt tolerances and show good results as adjuvants in several post-applied herbicides, such as the control of giant foxtail in soya beans with Assure II (DuPont) and control of common lambsquarters in soya bean with Pursuit (American Cyanamid). The short-chain products ($C_{8/10}$ and C_{9-11}) have now been approved as inert ingredients by the United States Environmental Protection Agency (EPA) [48].

4.3.4.3 Fatty Acid Glucamides

The synthesis to produce the fatty acid glucamides involves the reaction of glucose with methyl amine under reductive conditions to form the corresponding *N*-methyl glucamine. In a subsequent reaction step this intermediate is converted with fatty acid methyl ester to the corresponding fatty acid amide. Compared to the alkyl polyglycosides, fatty acid glucamides are composed of only one single carbohydrate molecule attached to the fatty acid chain. This is one reason why fatty acid glucamides are less soluble and tend to crystallize more easily from aqueous solutions. Scheme 4.5 shows the manufacturing scheme for the production of fatty acid glucamides. To avoid significant amounts of unreacted *N*-methyl glucamine, which could be considered as potential precursors for nitrosamines, Procter & Gamble has developed an optional reaction with acetic anhydride in the product finishing. Free secondary amines can be acetylated in this step and the resulting acetates can remain in the final product [49, 50].

For several years fatty acid glucamides with $C_{12/14}$ and $C_{16/18}$ alkyl chains have been used exclusively in powdered and liquid detergents and liquid dishwashing agents.

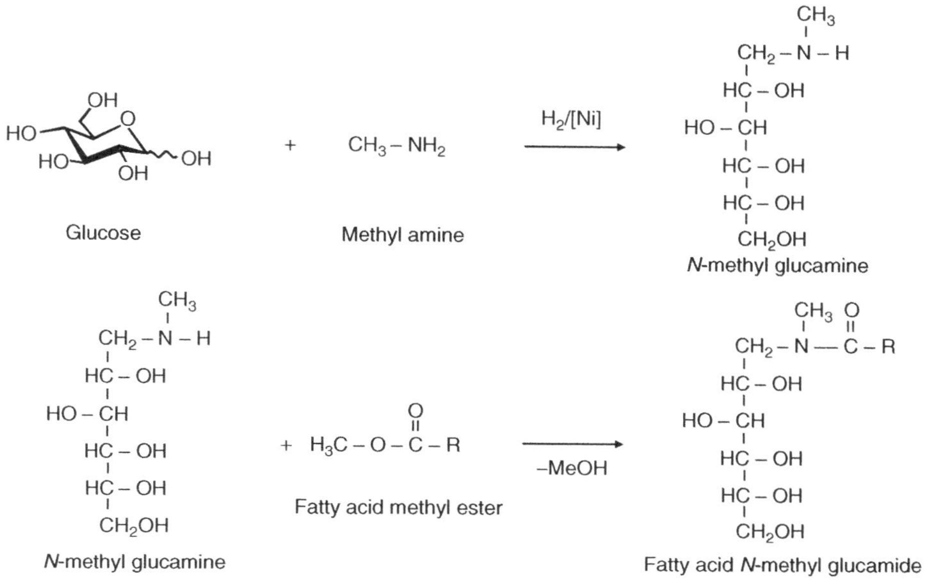

Scheme 4.5 Two-step synthesis of fatty acid glucamides by reductive alkylation of methyl amine with glucose to obtain N-methyl glucamine, which is acylated by a base-catalysed reaction with fatty acid methyl ester in a second step.

A couple of years ago their use was limited to a few applications only and at present they are rarely found in any application. Basically they have disappeared from the market. Derivatives of glucamides, such as glucamine oxides and betaines or anionic glucamides, as well as bifunctional glucamides, have also been described in the literature [51], but are hardly found in any market product so far.

4.3.5 Derivatives of Alkyl Polyglycosides

Since alkyl polyglycosides are available in sufficient quantities their use as a raw material for the development of specialty surfactants is arousing considerable interest. The derivatisation of alkyl polyglycosides is currently being pursed with great commitment, with the goal to modify the surfactant properties [52]. A broad range of alkyl polyglycoside derivatives can be obtained by using relatively simple methods, e.g. nucleophilic substitution. Besides the reaction to esters or ethoxylates, ionic products, such as sulfates and phosphates, can also be synthesized (Scheme 4.6). However, until now there have only been a few products that are established in the market: methyl glucoside esters and selected anionic derivatives of alkyl polyglycosides.

4.3.5.1 Methyl Glucoside Esters

Esterification of methyl glucoside (obtained from starch or glucose and methanol) with methyl esters of stearic or oleic acid leads to the desired products (Scheme 4.7). By

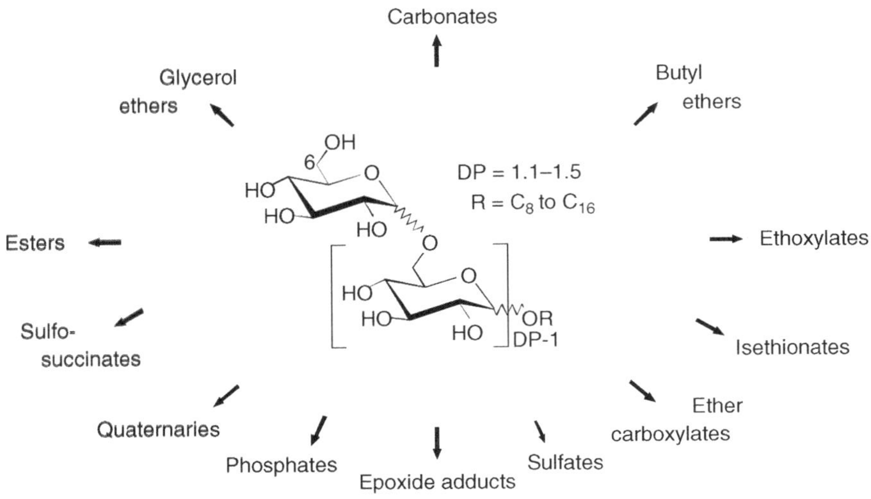

Scheme 4.6 Overview of alkyl polyglycoside derivatives.

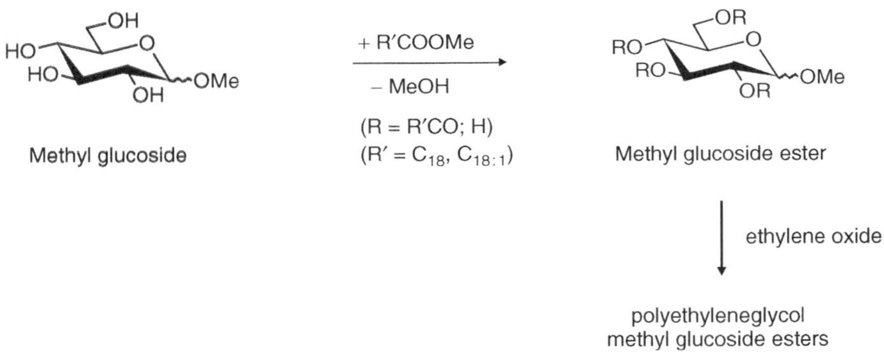

Scheme 4.7 Synthesis of methyl glucoside ester by base-catalysed (K_2CO_3) trans-esterification of methyl glucoside with fatty acid methyl ester (R′COOMe) at 120–160 °C.

subsequent ethoxylation the respective polyethyleneglycol methyl glucoside esters are obtained. Due to their relatively lipophilic character, methyl glucoside esters are, in contrast to alkyl polyglycosides with the same hydrophobic chain length, hardly soluble in water, but exhibit excellent emulsification properties [16, 17, 53]. They have found application as emollients, moisturizing and emulsifying agents and thickeners for cosmetics. The hydrocarbon length and DS can be varied to adjust the rheological properties and obtain specific water-in-oil emulsification behaviour. A major manufacturer for methyl glucoside esters and ethoxylated derivatives is Lubrizol with its Noveon® line. The total market size, including the ethoxylated products, is estimated to be less than 10 000 tonnes/year (Table 4.3).

4.3.5.2 *Anionic Derivatives of Alkyl Polyglycosides*

Cesalpina Chemicals, a subsidiary of Lamberti Spa., Italy, introduced some years ago three nonionic alkyl polyglycoside esters, namely citrates, sulfosuccinates and tartrates, that can be used in personal care applications [54]. The synthesis starts with alkyl polyglycoside (typically $C_{12/14}$), which is esterified with citric acid, maleic anhydride and tartaric acid, respectively. The succinate is further sulfonated. Structures are shown in Figure 4.5 [55].

More recently, alkyl polyglucoside carboxylate (International Nomenclature of Cosmetic Ingredients (INCI) name sodium lauryl glucose carboxylate (and) lauryl glucoside) has been introduced to the market by Cognis as a new anionic surfactant with an excellent performance profile for personal care cleansing applications. In shampoo and shower bath formulations the anionic surfactant shows a clearly better foaming behaviour compared to nonionic sugar surfactants. In body wash applications it improves the sensorial effects. These properties make the product suitable for several cosmetic applications, e.g. mild facial wash gel, mild baby shampoo, mild body wash for sensitive skin, wet wipes and special sulfate-free shampoo applications. A new industrial process based on the reaction of sodium monochloroacetate with

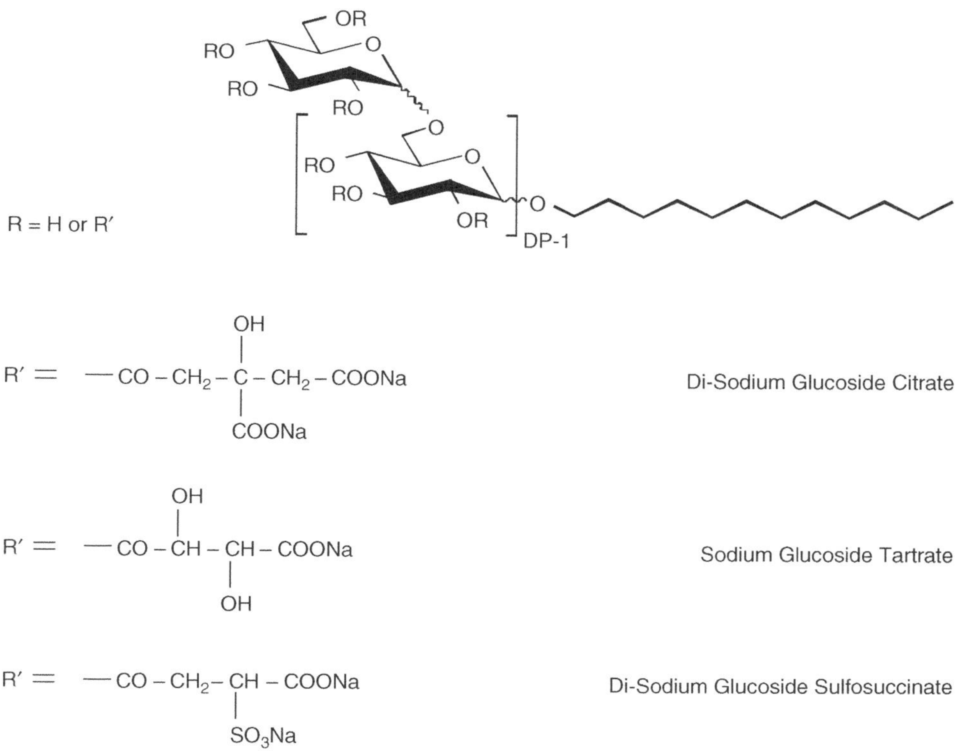

Figure 4.5 *Examples of anionic alkyl polyglycoside derivatives.*

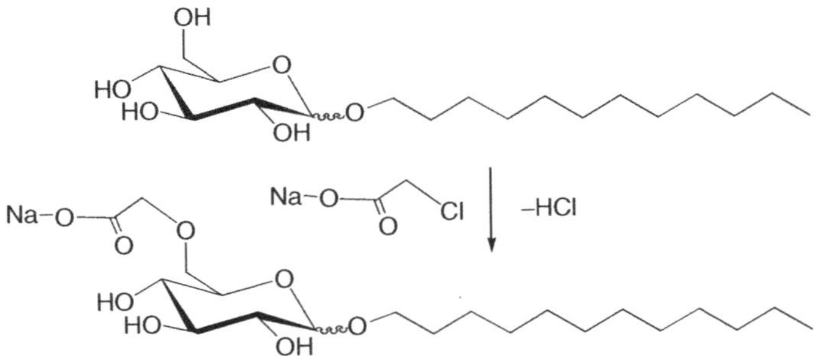

Scheme 4.8 Synthesis of alkyl polyglycoside carboxylate.

aqueous alkyl polyglycoside (without additional solvents) enables the manufacturing of this product in an economically and ecologically favourable way (Scheme 4.8) [56].

4.3.6 Other Carbohydrates and Proteins as Raw Materials in the Production of Plant-Based Surfactants

4.3.6.1 Hemicellulose

Recently, development activities for alkyl polypentosides have been described. The products are based on hemicellulose, which, after hydrolysis, leads to a mixture of glycosides (mainly pentoses such as xylose and arabinose). These pentoses are transformed into the desired alkyl polypentosides via the reaction with fatty alcohols under standard reaction conditions. The products are reported to have similar properties to alkyl polyglycosides. These developments have been carried out by Wheatoleo, a joint venture between Oleon and A.R.D./Soliance [57].

4.3.6.2 Acylated Proteins (Protein Fatty Acid Condensates)

The use of proteins as raw material for personal and home care products has been known for more than 60 years [58]. Both animal (e.g. leather waste) and plant-based proteins are available as raw materials. Nowadays mostly plant-based proteins (e.g. from wheat, soya bean, rice, peas) are used. Typically the natural proteins are degraded by hydrolysis (either chemically or enzymatically) and the respective protein hydrolysates are obtained. After purification and work-up they are used for skin and hair care formulations. The main claims are protection and care for hair and skin. Subsequent quarternization of the protein hydrolysates leads to products with high substantivity and conditioning effects, whereas acylation of the protein hydrolysates with fatty acids results in protein surfactants, the so-called protein fatty acid condensates. In the latter case, as already described for the sugar-based surfactants, the products are based completely on natural raw materials, namely fatty acids (from vegetable oil) and the protein or protein hydrolysate as the hydrophilic part in the surfactant molecule (Figure 4.6). The

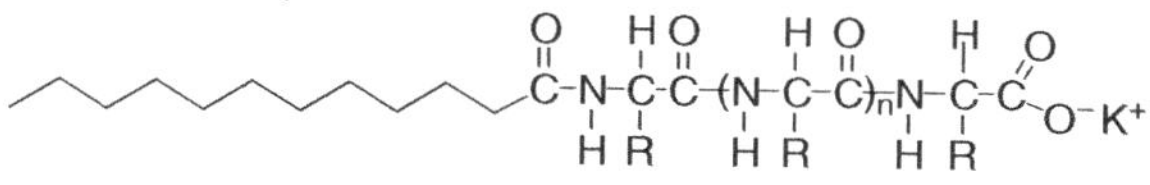

Figure 4.6 *Structure of protein hydrolysate fatty acid condensates.*

condensation is carried out by reaction of the protein hydrolysate with fatty acid chloride under 'Schotten–Baumann' conditions using water as the solvent. Products are obtained that have an excellent skin compatibility and additionally a good cleaning effect (particularly on the skin) and, in combination with other surfactants, lead to an increase in performance (synergy). For instance, even small additions of the acylated protein hydrolysate improve the skin compatibility of other surfactants. An explanation for this protective effect could be the amphoteric behaviour of the product and, as a result, an interaction of the protein fatty acid condensate with skin collagen. This could lead to the formation of a protective layer, which reduces the excessive attack of other surfactants on the upper layers of the skin, their degreasing effect and the direct interaction of anionic surfactants with the skin.

In the personal care market, fatty acid derivatives of proteins and amino acids (in particular acyl glutamates) have found broad uses in recent years and are mainly used in mild shower and bath products, mild shampoos, surfactant-based face cleansers, cold-wave preparations and fixatives, baby wash formulations, as well as special emulsifiers for leave-on products.

4.4 Conclusion

Comparing the existing carbohydrate-based surfactants it is clear from a chemical point of view that the generation of an amphiphilic structure can be perfectly accomplished by using glucose as the carbohydrate source – so far no methods for the selective derivatisation of sorbitol and sucrose have been developed on an industrial scale. This, combined with outstanding performance, multifunctionality, competitive price, high product safety and an environmentally favourable manufacturing process, could be one reason why alkyl polyglycosides are the most successful carbohydrate-based surfactants nowadays.

Acknowledgements

Ryoto is a registered trademark of Mitsubishi-Kagaku Food; APG surfactants, Glucopon, Plantacare, Plantaren are registered trademarks of the Cognis group; and Noveon is a registered trademark of Lubrizol.

References

1. Falbe, J. (ed.) (1986) *Surfactants in Consumer Products – Theory, Technology and Application*, Springer, Berlin.

2. Baumann, H. and Biermann, M. (1993) Neue Tenside aus nativen Rohstoffen, in *Nachwachsende Rohstoffe, Perspektiven für die Chemie* (eds M. Eggersdorfer, S. Warwel and G. Wulff), Wiley-VCH Verlag GmbH, Weinheim, pp. 33–55.
3. Hill, K. and Rhode, O. (1999) Sugar-based surfactants for consumer products and technical appllications. *Fett/Lipid*, **101** (101), 25–33.
4. Hill, K. (2007) Industrial development and application of biobased oleochemicals. *Pure Appl. Chem.*, **79** (11), 1999–2011.
5. Sauthoff, H. (2005) Bioenergy and Biofuels – Opportunity or Threat for Oils and Fats Trading? Handout FOSFA 'Contact Day', London, September 8, 2005.
6. Noweck, K. (2007) Die Energie der Zukunft kommt aus Pflanzen/Verfahrenstechnik und Anlagenkonzepte. *CHEManager*, **2**, 11.
7. Schulz, P. (1992) Chemistry and technology of alkyl glycosides. *Chimica Oggi*, **10** (8–9), 33–38.
8. Biermann, M., Schmid, K. and Schulz, P. (1993) Alkylpolyglucoside – Technologie und Eigenshaften. *Starch/Stärke*, **45**, 281–288.
9. Ruback, W. and Schmidt, S. (1996) Alkyl polyglucoside, a carbohydrate-based surfactant, in *Carbohydrates as Organic Raw Materials III* (eds H. van Bekkum, H. Röper and A. G. J. Vorhagen), Wiley-VCH Verlag GmbH, Weinheim, pp. 231–253.
10. Brancq, B. (1992) Development and trends of sugar derived surfactants. *SÖFW-Journal*, **118**, 905–908.
11. Lichtenthaler, F. W. (2007) Carbohydrates as renewable raw materials: a major challenge of green chemistry, in *Methods and Reagents for Green Chemistry* (eds P. Tundo, A. Perosa and F. Zecchini), Chapter 2, John Wiley & Sons, Ltd., Hoboken, pp. 23–63.
12. Biermann, M., Lange, F., Piorr, R. *et al.* (1987) Synthesis of surfactants, in *Surfactants in Consumer Products – Theory, Technology and Application* (ed. J. Falbe), Chapter 3, Springer-Verlag, Berlin.
13. Ropuszynski, S. and Sczesna, E. (1990) Dehydration of D-sorbitol in the presence of sodium phosphates. *Tenside Surfactant Deterg.*, **27**, 350–351.
14. Ropuszynski, S. and Sczesna, E. (1985) The effect of catalysts on the synthesis of fatty acid esters of anhydrosorbitols (Spans). *Tenside Deterg.*, **22**, 190–192.
15. Rupp-Dahlem, C. (2007) Bio refinery concept from cereals: The BioHub® Program, in Renewable Raw Materials for Industry: Contribution to Sustainable Chemistry, Brussels .
16. Desai, N. B. (1990) Esters of sucrose and glucose as cosmetic materials. *Cosmet. Toiletries*, **105**, 99–107.
17. Nakamura, S. (1997) Using sucrose esters as food emulsifiers. *Inform*, **8**, 866–874.
18. Cecchin, M. (2001) Domaines d'Application des Sucroesters et Sucroglycérides, DESS Ingénierie Documentaire, ENSSIB de Lyon, p. 87.
19. Osipow, L. I., Snell, F. D. and Finchler, W. C. (1956) Methods of preparation of fatty acid esters of sucrose. *Ind. Engng Chem.*, **48**, 1459.
20. Hass, H. B. (1968) In *Sugar Esters* (ed. F. D. Snell), Chapters 1–7, Sugar Research Foundation, New York.
21. Osipow, L. I. and Rosenblatt, W. (1967) Micro-emulsion process for the preparation of sucrose esters. *J. Am. Oil. Chem. Soc.*, **44** (5), 307.
22. Claverie, V. (1998) Mise au Point d'un Nouveau Procédé de Synthèse d'Esters de Saccharose de Degré de Substitution Contrôlé, Thèse ENCT de Toulouse.
23. Pian, V., Le Hen-Ferrenbach, C., Beuché, M. and Roussel, M. (2005) *New Sucrose Polystearate Effective Emulsifier Combining Skin Touch with Moisturizing Properties*, ISFCC Florence.
24. Le Hen-Ferrenbach, C., Beuché, M. and Roussel, M. (2005) 054432, Cognis.
25. Mouloungui, Z. (1987) Réactions en Milieu Hétérogène Solide/Liquide Faiblement Hydraté, Thèse d'état INPT de Toulouse.

26. Drummond, C. J., Fong, C., Krodkiewska, I. *et al.* (2003) Sugar fatty acid esters, in *Novel Surfactants: Preparation, Applications, and Biodegradability*, 2nd edn, Revised and Expanded (ed. K. Holmberg), Surfactant Science Series 114, Chapter 3, Marcel Dekker, New York, pp. 95–128.
27. Fischer, E. (1893) Über die Glucoside der Alkohole, *Ber. Deutsch. Chem. Ges.*, **26**, 2400–2412.
28. Jürges, P. and Turowski, A. (1996) Vergleichende Untersuchung von Zuckerestern, N-Methylglucamiden und Glycosiden am Beispiel von Reinigungsprodukten, in *Perspektiven Nachwachsender Rohstoffe in der Chemie* (ed. H. Eierdanz), Wiley-VCH Verlag GmbH, Weinheim, pp. 61–70.
29. Kelkenberg, H. (1988) Detergenzien auf Zuckerbasis, *Tenside Surfactant Deterg.*, **25**, 8–13.
30. Hill, K., von Rybinski, W. and Stoll, G. (eds) (1997) *Alkyl Polyglycosides – Technology, Properties and Applications*, Wiley-VCH Verlag GmbH, Weinheim.
31. von Rybinsky, W. and Hill, K. (1998) Alkyl polyglycosides – properties and applications of a new class of surfactants. *Angew. Chem. Int. Ed.*, **37** (10), 1238–1345.
32. von Rybinski, W. and Hill, K. (1998) Alkyl polyglycosides, in *Novel Surfactants, Preparations, Applications, and Biodegradability* (ed. K. Holmberg), Chapter 2, Marcel Dekker, New York, pp. 31–85.
33. Koeltzow, D. E. and Urfer, A. D. (1984) Preparation and properties of pure alkyl glucosides, maltosides and maltotriosides, *J. Am. Oil Chem. Soc.*, **61**, 1651–1655.
34. Straathof, A. J. J., van Bekkum, H. and Kieboom, A. P. G. (1988) Solid state and solution properties of octyl D-glucopyranosides, *Starch/Stärke*, **40**, 438–440.
35. Böcker, Th. and Thiem, J. (1989) Synthese und Eigenschaften von Kohlenhydrattensiden, *Tenside Surfactant Deterg.*, **26**, 318–324.
36. Anastas, P. T. and Warner, J. C. (1998) *Green Chemistry: Theory and Practice*, Oxford University Press, New York.
37. Steber, J., Guhl, W., Stelter, N. and Schröder, F. R. (1997), Ecological evaluation of alkyl polyglycosides, in *Alkyl Polyglycosides – Technology, Properties and Applications*, Chapter 11, Wiley-VCH Verlag GmbH, Weinheim, pp. 177–188.
38. Stalmans, M., Matthijs, E., Weeg, E. and Morris, S. (1993) The environmental properties of glucose amide – a new nonionic surfactant, *SÖFW-Journal*, **119**, 794–808.
39. Willing, A., Messinger, H. and Aulmann, W. (2004) Ecology and toxicology of alkyl polyglycosides, in *Handbook of Detergents* (ed. U. Zoller), Part B, Chapter 18, Marcel Dekker, New York, pp. 487–521.
40. Aulmann, W. and Sterzel, W. (1997) Toxicology of alkyl polyglycosides, in *Alkyl Polyglycosides – Technology, Properties and Applications*, Chapter 9, Wiley-VCH Verlag GmbH, Weinheim, pp. 151–160.
41. Matthies, W., Jackwerth, B. and Kraechter, H.-U. (1997) Dermatological properties of alkyl polyglycosides, in *Alkyl Polyglycosides – Technology, Properties and Applications*, Chapter 10, Wiley-VCH Verlag GmbH, Weinheim, pp. 169–176.
42. Mehling, A., Kleber, M. and Hensen, H. (2007) Comparative studies on the ocular and dermal irritation potential of surfactants. *Food Chem. Toxicol.*, **45** (5), 747–758.
43. Schmid, K. (1996) Tenside aus nachwachsenden Rohstoffen für Wasch- und Reinigungsmittel, in *Perspektiven Nachwachsender Rohstoffe in der Chemie* (ed. H. Eierdanz), Wiley-VCH Verlag GmbH, Weinheim, pp. 41–60.
44. Andree, H., Hessel, J. F., Krings, P. *et al.* (1997) Alkyl polyglycosides in hard surface cleaners and laundry detergents, in *Alkyl Polyglycosides – Technology, Properties and Applications*, Chapter 6, Wiley-VCH Verlag GmbH, Weinheim, pp. 99–130.
45. Tesmann, H., Kahre, J., Hensen, H. and Salka, B. A. (1997) Alkyl polyglycosides in personal care products, in *Alkyl Polyglycosides – Technology, Properties and Applications*, Chapter 5, Wiley-VCH Verlag GmbH, Weinheim, pp. 71–88.

46. Ansmann, A., Busch, B., Hensen, H. *et al.* (2005) Personal care formulations, in *Handbook of |Detergents, Part D: Formulations* (ed. M. S. Showell), Chapter 7, CRC Press and Taylor & Francis Group, Boca Raton, Florida, p. 207–260.
47. Weuthen, M., Kawa, R., Hill, K. and Ansmann, A. (1995) Long chain alkyl polyglycosides – a new generation of emulsifiers. *Fat Sci. Technol.*, **97**, 209–211.
48. Garst, R. (1997) Alkyl polyglycosides – new solutions for agricultural applications, in *Alkyl Polyglycosides – Technology, Properties and Applications*, Chapter 7, Wiley-VCH Verlag GmbH, Weinheim, pp. 131–138.
49. Laughlin, R. G., Fu, Y.-C., Wireko, F. C. *et al.* (1998) The physical science of *N*-dodanoyl-*N*-methlyglucamine, in *Novel Surfactants, Preparations, Applications, and Biodegradability* (ed. K. Holmberg), Chapter 1, Marcel Dekker, New York, pp. 1–30.
50. Scheibel, J. J., Connor, D. S., Shumate, R. E. and St. Laurent, J. C. T. R. B. (1990) Process for preparing N-alkyl polyhydroxy amines and fatty acid amides therefrom in hydroxy solvents, EP-B 0558515, US-A 598462; Procter & Gamble, *Chem. Abstr.*, **117** (1992), 114045.
51. Beck, R. (2002) Glucamide and glucamine surfactants. *Chimica Oggi*, May, 45–47.
52. Rhode, O., Weuthen, M. and Nickel, D. (1997) New nonionic derivatives of alkyl polyglycosides – synthesis and properties, in *Alkyl Polyglycosides – Technology, Properties and Applications*, Chapter 8, Wiley-VCH Verlag GmbH, Weinheim, pp. 139–149.
53. Anonymous (1998) Tensioactifs naturel: sucroesters et APG en forte croissance. *Parfums Cosmet. Actual.*, **139**, 44–48.
54. Bernardi, P., Fornara, D., Paglino L. and Verzotti, T. (1996) Eucarol APG series: a new generation of surfactants derived from glucose, in The 1st Concise Surfactants Directory, p. 15.
55. Anonymous (1996) *Chem. Market. Rep.*, June 17, 23.
56. Behler, A., Hensen, H. and Seipel, W. (2004) Proceedings of the 6th World Surfactant Congress, CESIO June, 2004, Berlin.
57. ICIS (2007) ICIS Chemical Business.
58. Sander, A., Eilers, E., Heilemann, A. and von Kries, E. (1997) Production and appllication of acylated protein hydrolyzates. *Fett/Lipid*, **99** (4), 115–120.

5

Amino Acids, Lactic Acid and Ascorbic Acid as Raw Materials for Biocompatible Surfactants

Carmen Moran[1], Lourdes Perez[2], Ramon Pons[2], Aurora Pinazo[2] and Maria Rosa Infante[2]

[1]*Chemistry Department, Coimbra University, Coimbra, Portugal*
[2]*Instituto de Química Avanzada de Cataluña, Barcelona, Spain*

5.1 Introduction

Society is becoming progressively more worried with safety issues while wishing for a sustainable future. Sustainable chemistry represents an area of innovation, which not only preserves resources but also stands for a development process in the chemical industry. The most important goals in fine chemicals, skin care products and pharmaceuticals including biomedical compositions are the development of nontoxic and biodegradable compounds, the advance on new reaction conditions (cleaner solvents, biotechnological processes, etc.) and the use of raw materials from renewable feedstock [1].

There is today a strong trend to replace conventional surfactants with more environmentally benign compounds. Manufacturers and consumers demand novel environmentally friendly surfactants from renewable resources produced by clean and sustainable technologies (bio-based surfactants). The challenge is to find surfactants that are mild and biodegradable but meet performance and cost–benefit requirements.

Nature offers a complete renewable and sustainable source of a wide range of raw materials of different structure, polarity and size. Cellulose, chitin, starch, fruits, lignin, waste proteins, fats and oils are important bio-based raw materials for sugars, peptides,

Surfactants from Renewable Resources Edited by Mikael Kjellin and Ingegärd Johansson

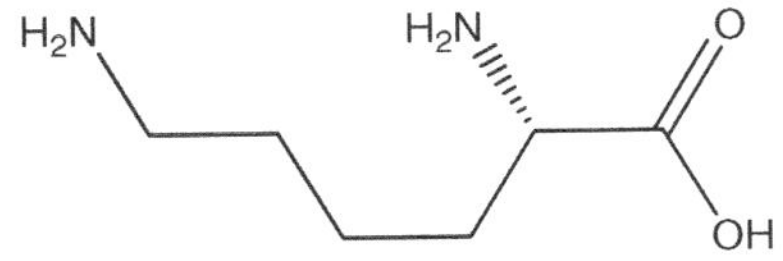

lysine

1

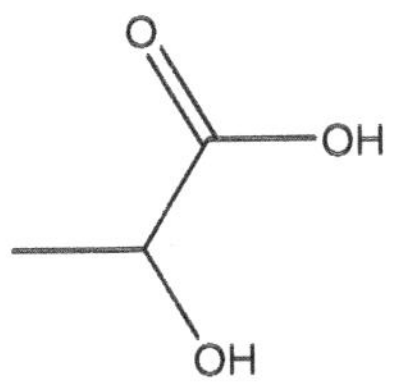

lactic acid

2

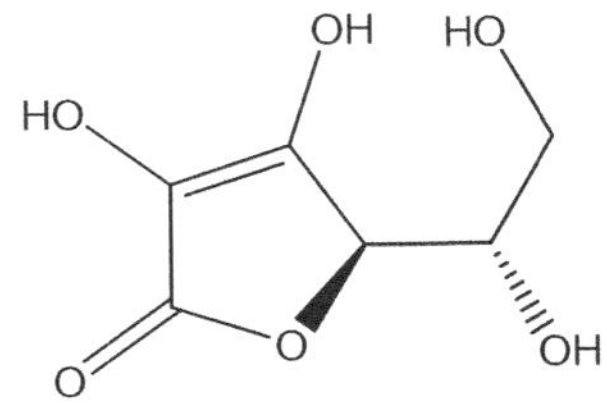

ascorbic acid

3

Figure 5.1 *Chemical structure of lysine (1), lactic acid (2) and ascorbic acid (3).*

amino acids, fatty acids, hydroxyacids, lipids, and so on. The use of hydrophilic renewable raw materials to prepare novel 'natural' surfactants is an exciting and attractive research activity to conciliate sustainable issues with industrial development.

Small and multifunctional structures such as amino acids, hydroxyl and polyhydroxy acid molecules that can act as the hydrophilic entities are an interesting option to be investigated. The interest lies in the ready availability of cheap raw materials that are not derived from petrochemicals. These biodegrade and are easily accepted by the consumer. In this chapter we will describe significant advances that have been made in the field of surfactants derived from three types of renewable and biodegradable sources: amino acids, in particular lysine (Figure 5.1, **1**), lactic acid (Figure 5.1, **2**) and ascorbic acid (Figure 5.1, **3**).

5.2 Production of Raw Materials

5.2.1 Lysine

The economic importance of amino acids is noteworthy; consequently, demand is continually growing for their utilization as food additives, feed supplements, therapeutic agents and precursors for the synthesis of peptides or agrochemicals. L-Glutamic acid is the highest produced amino acid (approximately 900 000 tonnes per year) followed by L-lysine (420 000 tonnes per year). Lysine is an essential amino acid and belongs to the aspartate biosynthetic pathway. Lysine can be produced by fermentation of bacteria or their mutants [2, 3]. For nearly 50 years, constant efforts to increase production performance have been carried out, both directed towards the microorganisms themselves and towards technical improvements of the respective processes [4]. *Corynebacterium glutamicum* has traditionally occupied a special position within the lysine-producing microorganisms [5]. Another important route for the large-scale production of L-lysine is from methanol using auxotrophic mutants of thermotolerant *Bacillus methanolicus* [6, 7]. The advantages of this approach are a readily available, stable, inexpensive substrate and existing large-scale fermentation technology for methylotrophic organisms [8, 9].

5.2.2 Lactic Acid

Lactic acid, a naturally occurring hydroxy acid, alpha hydroxyl acid (AHA, is well known as a food preservative, acidulant and flavouring agent. The current worldwide market is 150 000 tonnes per year. It also has many pharmaceutical and cosmetic applications [10]. Its biodegradable polymer has medical uses as sutures, orthopaedic implants, controlled drug release and so on. Lactic acid esters like ethyl/butyl lactate can be used as green solvents, which are high boiling, nontoxic and degradable [11]. Lactic acid can be manufactured by chemical synthesis (production of racemic mixture) or carbohydrate fermentation (racemic or stereo specific acid) [12]. Other possible routes are base-catalysed degradation of sugars, oxidation of propylene glycol, reaction of acetaldehyde, carbon monoxide and water at elevated temperature and pressures, hydrolysis of chloropropionic acid, carbohydrate fermentation and nitric acid oxidation of propylene.

5.2.3 Ascorbic Acid

L-Ascorbic acid (l-AA) is a hydrophilic vitamin (vitamin C) with high reducing activity due to its enediol-lactone resonant structure. With an estimated world production of about 110 000 tonnes per year, the vitamin C market is becoming increasingly competitive [13]. The synthetically produced l-AA is used for different applications in the pharmaceutical, food, beverage and feed industries [14]. Originally isolated from lemons, the first chemical l-AA synthesis process from *l*-xylosone was achieved in 1933 [15, 16]. During the past 70 years, the Reichstein process (from *d*-glucose) has gained all the advantages that would be expected after such a long period of application and development [16]. Nevertheless, the process is still mostly energy consuming; high temperatures and pressures are necessary for some steps of the process. These economic factors gave rise to tremendous interest in the use of alternative processes. As a result, microbiological biotransformations using reasonable raw materials became the focus of particular attention. The modern fermentation processes as well as cell-free biocatalysis systems combined with recent innovations in biochemistry and molecular biology have raised the chance of developing a successful process [17].

5.3 Lysine-Based Surfactants

5.3.1 Structure and Synthetic Aspects

Amino acid-based surfactants constitute an important class of natural surface-active biomolecules of great interest to organic and physical chemists as well as to biologists with an unpredictable number of basic and industrial applications [18]. Structurally, lipo amino acids are a very heterogeneous group of compounds but with a common advantage, in that they are relatively easy to design and synthesize. Often these molecules combine charged, or noncharged residues (i.e. glutamic acid (Glu), lysine (Lys), arginine (Arg), serine (Ser), leucine (Leu), phenylalanine (Phe), alanine (Ala)) as the hydrophilic head group with a hydrophobic tail of different structures, lengths and numbers (i.e. fatty acids, fatty alcohols, fatty amines) as synthons for the amphiphilic structure [19–22]. This fact explains the diversity of amino acid/peptide-based surfactants and the variety of their physicochemical and biological properties [23–27]. Since the middle of the last century a lot of lysine surfactant molecules of different structures have been described although only a few have been commercialized, that is Ajinomoto Co., Inc. sells Amihope LL, a salt of N^{α}-lauroyl lysine, or Sigma-Aldrich, the N^{ε}-lauroyl lysine analogue. The presence of one carboxylate (anionic) and two amino (cationic) moieties in lysine makes it possible to design various types of surfactants of different ionic character (anionic, cationic, nonionic and amphoteric derivatives) by introducing hydrophobic groups (fatty acid/fatty amine or fatty alcohol) into the molecule. The ionic nature of the amphiphile with lysine in the head group depends on pH and the specific structure modification of the surfactant.

Different structure amphiphiles can be designed by hydrophobic modulation. Thus, monocatenary, one lysine residue bearing one hydrophobic tail (Figure 5.2, **1**), dicatenary, one lysine residue bearing two hydrophobic chains (Figure 5.2, **2**), geminal, two lysine polar heads and two hydrophobic tails per molecule (Figure 5.2, **3**), and glycerolipid derivatives, one lysine polar head and one or two hydrophobic moieties linked together through a glycerol skeleton (Figure 5.2, **4**), characterized by the presence of weak amide and/or ester bonds anywhere in the molecule have been described [28–33].

Most of the lysine-based surfactants are N^{α}-acyl lysine (Figure 5.3, **1**) or N^{ε}-acyl lysine (Figure 5.3, **2**) salt or ester derivatives.

The structure is characterized by a fatty acyl radical linked to the α- or ε-amino group of lysine through an amide linkage, which has a strong hydrogen bonding ability [19]. The synthesis of these compounds has been carried out following the methodologies for the formation of amide functions in the liquid phase, including peptide chemistry [34]. Ideal chemical conditions for the preparation of N-acyl amino acids would allow acyl bond formation to be carried out rapidly and quantitatively under mild conditions, avoiding side-reactions, while maintaining all the adjacent chiral centres. In practice, however, diverse methodologies have been devised to overcome the problems related to the reactivity and purity processes.

Symmetrical (two equal fatty acid chains) (Figure 5.4, **1**) and asymmetrical (two different fatty acid chains) (Figure 5.4, **2**) nonionic double-chain surfactants of the type N^{α},N^{ε}-diacyl lysine polyoxyethylene glycol amide compounds, with a structural resemblance to natural lecithin phospholipids, have been reported by the authors' lab [35–38] to determine the effect of several structural parameters (hydrophobic chain length, polyoxyethylene (POE) chain length and number of polyoxyethylene chains) on the physicochemical properties and biological performance of these natural mimics.

The structure of these nonionic amphiphilic compounds is based on N^{α},N^{ε}-diacyl lysine and contains no active hydrogen in the ethylene oxide (EO) hydrophilic head (they possess a methoxy capped EO chain), with the result that they are more chemically

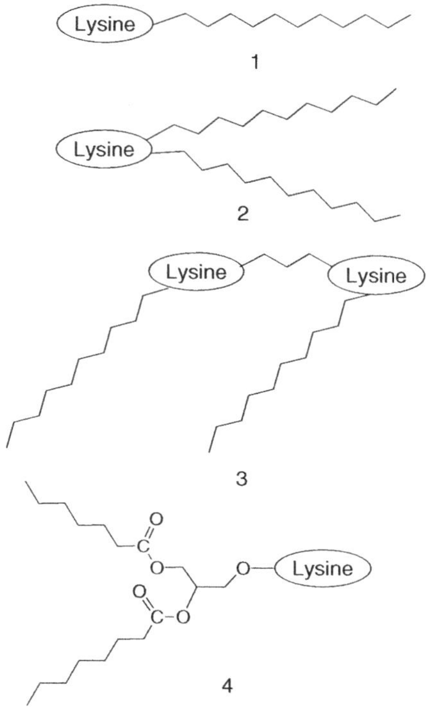

Figure 5.2 *Schematic structure of lysine-based surfactants.*

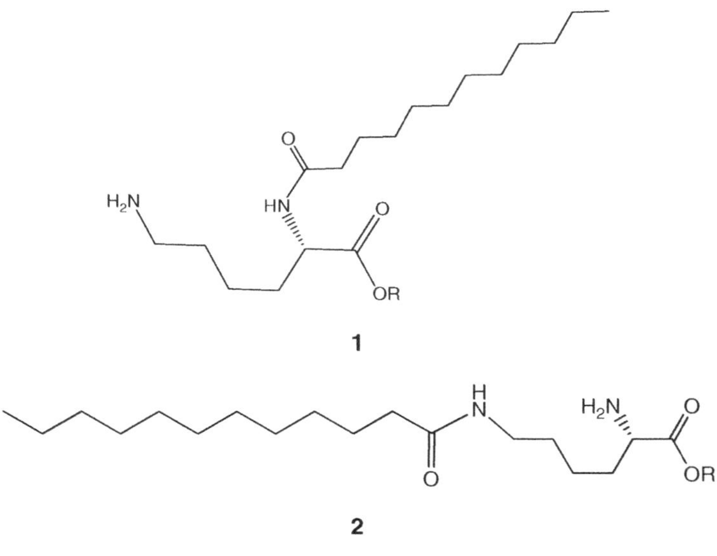

Figure 5.3 *Chemical structure of N^{α}-acyl lysine **(1)** and N^{ε}-acyl lysine **(2)** derivatives where R is Me, Et, H, Na^+.*

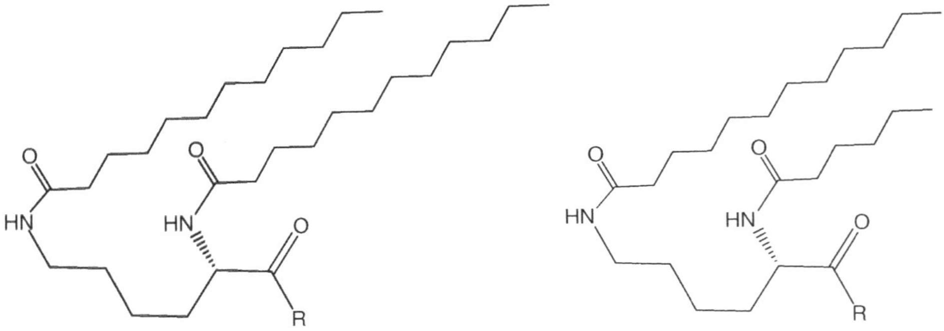

1 2

Figure 5.4 Chemical structure of symmetrical N^{α},N^{ε}-dilauroyl lysine (**1**) and asymmetrical N^{α}-butyl, N^{ε}-lauroyl lysine (**2**) derivatives where R is $-NH-(CH_2-CH_2-O)_nMe$; $-N-[(CH_2-CH_2-O)_nMe]_2$; $-ONa^+$, $-O$ Lysine$^+$, $-OK^+$, $-OLi^+$.

inert. As with the lecithins these compounds have two hydrophobic tails of identical fatty acid groups and one hydrophilic head. However, in the latter compounds the polar head is of the nonionic type (one or two chains of monomethyl ether ethylene oxide glycol with different EO units) whereas in the lecithins it is of the zwitterionic type. The central pivot in the structure of lecithins, that is the glycerol, is mimicked by the natural trifunctional amino acid lysine. The fatty acids and methyl ether oxyethylene glycol amine residues are introduced into the α- and ε-amino or α-carboxylic functions of the lysine through amide bonds in place of the ester bonds in the lecithins. These amide bonds offer greater resistance to hydrolysis than the ester of lecithins. They have been prepared by condensation of fatty acid compounds to the conveniently protected lysine in organic media. The formation of N-acyl lysine derivatives takes place by reaction of the corresponding fatty acyl chloride with lysine, but this procedure requires prior protection of the second free amino group of the amino acid. The principal protecting groups for the N-amino function of Lys and homologues are *tert*-butoxycarbonyl (Boc) and benzyloxycarbonyl (Z), which can be easily removed by acidolysis and hydrogenolysis respectively. Similarly, symmetrical anionic double-chain surfactants of the type N^{α},N^{ε}-diacyl lysine have been chemically synthesized by our group and later by others [39].

The rapid advance of biotechnology has led to the production of interesting structures via mediation of enzymatic or chemoenzymatic catalysts. The main advantages associated with the use of biocatalysts are mild reaction conditions and high enzymatic specificity, which often eliminate the need for regioselective protection of multifunctional starting materials. This point has been clearly demonstrated by a number of recent reports dealing with the application of enzymes to synthesis and/or modification of amino acids, sugar fatty acid esters, phospholipids and alkyl glycosides [40–44].

5.3.2 Physicochemical Properties

Our group has already reported the physicochemical properties of N^{α}-acyl lysine salts both as cationic surfactants and amphoteric derivatives [45] (Figure 5.3, **1**). N^{ε}-acyl esters of lysine (Figure 5.3, **2**) have been studied recently [46, 47]. The critical micelle concentration (CMC) values of these cationic surfactants are slightly smaller than those of the corresponding quaternary ammonium surfactants, 5.5 mM for lauroyl lysine compared to 16 mM for dodecyl trimethylammonium bromide. The lauroyl derivative was found to form spherical micelles while the myristoyl form elongated micelles and the palmitoyl derivative has low solubility at 25 °C but also forms spherical micelles at 50 °C. The N^{α}-lauroyl derivative presented a CMC very close to that of the N^{ε}-lauroyl derivative (6–8 mM) [48].

N^{α},N^{ε}-diacyl lysine compounds have been studied recently both as nonionic compounds and as salts. Pinazo *et al.* [49] and Brito *et al.* [50] studied the effect of alkyl chain length on the dry product and a cubic to lamellar transition was observed as the hydrophobic chain length increases. The shorter (N^{α},N^{ε}-dihexanoyl lysine) product did not follow this trend and crystallized in an inverse hexagonal phase. The reason for this difference was discussed in terms of the difficulty of bringing together two short chains. The thermotropic phase behaviour of the nonionic methyl esters was also studied as a function of hydrophobic chain length. Anomalies in the crystallization of the shorter chain analogues were found. Diacyl salts in the dry state showed a similar trend with chain length as the nonionic compounds (both free acid and ester derivatives). The influence of the counterion volume was studied for the dioctanoyl derivative; big counterions induce the formation of lamellar phases while small counterions induce the formation of *pn3m* cubic phase [49].

Seguer *et al.* [36, 37] studied the micellization properties of the same basic structure, N^{α},N^{ε}-diacyl lysine oxyethylene derivatives linked to the acid group via an amide bond (Figure 5.4, **1**). The EO chains increase the water solubility of the products and allow for CMC determination. Two series of compounds were studied, a single EO chain (Figure 5.4, **1**; R = $-NH-(CH_2-CH_2-O)_nMe$) and two EO chains (Figure 5.4, **1**; R = $-N-[(CH_2-CH_2-O)_nMe]_2$) bonded to the same amino group. The CMC and surface tension reduction for the two EO derivatives was similar to that of single-chain alcohol ethoxylates with the same hydrocarbon chain length and a similar number of EO units in the head group and much bigger than the corresponding lecithin with the same chain length. The CMC of the single EO chain per head group derivatives was smaller than those with two chains in the head group.

In Figure 5.5 data for the EO derivatives with a single EO chain, two EO chains, lecithins and salt derivatives are shown as a function of total carbon number. From this figure it is clear that the slope for lecithins is almost double than that of the two EO chains family and that of the single-chain family is intermediate. This implies that the energy of transfer of a methylene group ($-CH_2$) from a hydrocarbon to an aqueous environment is smaller for those lysine derivatives than for the usual molecules (the energy transfer per methylene can be evaluated as 0.54 for the lecithins and 0.28 for the double EO chain surfactants. This is probably related to some enhanced interaction between the hydrophobic chains for those later surfactants that would be responsible for a lower interaction with water. It was found that the short chain N^{α},N^{ε}-dioctyl lysine derivatives formed small spherical micelles in aqueous solution [49]. The area per molecule as obtained from surface tension measurements gives a high value for the dioctyl lysine derivatives [49, 51], agreeing with the small micellar sizes. Longer derivatives

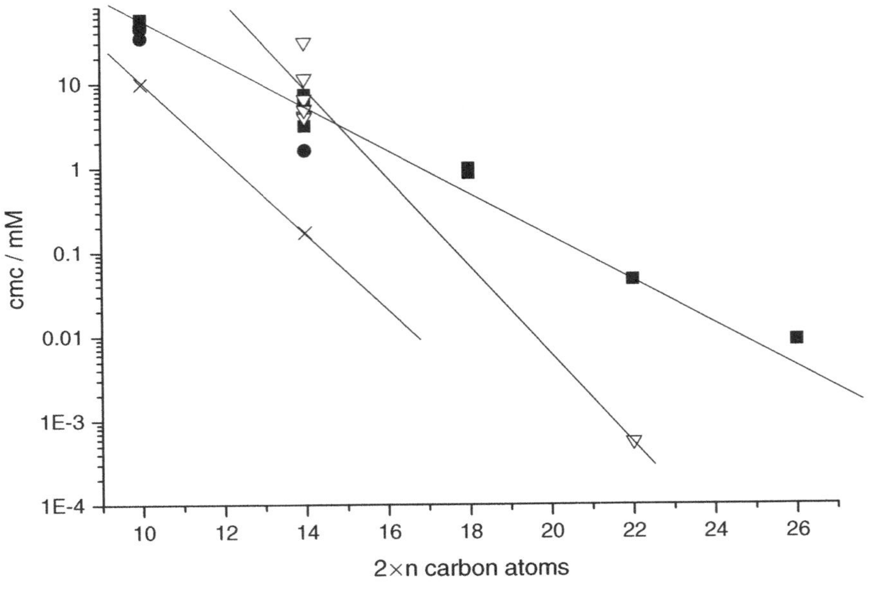

Figure 5.5 CMC of several diacyl lysine derivatives compared to short-chain lecithins, with circles: N^{α},N^{ε}-diacyl lysine dimethyl ether monopolyoxyethylene glycol amides [36, 37]; squares: N^{α},N^{ε}-diacyl lysine dimethyl ether dipolyoxyethylene glycol amides [36, 37]; triangles N^{α},N^{ε}-diacyl lysine salts with several counterions [37, 49, 50, 53], compared to short-chain lecithins (crosses) (dihexanoyl lecithin and dioctanoyl lecithin) [37].

like N^{α},N^{ε}-dilauroyl lysine salt, however, show a very small area per molecule, which could indicate not the formation of micelles but of lamellar aggregates. It was also shown that the homologues containing two short alkyl chains and two methyl ether oxyethylene glycol chains were interesting nonhemolytic molecules that could be applied as water-soluble surfactants in the biological field. By contrast, the ones with higher alkyl chain lengths (even with two EO chains) showed no water solubility, yielding compounds, in principle, with little technological interest. The asymmetrical homologues resulted in higher surface-active properties and in a greater capacity for micellization when compared with their symmetrical analogues, albeit with a slight increase in toxicity [52].

Suzuki *et al.* have extensively studied the ability to form hydrogels and organogels with lysine-derived surfactants. The structures studied correspond to a wide range of structures: N^{ε}-lauroyl lysine hydrocarbon terminated with a positive group and with the carboxylic group esterified [53, 54] as hydrogelators, Gemini surfactants from N^{ε}-lauroyl lysine with free carboxylic acid groups, partial salts or with these groups esterified with fatty alcohols as organogelators [55, 56], asymmetric N^{α},N^{ε}-diacyl lysine as mixtures of free carboxylic derivatives and salts as both hydrogelators and organogelators [57, 58] and N^{α}-glucoheptonamide-N^{ε}-lauroyl lysine as hydrogelator [59]. α-Lysine ω-amino bolaamphiphiles were shown to aggregate, forming monolayer nanotubes [60]. Tubular (short chain) or vesicular structures (long chain) were obtained in aqueous dispersions of lysine-glutamine diacyl surfactants [61].

5.3.3 Applications

In general, the amino acid compounds can be classified as biocompatible surfactants; they are readily biodegradable compounds and have low toxic effects in the aquatic environment. Depending on their chemical structure, they can act as emulsifiers, detergents, wetting agents, foaming or dispersing compounds. They can therefore be used in a wide spectrum of industrial and biomedical applications, so the market place projection of these surfactants is encouraging. Salts of long-chain *N*-acyl amino acid are currently used as detergents, foaming agents and shampoos because they are nonirritating to the human skin and highly biodegradable. Among the principal producers of commercially available amino acid and protein fatty acid condensate surfactants can be mentioned Ajinomoto Co., Inc. (glutamic, alanine, aspartic, glycine, arginine and lysine derivatives); Cognis GmbH and Inolex Chemical (protein condensates sold as Lamepon® and Maypon® 4C, respectively); Croda, Stepan and Hoechst, which supply acyl sarcosinates. For more information consult the chapters in the book *Protein-Based Surfactants* cited in References [62] and [63].

Biocompatible cationic surfactants from the amino acid lysine (hydrochloride salts of N^{ε}-lauroyl-lysine methyl ester, N^{ε}-miristoyl-lysine methyl ester and N^{ε}-palmitoyl-lysine methyl ester) show moderate antimicrobial activity against the Gram-positive bacteria. The haemolytic activity of these compounds is considerably lower than those reported for other cationic N^{α}-acyl amino acid analogues [64]. Taking into account the high biodegradation level and the low haemolytic activity, these compounds could be considered safe surfactants in relation to the cell of the human body. These properties make them suitable candidates for biological and medical applications [65].

Diacyl lysine anionic surfactants, N^{α}-octanoyl-N^{ε}-octanoyl lysine with different counterions (lysine, Na, K, Li) (Figure 5.4, **1**; R = $-ONa^{+}$, $-OLysine^{+}$, $-OK^{+}$, $-OLi^{+}$) showed less cytotoxicity than sodium dodecyl sulfate (SDS). Moreover, they were less eye-irritating than SDS and none showed phototoxic effects. These surfactants are a promising alternative to commercial anionic surfactants given their low ocular and dermal irritancy. These properties offer great potential for topical preparations [66].

The efficient delivery of deoxyribonucleic acid (DNA) to cells *in vivo* has been a major goal for some years and there is a continuing need to develop novel low-toxicity surfactant molecules to facilitate the effective transfer of polynucleotides into cells. We studied the interaction of the single-chain arginine-based surfactant, the salt of lauryl amide of arginine (arginine-*N*-lauroyl amide, or ALA) with DNA [67]. The ability of this surfactant alone to compact DNA is compared by fluorescence microscopy studies to classical cationic surfactants. Furthermore, toxicity studies revealed that the incorporation of ALA in the catanionic vesicle system transformed them into cell-viable systems, therefore extending their use to drug and gene delivery systems. Nontoxic catanionic vesicles can also be obtained from diacyl lysine derivative compounds [28]. Cationic diether lipids based in arginine, tryptophan, histidine and lysine have also been proposed as gene transfer agents [68].

Gemini surfactants show greatly enhanced surfactant properties relative to the corresponding monovalent compounds, which makes them of special interest for biomedical applications. In the last 10 years different gemini cationic surfactants with amino acids in the polar head have been proposed as new synthetic vectors for gene transfection [69–72]. Amino acid surfactants are chiral compounds that can form gels containing helical fibres stabilized by amide hydrogen bonds. Gemini surfactants with two L-lysine derivatives linked by different alkylene chain lengths through the amide bond are good organogelators that gel most organic solvents such as alcohols, cyclic ethers, aromatic solvent and acetonitrile [55].

5.4 Lactic Acid-Based Surfactants

5.4.1 Structure and Synthetic Aspects

Within this group of surfactants we can mention the fatty acid esters of lactic acid, lactylates (esters of the fatty acid with the hydroxyl group of lactic acid and lactylates) and the alkyl lactates (esters of the lactic acid with a fatty alcohol) (Figure 5.6), all of them commercially available as safe products for food and personal care formulations. Some examples of manufacturers are Eurolabs Limited, Stepan Co., Cognis GmbH and Spectrum Chemical Mfg Corp.

The chemical synthesis of fatty acid esters of lactic acid and lactylates involve protection and deprotection of lactic acid. These methods use strong acid catalysts and high temperatures [73, 74], leading to moderate conversions along with many unwanted

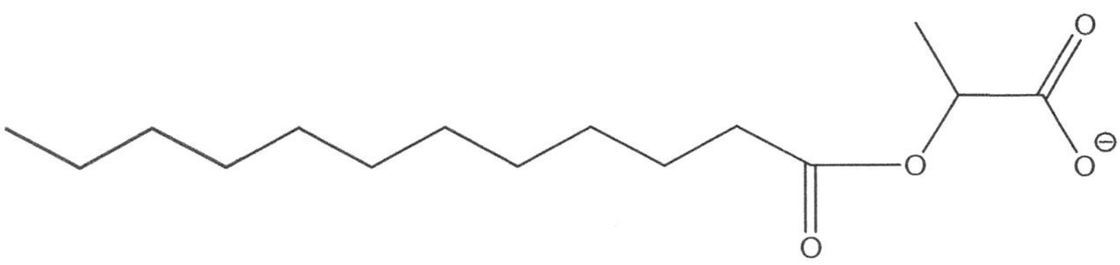

1

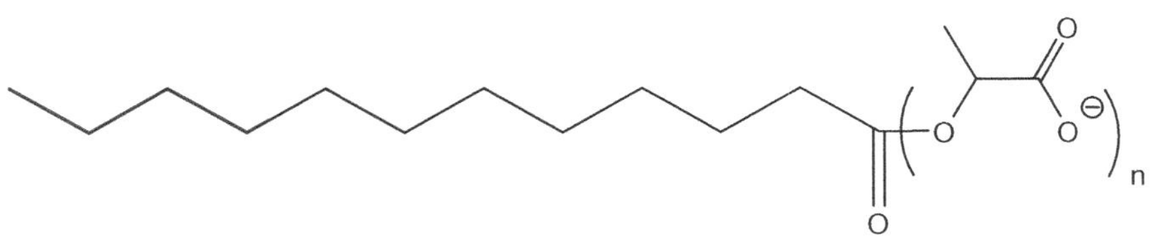

2

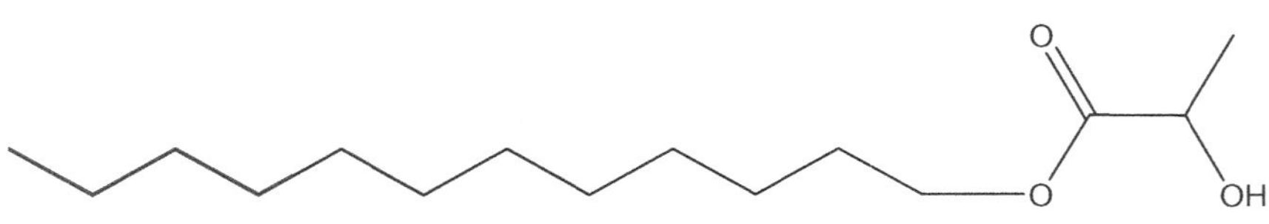

3

Figure 5.6 Sodium lauroyl lactate salt (1), sodium lauroyl lactylate salt (2), lauryl lactate ester (3).

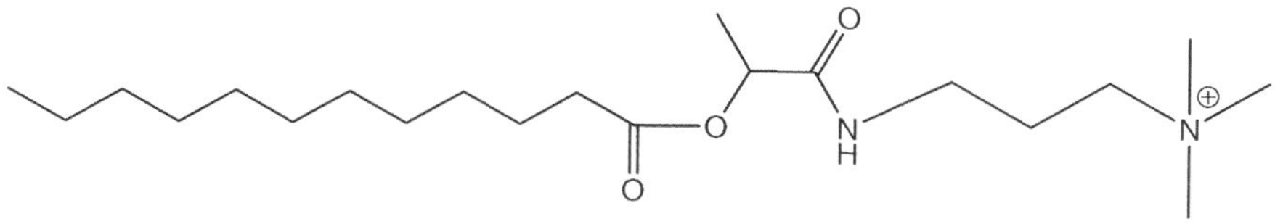

Figure 5.7 Chemical structure of fatty amine salts of lactic acid.

by-products. Lipase catalysed synthesis of fatty acid esters and the hydroxyl group of lactic acid in nonaqueous media is available in the literature [75]. Very few reports are described using enzymatic catalysts for the preparation of alkyl lactates. The enzymatic catalyst procedure with yields of about 80 % using a solid-supported lipase B from *Candida Antarctica* in the absence of solvent is also described [76–78]. Lactic acid-based cationic surfactants (Figure 5.7) were prepared by condensation of lactic acid with *N*,*N*-dimethyl propylamine, followed successively by acetylation with fatty acid chloride or fatty acid ester and by quaternization with dimethylsulfate [79].

5.4.2 Physicochemical Properties

In general the physicochemical properties of lactic acid-based surfactants have been studied with not very well defined and pure products. Since the first paper dealing with *N*-substituted alkyl lactamides synthesized by Ratchford [80] was published in 1950 there was not very much activity until the year 2000. In their works Ratchford and Fisher described some physical properties (melting and boiling points, density and refractive index), but they did not explicitly identify their products as surfactant or amphiphiles and therefore did not investigate the micellization and adsorption properties. The micellization properties of two series of homologues, one nonionic and one cationic, with varying hydrocarbon chain lengths have been recently described [79, 81]. However no clear trends were obtained with varying hydrocarbon chain lengths and even the units used for giving the obtained CMCs were not clear. Winter *et al.* [82] studied the thermotropic phase behaviour of several fractions of a commercial surfactant (Durlac 300, Loders Croklaan, Channahon, Illinois) with varying lactic acid units at the head group by differential scanning calorimetry (DSC) and X-ray diffraction. Formation of L_α and L_β lamellar phases was found in the dry state. These authors also studied the formation of mixed bilayers with colipids in aqueous systems [82, 83]. An important part of recent literature is devoted to the study of the stabilization of foams and air microbubbles in complex systems where sodium or calcium steaoryl lactylate (Figure 5.6, **2**) is one of the major components [84–88]. These studies try to set the fundamentals of the action of some commercial mixtures in the food industry. Some fundamentals of the adsorption have been studied by Grigoriev *et al.*, who found that the shorter soluble dilactylates in a mixture of mono- and dilactylates were squeezed out of the surface of water monolayers [89].

Also the interaction of lactic acid ester and monoglyceride with proteins has been studied. The complexity of the systems makes it difficult to extract general trends. There seems to exist interactions both with the hydrophilic residues of the proteins via the head group and with the hydrophobic residues via the surfactant hydrocarbon tail [90–92].

5.4.3 Applications

Long-chain fatty acid esters of lactic acid, such as palmitoyl or stearoyl lactate (Figure 5.6, **1**), are ionic biocompatible surfactants widely used in the food industry [93]. Sodium stearoyl lactate (and the similar calcium stearoyl lactate) is a high efficient emulsifier used in a wide variety of modern food [94–98], pharmaceutical [99] and personal care [100, 101] technologies. Liquid systems are increasingly desired in the food industry and the use of stearoyl lactylates improves the liquid systems. Liquid systems are increasingly desired in the food industry as the traditional powder form is difficult to handle. Liquid pumpable bread improver systems, based on an oil fraction, have been developed. The drawback of these systems is that emulsifiers can be added only in low percentages as otherwise very viscous systems are obtained, which cannot be pumped. Stearoyl lactylates can be added to the oil fraction in large amounts, resulting in a liquid pumpable system, without thickening of the system [102].

Nonionic surfactants from lactic acid can be obtained by the esterification of the free hydroxyl group of mono- and diglycerides with lactic acid. These compounds are commercially available with the name of LACTEM (Figure 5.8).

These mono- and diglyceride derivatives are used as an oil-soluble emulsifier in special fats for dessert products such as nondairy creams, chocolates [103] and cakes [104, 105]. Another commercially available mixture of lactic-based surfactants is LFEGPG, composed of fatty acid esters of glycerol and fatty acid esters of propylene glycol in which the free hydroxyl group has been esterified with lactic acid. The introduction of the lactic acid in the polar head of mono- and diglycerides changes the polar/apolar interface of the lipid and reduces the melting temperature by at least 10 °C [106]. These changes make these compounds good alternatives to the original ones used in the food industry as emulsifying agents. Nonionic biocompatible and biodegradable lactide-based surfactants with the general structure $LA_2EO_mLA_n$ (m and n are the number of LA and EO repeat units) can be used to stabilize emulsions of water and CO_2 as well as CO_2-based dispersions [107].

Lactic acid is also used in cosmetic industries as skin peels and emollients. Cationic surfactants derived from lactic acid have been reported for increasing the skin absorbance of the lactic acid and decreasing irritating side effects [81].

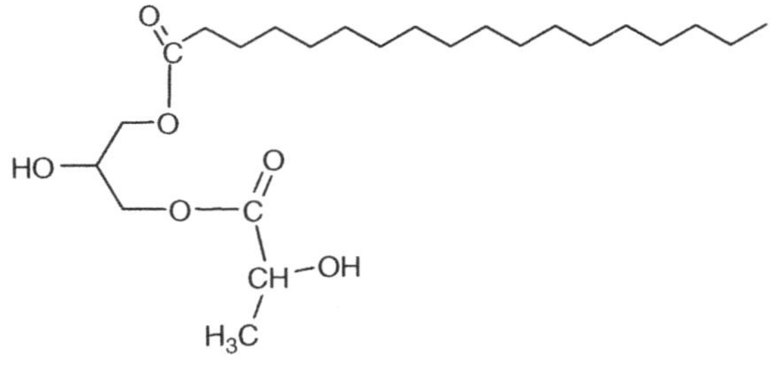

Figure 5.8 *Chemical structure of lactic acid glycerol monostereate (LACTEM).*

5.5 Ascorbic Acid-Based Surfactants

5.5.1 Structure and Synthetic Aspects

Ascorbyl fatty acid esters of lauric, myristic, palmitic or stearic acid were first synthesized by Swern *et al.* in 1943 [108], aiming at combining the reductive properties of ascorbic acid and the lipid solubility of a fatty acid. Several methods were used: treating the ascorbic acid with the appropriate fatty acid chloride in the absence of a solvent or in pyridine and heating the ascorbic acid with the fatty acid in the presence of basic catalyst and in the presence of sulfuric acid solution. The reaction involves the condensation of the carboxylic fatty acid and the primary –OH group in position 6 of the l-AA.

Lipase-catalysed synthesis of vitamin-C fatty acid esters of ascorbic acid with emulsifying and antioxidant properties has received considerable attention due to the mild reaction conditions and high regioselectivity [109, 110]. 6-*O*-acyl *l*-ascorbates (Figure 5.9) with long chains between 8 and 18 have been obtained in organic solvents with a low water content. The ascorbates with short acyl chains would be more water soluble and have more potential as emulsifiers for preparing oil-in-water emulsions than those with long acyl chains [111–113]. L-Ascorbyl linoleate was successfully prepared by enzymatic esterification and transesterification in a nonaqueous medium using immobilized lipase as the biocatalyst [114]. Recently, the synthesis of new amphiphilic ascorbate salts from ascorbic acid with surfactant and antioxidant properties was carried out using different fatty amines of 8, 12, 16 and 18 carbon atoms [115] (Figure 5.10).

Characterization of physicochemical properties of these new compounds indicated a classical behaviour of cationic surfactants.

5.5.2 Physicochemical Properties

Alcanoyl-6-*O*-ascorbic acid esters with different chain lengths (8–18) possess three free –OH groups in positions 2, 3 and 5 and behave as anionic single-chain surfactants. The

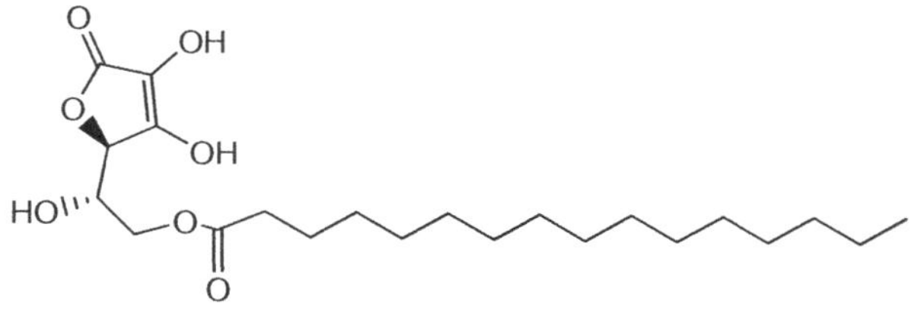

Figure 5.9 Chemical structure of 6-O-palmitoyl- ascorbate.

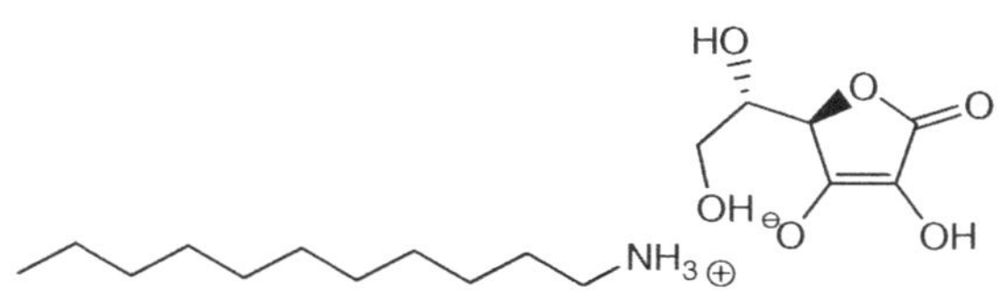

Figure 5.10 Chemical structure of fatty amine salts of ascorbic acid.

shorter homologue, octanoyl-6-*O*-ascorbic acid forms nearly spherical micelles in water solution above a CMC of about 6 mM at 30 °C while decanoyl and dodecyl derivatives are poorly soluble in water. Ascorbyl palmitate and its alkali salts are amphiphilic molecules practically insoluble in water at room temperature, but their solubility increases rapidly with temperature. Once a given temperature, which depends on each specific alkali salt, is reached a clear solution is formed. At this temperature, known as the critical micellar temperature (CMT) or Krafft temperature, ascorbyl palmitates and their salts form micelles and become soluble. The CMT was determined by Mantsch and co-workers [116, 117] using infrared spectroscopy (Table 5.1). The transition temperature of these alkali salts depends on the critical balance between the energy required to disrupt the crystal structure and the energy gained by micelle formation.

At temperatures higher than the CMT, ascorbyl palmitate solutions are transparent. As the temperature decreases, a crystalline mesofase, called coagel, forms [118, 119] Ascorbic acid esters with six or eight carbon atoms in the side chain evolve from a coagel to a micellar dispersion when the temperature and water volume increase. For homologues with larger chains ($10 \leq n \leq 16$) the coagel turns into a stable transparent gel phase [120, 121]. The coagel phase is usually thought to be formed of surfactant lamellae separated by thin interlayers of strongly bound, essentially 'frozen' molecules. Ambrosi *et al.* [122] conducted measurements to explore the interactions between water and surfactant molecules in coagels. Using different techniques they detected two kinds of water: interlayer hydration water and bulk water. In bulk water hydrated surfactant islands can be considered to be dispersed. Also they found that the Krafft temperature was determined predominately not by the hydrocarbon chain phase transition but by a change in the hydration water.

The Hofmeister effect in coagels of ascorbyl alkanoates has been studied by Lo Nostro *et al.* [123]. They found that the presence of 0.5 M salt solutions has a large impact on the coagel properties due to the adsorption of the different anionic species at the surface of ascorbyl-alkanoate aggregates. The phase transition temperature changes significantly with the anions, increasing in the following sequence: $SCN^- > I^- > NO_3^- > Br^- > Cl^- > H_2O > BF_4^- > H_2PO_4^- > SO_4^- > HCOO^- > F^- > AcO^-$. Anomalous results that do not fit the pattern were observed for formate and acetate ions. This is due to the hydrolysis equilibrium that they generate in the presence of the ascorbic acid esters [83].

Table 5.1 *Critical micellization temperatures of aqueous ascorbyl palmitate (AP) and its alkali salts*

Solute	CMT (°C)
AP	65
$Li^+ AP^-$	27
$Na^+ AP^-$	28
$K^+ AP^-$	48
$Rb^+ AP^-$	44
$Cs^+ AP^-$	32

Reproduced with permission from U. Köhler, P.W. Yang, H. H. Mantsch, Structure and Polymorphic Phase Behavior of Ascorbyl Palmitate in Water, *Canadian J. Spectros.* 1988, 33, 122–127.

An important category of properties of the ascorbic acid derivatives and their mixtures with other compounds refers to the air/water interface behaviour. Long-chain derivatives of ascorbic acid readily produce monomolecular films at the air/water interface, and give stable mixed monolayers with some vitamins that possess an amphiphilic structure as well. This feature is particularly important in order to determine the mutual miscibility of ascorbic acid derivatives and some relevant natural compounds with a perspective of producing stabilized systems where the ascorbyl derivative protects the other components against oxidation. Several studies [124–127] have shown that the association of ascorbic acid derivatives and tocopherols enhances the antioxidant capability of the two vitamins. The ascorbic acid derivatives/tocopherol mixtures are therefore a very promising and important tool for the stabilization and protection of many different systems, such as biological materials, food and artificial membranes used as drug carriers, from both the scientific and industrial points of view. Capuzzi *et al.* [128] studied the monolayer behaviour of ascorbic acid derivatives, the tocopherol and their mixtures. Ascorbyl stearate and tocopherol are miscible in all proportions in a monolayer at the air/water interface at 20 °C. The collapse pressure of the mixed film depends on the composition of the mixture. The Gibbs free energy of mixing, the molecular area at fixed pressure and the limit area indicated the presence of repulsive interactions among the molecules of the two surfactants, probably because of the branched methyl groups of the phytyl tail in tocopherol.

Cappuzzi *et al.* [129] studied the mixtures of stearoyl ascorbic acid and vitamin K_1. These two compounds form stable Langmuir films at the air/water interface for temperatures ranging from 20 to 35 °C and are completely miscible in the monolayer in all ratios. The mixed films are more expanded than the pure monolayer components and, similarly to tocopherol, repulsive intermolecular interactions between the two components are present. In this case the repulsions can be attributed to the presence of five branched methyl groups in the vitamin K_1 and a negative charged polar head in the stearoylascorbic acid. Mixtures of ascorbyl-stearate and vitamin D_3 form a stable monolayer at the air/water interface at different mixture compositions and temperatures. The presence of calcium ions in the subphase results in a more stable and rigid monolayer of the pure compounds than that of mixtures where the divalent cations strongly interact with the negative ascorbate head group, causing a significant compression of the monolayer [130]. Recently new ascorbic acid/fatty amine salts (alkyl ammonium ascorbates) have been described [131]. It was found that surfactant properties of these new compounds are strongly influenced by the length of the fatty alkyl amine used, which is also the case in traditional cationic surfactants.

5.5.3 Applications

Ascorbyl palmitate, a derivative of ascorbic acid, is actually a lipophilic vitamin C and has been shown to be as biologically active as its original hydrophilic counterpart. Since both vitamin C and palmitate esters are natural food ingredients, ascorbyl palmitate is widely used in the food industry as a natural preserver of oils and fats.

Vitamin C (l-AA) is a powerful, water-soluble antioxidant. It is one of the few topical agents whose effectiveness against wrinkles and fine lines is backed by a fair amount of reliable scientific evidence. Unfortunately, the use of the vitamin C in finished products is limited by its chemical instability: this molecule easily undergoes oxidation in aqueous solutions and is poorly adsorbed through the skin. Moreover, since ascorbic acid is poorly soluble in organic solvents, it can only be used in an aqueous environment. Surfactants from vitamin C increase the stability of the ascorbic acid and allow the use of these compounds in different environments. Alkanoyl ascorbate surfactants are nonirritating compounds more stable than vitamin C and provide antioxidant properties to the skin [132]. Ester derivatives of ascorbic acid produce self-assembled aggregates in water solutions that can solubilise hydrophobic drugs for pharmaceutical formulations These derivatives keep the same extraordinary radical scavenging as the ascorbic acid so their micellar solutions can be the ideal medium for the solubilisation of poorly water soluble chemicals such as sensitive drugs and at the same time they provide a suitable shield against oxidative deterioration [133]. Dispersed in water, 1,12-diascorbyl dodecanoate forms nanotubes with ascorbic acid rings in the molecular architecture. These structures can be exploited for the production of nanosized metallic materials [134]. Double-chain surfactants from ascorbic acid can produce organogels. The presence of a redox-active polar head group in the surfactant adds an important and new functionality to the final organogel that extends their application and uses to the solubilisation, storage and protection of valuable but oxidable materials [135]. Nanostructures from alkyl vitamin C derivatives possess very interesting properties that make these compounds promising pharmaceutical derivatives for drug delivery systems [136, 137].

Recently, alkyl ammonium ascorbates have been investigated in the formulation of wood preservation mixtures containing propiconazole, a widely used fungicide. The protection achieved with these antioxidant surfactants was slightly superior to that obtained with conventional surfactants. The use of these aqueous solutions allows the quantity of biocide to be reduced [131].

References

1. Vemula, K. and John, G. (2008) Crops: a green approach toward self-assembled soft materials. *Acc. Chem. Res.*, **41**, 769–782.
2. Kinoshita, S., Udaka, S. and Shimono, M. (1957) Studies on the amino acid fermentation. Part I. Production of L-glutamic acid by various microorganisms. *J. Gen. Microbiol.*, **3**, 193–205.
3. Parekh, S., Vinci, V. A. and Strobel, R. J. (2000) Improvement of microbial strains and fermentation processes. *Appl. Microbiol. Biotechnol.*, **54**, 287–301.
4. Eggeling, L. and Sahm, H. (1999) L-Glutamate and lysine: traditional products with impetous developments. *Appl. Microbiol. Biotechnol.*, **52**, 146–153.
5. Tryfona, T. and Bustard, M. T. (2005) Fermentative production of lysine by *Corynebacterium glutamicum*: transmembrane transport and metabolic flux analysis. *Process Biochem.*, **40**, 499–508.
6. Arfman, N., Dijkhuizen, L., Kirchholf, G. *et al.* (1992) *Bacillus mefhanolicus* sp. nov., a new species of thermotolerant, methanol-utilizing, endospore-forming bacteria. *Int. J. Syst. Bacteriol.*, **42**, 439–445.
7. Schendel, F. J., Bremmon, C. E., Flickinger, M. C. *et al.* (1990) L-Lysine production at 50 °C by mutants of a newly isolated and characterized methylotrophic *Bacillus* sp. *Appl. Environ. Microbiol.*, **56**, 463–470.

8. Ekeroth, L. and Villadsen, J. (1991) Single cell protein production from C compounds, in *Biology of Methylotrophs* (eds I. Goldberg and J. S. Rokem), Butterworth-Heineman, Boston, Massachusetts, pp. 205–232.
9. Leak, D. J. (1992) Biotechnological and applied aspects of methane and methanol utilizers, in *Methane and Methanol Utilizers* (eds J. C. Murrell and H. Dalton), Plenum Press, New York, pp. 245–279.
10. Wee, Y. J., Kim, J. N. and Ryu, H. W. (2006) Biotechnological production of lactic acid and its recent applications. *Food Technol. Biotechnol.*, **44**, 163–172.
11. Datta, R. (1995) Technological and economical potential of polylactic acid and lactic acid derivatives. *FEMS Microbiol. Rev.*, **16**, 221–231.
12. Narayanan, N., Roychouhury, P. and Srivastava, A. (2004) Lactic acid fermentation and its product polymerization. *Electron. J. Biotechnol.*, **7**, 167–179.
13. Macauley, S., McNeil, B. and Harvey, L. M. (2001) The genus *Gluconobacter* and its applications in biotechnology. *Crit. Rev. Biotechnol.*, **21**, 1–25.
14. Lupo, M. P. (2001) Antioxidants and vitamins in cosmetics. *Clin. Dermatol.*, **19**, 467–473.
15. Haworth, W. N. and Hirst, E. L. (1933) Synthesis of ascorbic acid. *J. Soc. Chem. Ind.*, **52**, 645–647.
16. Reichstein, T., Grüssner, A. and Oppenauer, R. (1933) Die Synthese der *d*-Ascorbinsäure (*d*-Form des C-Vitamins). *Helv. Chim. Acta*, **16**, 561–565.
17. Bremus, C., Herrmann, U., Bringer-Meyer, S. and Sahm, H. (2006) The use of microorganisms in *l*-ascorbic acid production. *J. Biotechnol.*, **124**, 196–205.
18. Xia, J. and Nnanna, I. A. (2001) An overview of the basis, technology, and surface phenomena of protein-based surfactants, in *Protein-Based Surfactants. Synthesis, Physicochemical Properties, and Applications* (eds I. A. Nnanna and J. Xia), Marcel Dekker, New York, pp. 1–14.
19. Takehara, M. (1989) Properties and applications of amino-acid based surfactants. *Colloids Surf.*, **38**, 149–167.
20. Boyat, C., Rolland-Fulcrand, V., Roumestant, M.-L. *et al.* (2000) Chemo-enzymatic synthesis of new nonionic surfactants from unprotected carbohydrates. *Prep. Biochem. Biotechnol.*, **30**, 281–294.
21. Gerova, M., Rodrigues, F., Lamere, J. F. *et al.* (2008) Self-assembly properties of some chiral *N*-palmitoyl amino acid surfactants in aqueous solution. *J. Colloid Interface Sci.*, **319**, 526–533.
22. Vijay, R., Angayarkanny, S. and Bhasker, G. (2008) Amphiphilic dodecyl ester derivatives from aromatic amino acids: significance of chemical architecture in interfacial adsorption characteristics. *Colloids Surf. A*, **317**, 643–649.
23. Roy, S. and Dey, J. (2006) Self-organization properties and microstructures of sodium *N*-(11-acrylamidoundecanoyl)-L-valinate and -L-threoninate in water. *Bull. Chem. Soc. Jpn*, **79**, 59–66.
24. Das, D., Dasgupta, A., Roy, S. *et al.* (2006) Water gelation of an amino acid-based amphiphile. *Chem. Eur. J.*, **12**, 5068–5074.
25. Varka, E. M., Heli, M. G., Coutouli-Argyropoulou, E. and Pegiadou, S. A. (2006) Synthesis and characterization of nonconventional surfactants of aromatic amino acid–glycerol ethers: effect of the amino acid moiety on the orientation and surface properties of these soap-type amphiphiles. *Chem. Eur. J.*, **12**, 8305–8311.
26. Ohta, A., Toda, K., Morimoto, Y. *et al.* (2008) Effect of the side chain of *N*-acyl amino, acid surfactants on micelle formation: an isothermal titration calorimetry study. *Colloids Surf. A*, **317**, 316–322.
27. Capone, S., Walde, P., Seebach, D. *et al.* (2008) pH-sensitive vesicles containing a lipidic beta-amino acid with two hydrophobic chains. *Chem. Biodivers.*, **5**, 16–30.

28. Brito, R. O., Marques, E. F., Gomes, P. *et al.* (2006) Self-assembly in a catanionic mixture with an amino acid-derived surfactant: from mixed micelles to spontaneous vesicles. *J. Phys. Chem. B*, **110**, 18158–18165.
29. Tan, H. and Xiao, H. (2008) Synthesis and antimicrobial characterization of novel L-lysine gemini surfactants pended with reactive groups. *Tetrahedron Lett.*, **49**, 1759–1761.
30. Infante, M. R., García Domínguez, J. J., Erra, P. *et al.* (1984) Surface active molecules: preparation and properties of long chain *N*[alpha]acyl-*l*-[alpha]-amino-[omega]-guanidine alkylic acid derivatives. *Int. J. Cosmet. Sci.*, **6**, 275–282.
31. Pérez, L., Torres, J. L., Manresa, A. *et al.* (1996) Synthesis, aggregation, and biological properties of a new class of gemini cationic amphiphilic compounds from arginine, bis(Args). *Langmuir*, **12**, 5296–5301.
32. Clapés, P., Morán, M. C. and Infante, M. R. (1999) Enzymatic synthesis of arginine-based cationic surfactants. *Biotechnol. Bioengng*, **63**, 333–343.
33. Morán, M. C., Pinazo, A., Pérez, L. *et al.* (2004) 'Green' amino acid-based surfactants. *Green Chem.*, **6**, 233–240.
34. Bodarszky, M. and Bodanszky, A. (1984) *The Practice of Peptide Synthesis*, Springer, Berlin, p. 21.
35. Seguer, J., Infante, M. R., Allouch, M. *et al.* (1994) Synthesis and evaluation of nonionic amphiphilic compounds from amino-acids –molecular mimics of lecithins. *New J. Chem.*, **18** (6), 765–774.
36. Seguer, J., Selve, C. and Infante, R. (1994) New nonionic surfactants from lysine and their performance. *J. Dispersion Sci. Technol.*, **15** (5), 591–610.
37. Seguer, J., Selve, C., Allouch, M. and Infante, M. R. (1996) Nonionic amphiphilic compounds from lysine as molecular mimics of lecithins. *J. Am. Oil Chem. Soc.*, **73** (1), 79–86.
38. Macian, M., Seguer, J., Infante, M. R. *et al.* (1996) Preliminary studies of the toxic effects of non-ionic surfactants derived from lysine. *Toxicology*, **106**, 1–9.
39. Gomes, P., Araujo, M. J. and Marques, E. F. (2008) Straightforward method for the preparation of lysine-based double-chained anionic surfactants. *Synth. Commun.*, **38**, 2025–2036.
40. Valivety, R., Jauregi, P., Gill, I. and Vulfson, E. (1997) Chemo-enzymatic synthesis of amino acid-based surfactants. *J. Am. Oil Chem. Soc.*, **74**, 879–886.
41. Zhang, X. M., Adachi, S., Watanabe, Y. and Matsuno, R. (2005) Lipase-catalyzed synthesis of *O*-lauroyl L-serinamide and *O*-lauroyl L-threoninamide. *Food Res. Int.*, **38**, 297–300.
42. Villeneuve, P. (2007) Lipases in lipophilization reactions. *Biotechnol. Adv.*, **25**, 515–536.
43. Soo, E. L., Salleh, A. B., Basri, M. *et al.* (2004) Response surface methodological study on lipase-catalyzed synthesis of amino acid surfactants. *Process Biochem.*, **39**, 1511–1518.
44. Wada, E., Handa, M., Imamura, K. *et al.* (2002) Enzymatic synthesis of *N*-acyl-L-amino acids in a glycerol–water system using acylase I from pig kidney. *J. Am. Oil Chem. Soc.*, **79**, 41–46.
45. Infante, M. R., Pinazo, A. and Seguer, J. (1997) Non-conventional surfactants from amino acids and glycolipids: structure, preparation and properties. *Colloids Surf. A*, **123-124**, 49–70.
46. Pérez, L., Pinazo, A., García, M. T. *et al.* (2009) Cationic surfactants from lysine: synthesis, micellization and biological evaluation. Eur. *J. Medicinal Chem.*, **44** (5), 1884–1892.
47. Pinazo, A., Pons, R., Angelet, M. *et al.* (2007) *Book of Proceedings, 2nd Iberic Meeting of Colloids and Interface* (eds A. Valente and J. Seixas de Melo), Coimbra, Portugal, pp. 321–329.
48. Infante, M. R., Molinero, J., Erra, P. *et al.* (1985) A comparative study on surface active and antimicrobial properties of some N^{α}-lauroil-L-α,ω-dibasic amino acid derivatives. *Fette Seifen Anstrichm.*, **8**, 309–313.

49. Pinazo, A., Pérez, L., Lozano, M. *et al.* (2008) Aggregation properties of diacyl lysine surfactant compounds: hydrophobic chain length and counterion effect. *J. Phys. Chem. B*, **112**, 8578–8585.
50. Brito, R. O., Marques, E. F., Gomes, P. *et al.* (2008) Structure/property relationships for the thermotropic behavior of lysine-based amphiphiles: from hexagonal to smectic phases. *J. Phys. Chem. B*, **112** (47), 14877–14887. DOI: 10.1021/jp8042494
51. George, A., Modi, J., Jain, N. and Bahadur, P. (1998) A comparative study on the surface activity and micellar behaviour of some *N*-acylamino acid based surfactants. *Indian J. Chem.*, **37A**, 985–992.
52. Infante, M. R., Seguer, J., Pinazo, A. and Vinardell, M. P. (1999) Synthesis and properties of asymmetrical nonionic double chain surfactants from lysine. *J. Dispersion Sci. Technol.*, **20**, 621–642.
53. Suzuki, M., Yumoto, M., Shirai, H. and Hanabusa, K. (2005) L-Lysine-based supramolecular hydrogels containing various inorganic ions. *Org. Biomol. Chem.*, **3**, 3073–3078.
54. Suzuki, M., Yumoto, M., Kimura, M. *et al.* (2002) Novel family of low molecular weight hydrogelators based on L-lysine derivatives. *Chem. Commun.*, 884–885.
55. Suzuki, M., Nigawara, T., Yumoto, M. *et al.* (2003) L-Lysine based gemini organogelators: their organogelation properties and thermally stable organogels. *Org. Biomol. Chem.*, **1**, 4124–4131.
56. Suzuki, M., Nanbu, M., Yumoto, M. *et al.* (2005) Novel dumbbell-form low-molecular-weight gelators based on L-lysine: their hydrogelation and organogelation properties. *New J. Chem*, **29**, 1439–1444.
57. Suzuki, M., Yumoto, M., Shirai, H. and Hanabusa, K. (2008) Supramolecular gels formed by amphiphilic low-molecular-weight gelators of N^{α},N^{ε}-diacyl-L-lysine derivatives. *Chem. Eur. J.*, **14**, 2133–2144.
58. Suzuki, M., Yumoto, M., Kimura, M. *et al.* (2004) New low-molecular-mass gelators based on L-lysine: amphiphilic gelators and water-soluble organogelators. *Helv. Chim. Acta*, **87**, 1–10.
59. Suzuki, M., Owa, S., Shirai, H. and Hanabusa, K. (2007) Supramolecular hydrogel formed by glucoheptonamide of L-lysine: simple preparation and excellent hydrogelation ability. *Tetrahedron*, **63**, 7302–7308.
60. Furhop, J. H., Spiroski, D. and Boettcher, C. (1993) Molecular monolayer rods and tubules made of alpha-(L-lysine), omega-(amino) bolaamphiphiles. *J. Am. Chem. Soc.*, **115**, 1600–1601.
61. Sebyakin, Y. L. and Budanova, U. A. (2006) pH-sensitive cationic lipopeptides for the design of drug-delivery systems. *Russ. J. Bioorg. Chem.*, **32**, 407–412.
62. Nnanna, I. A., Cheng, G. Y. and Xia, J. (2001) Potential applications of protein-based surfactants, in *Protein-Based Surfactants, Synthesis, Physicochemical Properties and Applications* (eds I. A. Nnanna and J. Xia), Marcel Dekka, New York, pp. 227–260.
63. Sakamoto, K. (2001) Current market developments and trends in amino acid- and protein-based surfactants, in *Protein-Based Surfactants, Synthesis, Physicochemical Properties and Applications* (eds I. A. Nnanna and J. Xia), Marcel Dekka, New York, pp. 261–280.
64. Pérez, N., Pérez, L., Infante, M. R. and García, M. T. (2005) Biological properties of arginine-based glycerolipidic cationic surfactants. *Green Chem.*, **7**, 540–546.
65. Perez, L., Pinazo, A., Garcia, M. T. *et al.* (2007) Cationic surfactants from lysine, in *Book of Proceedings, 2nd Iberic Meeting of Colloids and Interfaces* (eds A. J. M. Valente and J. S. Seixas de Melo), Coimbra, Portugal.
66. Sanchez, L., Mitjans, M., Infante, M. R. and Vinardell, M. P. (2006) Potential irritation of lysine derivative surfactants by hemolysis and HaCaT cell viability. *Toxicol. Lett.*, **161**, 53–60.

67. Rosa, M., del Carmen Morán, M., da Graca Miguel, M. and Lindman, B. (2007) The association of DNA and stable catanionic amino acid-based vesicles. *Colloids and Surfaces A: Physicochem. Engng Aspects*, **301**, 361–375.
68. Heyes, J. A., Niculescu-Duvaz, D., Cooper, R. G. and Sprinter, C. J. (2002) Synthesis of novel cationic lipids: effect of structural modification on the efficiency of gene transfer. *J. Med. Chem.*, **45**, 99–114.
69. Buijnsters, P. J. J. A., García Rodriguez, C. L., Willighagen, E. L. *et al.* (2002) Cationic gemini surfactants based on tartaric acid: synthesis, aggregation, monolayer behaviour and interaction with DNA. *Eur. J. Org. Chem*, **8**, 1397–1406.
70. Kirby, A. J., Camilleri, P., Engberts, J. B. F. N. *et al.* (2003) Gemini surfactants: new synthetic vectors for gene transfection. *Angew. Chem. Int. Ed.*, **42**, 1448–1457.
71. Ronsin, G., Perrin, C., Guédat, P. *et al.* (2001) Novel spermine-based cationic Gemini surfactants for gene delivery. *Chem. Commun.*, **21**, 2234–2235.
72. Sen, J. and Chaudhuri, A. (2005) Gene transfer efficacies of novel cationic amphiphiles with alanine, β-alanine, and serine headgroups: a structure-activity investigation. *Bioconjugate Chem.*, **16**, 903–912.
73. Chaudhary, A. N. and Qadri, R. B. (1990) Higher fatty acids esters of lactic acid. *Pak. J. Sci. Ind. Res.*, **33**, 334–336.
74. Fein, M. L. and Fisher, C. H. (1948) Normal-alkyl beta-hydroxypropionates and beta-acetoxypropionates. *J. Org. Chem.*, **13**, 749–756.
75. Kiran, K. R. and Divakar, S. (2001) Lipase catalyzed synthesis of organic acid esters of lactic acid in non-aqueous media. *J. Biotechnol.*, **87** (2), 109–121.
76. From, M., Adlercreutz, P. and Mattiasson, B. (1997) Lipase catalyzed esterification of lactic acid. *Biotechnol. Lett.*, **19**, 315–317.
77. Torres, C. and Otero, C. (1999) Part I. Enzymatic synthesis of lactate and glycolate esters of fatty alcohols. *Enzyme Microb. Technol.*, **25** (8–9), 745–752.
78. Parida, S. and Nordick, J. S. (1991) Substrate structure and solvent hydrophobicity control lipase catalysis and enantioselectivity in organic media. *J. Am. Chem. Soc.*, **113** (6), 2253–2259.
79. Anoune, N., Nouiri, M., Berrah, Y. *et al.* (2002) Critical micelle concentrations of different classes of surfactants: a quantitative structure property relationship study. *J. Surfactants Deterg.*, **5**, 45–53.
80. Ratchford, W. P. (1950) Preparation and properties of *N*-n-alkylactamides. *J. Org. Chem.*, **15**, 326–332.
81. Anoune, N., Nouiri, M., Arnaud, C. *et al.* (2000) Synthesis and characterization of new cationic surfactants derived from lactic acid. *J. Surfactants Deterg.*, **3**, 381–386.
82. Winter, I., Lohner, K., Janssen, H. G. and Mel'nikov, S. M. (2004) Thermotropic polymorphism of a nonphospholipid mixture: lactylated fatty acid esters of glycerol and propylene glycol. *J. Am. Oil Chem. Soc.*, **81**, 961–969.
83. Winter, I., Pabst, G. and Mel'nikov, S. M. (2006) Miscibility of lactylated monoacylesters of propylene glycol with charged bilayer forming colipids in aqueous solution. *J. Molec. Liq.*, **123**, 86–91.
84. Bezelgues, J.-B., Serieye, S., Crosset-Perrotin, L. and Leser, M. E. (2008) Interfacial and foaming properties of some food grade low molecular weight surfactants. *Colloids Surf. A: Physicochem. Engng Aspects*, **331** (1–2), 56–62. DOI: 10.1016/ j.colsurfa.2008.07.022
85. Shen, Y., Powell, R. L. and Longo, M. L. (2008) Influence of the dissolution rate on the collapse and shedding behaviour of monostearin/monopalmitin-rich coated microbubbles. *Langmuir*, **24**, 10035–10040.
86. Shen, Y., Longo, M. L. and Powell, R. L. (2008) Stability and rheological behaviour of concentrated monodisperse food emulsifier coated microbubble suspensions. *J. Colloid Interface Sci.*, **327**, 204–210.

87. Shen, Y., Powell, R. L. and Longo, M. L. (2008) Interfacial and stability study of microbubbles coated with a monostearin/monopalmitin-rich food emulsifier and PEG40 stearate. *J. Colloid Interface Sci.*, **321**, 186–194.
88. Allen, K. E., Murray, B. S. and Dickinson, E. (2008) Whipped cream-like textured systems based on acidified caseinate-stabilized oil-in-water emulsions. *Int. Dairy J.*, **18**, 1011–1021.
89. Grigoriev, D. O., Leser, M. E., Michel, M. and Miller, R. (2007) Mixed micelles as delivery systems for enhanced emulsifier adsorption at the air/water interface: sodium stearoyl lactylate (SSL)/Tween80 solutions. *Colloids Surf. A: Physicochem. Engng Aspects*, **301**, 158–165.
90. Nuessli, J., Handschin, S., Conde-Petit, B. and Escher, F. (2000) Rheology and structure of amylopectin potato starch dispersions without and with emulsifier addition. *Starch/Stärke*, **52**, 22–27.
91. Chronakis, I. S., Fredholm, A., Öste-Triantafyllou, A. and Öste, R. (2004) Complex formation in aqueous medium of partially hydrolysed oat cereal proteins with sodium stearoyl-2 lactylate (SSL) lipid surfactant and implications for bile acids activity. *Colloids Surf. B: Biointerfaces*, **35**, 175–184.
92. Il'in, M. M., Semenova, M. G., Belyakova, L. E. *et al.* (2004) Thermodynamic and functional properties of legumin (11S globulin from *Vicia faba*) in the presence of small-molecule surfactants: effect of temperature and pH. *J. Colloid Interface Sci.*, **278**, 71–80.
93. St Angelo, A. J. and Vercellotti, J. R. (1989) Phospholipids and fatty acid esters of alcohols, in *Food Emulsifier Chemistry, Technology, Functional Properties and Applications* (eds G. Charalambous and G. Doxastakis), Elsevier, Amsterdam.
94. Giroux, H. J. and Britten, M. (2004) Heat treatment of whey proteins in the presence of anionic surfactants. *Food Hydrocolloids*, **18**, 685–692.
95. Thewlis, B. H. (2006) The fate of stearoyl lactylates when used in breadmaking. *J. Sci. Food Agric.*, **32** (2), 125–128.
96. Lugao, T. G. B. and Tadini, C. C. (2005) Influence of vegetable shortening and emulsifiers on the unfrozen water content and textural properties of frozen French bread dough. *Lebensm. Wiss. Technol.*, **38**, 275–280.
97. Shen, Y., Longo, M. L. and Powell, R. L. (2008) Stability and rheological behaviour of concentrated monodispersed food emulsifier coated microbubble suspensions. *J. Colloid Interface Sci.*, **327**, 204–210.
98. Grigoriev, D. O., Leser, M. E., Michel, M. and Miller, R. (2007) Mixed micelles as delivery systems for enhanced emulsifier adsorption at the air/water interface: sodium stearoyl lactylate(SSL)/Tween80 solutions. *Colloids Surf. A: Physicochem. Engng Aspects*, **301**, 158–165.
99. Castillo, E. J. and Ali, Y. (1994) Topical antimicrobial compositions/contains cationic microbial, polyelectrolyte and antimicrobial-enhancing amt. of anionic surfactant e.g. *N*-acyl sarcosinate or acyl-lactyc-lactate, Patent Nos EP590786-A; AU9344401-A; EP590786-A1; CA2104917-A; JP6211694-A; AU666957-B; JP2594877-B2; EP590786-B1; DE693154406-E; ES2111717-T3; MX184374-B;
100. Shapiro, I., Dado, G., Allen, D. and Sajic, B. (2008) Rinse-out skin/hair composition, e.g. shampoo, body wash, facial cleanser comprises acyl lactylate containing lactylic acid, non cationic surfactant, and water, Patent Nos WO2007100784-A2; WO2007100784-A3.
101. Dahms, G. H., Cook, J. W. (1999) Surfactant blends used for forming stable wet foams in personal care products, Patent No. US5911981-A.
102. Snijders, G. (2006) Stearoyl lactylates for liquid improver system. *Res. Disclosure*, **505**, 496–501.
103. Jovanovic, O. and Pajin, B. (2004) Influence of lactic acid ester on chocolate quality. *Trends Food Sci. Technol.*, **15**, 128–136.
104. Krog, N. (1997) Food emulsifiers, in *Lipid Technologies and Applications* (ed. F. D. Gunstone), Marcel Dekker, New York.

105. Allena, K. E., Murria, B. S. and Dickinson, E. (2008) Whipped cream-like textured systems based on acidified caseinate-stabilized oil-in-water emulsions. *Int. Dairy J.*, **18**, 1011–1021.
106. Winter, I., Lohner, K., Janssen, H.-G. M. and Melnikov, S. M. (2004) Thermotropic polymorphism of a nonphospholipid mixture: lactilated fatty acid esters of glycerol and propylene glycol. *J. Am. Oil Chem. Soc.*, **81** (10), 961–969.
107. Bharatwaj, B., Wu, L. and da Rocha, S. R. P. (2007) Biocompatible, lactide-based surfactants for the CO_2–water interface: high-pressure contact angle goniometry, tensiometry and emulsion formation. *Langmuir*, **23**, 12071–12078.
108. Swern, D., Stirton, A. J., Turner, J. and Wells, P. A. (1943) Fatty acid monoesters of *l*-ascorbic acid and *d*-isoascorbic acid. *Oil Soap*, **20**, 224–226.
109. Tang, L., Zhang, H., Shehate, M. M. *et al.* (2000) Study of the synthesis of ascorbate fatty acid esters catalyzed by immobilized lipase in organic media. *Biotechnol. Appl. Biochem.*, **32**, 35–39.
110. Yan, Y., Bornscheuer, U. T. and Schmid, R. D. (1999) Lipase-catalyzed synthesis of vitamin C fatty acid esters. *Biotechnol. Lett.*, **21**, 1051–1054.
111. Hemeau, C., Girardin, M., Rovel, B. and Miclo, A. (1995) Enzymatic synthesis of fatty acid ascorbyl esters. *J. Molec. Catal. B: Enzym.*, **5**, 19–23.
112. Watanabe, Y., Adachi, S., Nakanishi, K. and Matsuno, R. (1999) Condensation of *l*-ascorbic acid and medium-chain fatty acids by immobilized lipase in acetonitrile with low water content. *Food Sci. Technol. Res.*, **5**, 188–192.
113. Watanabe, Y., Adachi, S., Nakanishi, K. and Matsuno, R. (2001) Lipase-catalyzed synthesis of unsaturated acyl *l*-ascorbate and its ability to suppress the autoxidation of polyunsaturated fatty acid. *J. Am. Oil Chem. Soc.*, **78**, 823–826.
114. Song, Q., Zhao, Y., Xu, W. *et al.* (2006) Enzymatic synthesis of L-ascorbyl linoleate in organic media. *Bioprocess Biosyst. Engng*, **28** (4), 211–215.
115. Mounangaa, T., Gerardin, P., Poaty, B. *et al.* (2008) Synthesis and properties of antioxidant amphiphilic ascorbate salts. *Colloids Surf. A: Physicochem. Engng Aspects*, **318**, 134–140.
116. Sapper, H., Cameron, D. G. and Mantsch, H. H. (1981) The thermotropic behaviour of ascorbyl palmitate: an infrared spectroscopy study. *Can. J. Chem.*, **59**, 2543–2549.
117. Köhler, U., Yang, P. W. and Mantsch, H. H. (1988) Structure and polymorphic phase behaviour of ascorbyl palmitate in water. *Can. J. Spectrosc.*, **33** (5), 122–127.
118. Palma, S., Lo Nostro, P., Manzo, R. and Allemandi, D. (2002) Evaluation of the surfactant properties of ascorbyl palmitate sodium salt. *Eur. J. Pharm. Sci.*, **16**, 37–43.
119. Palma, S., Manzo, R. H., Allemandi, D. *et al.* (2002) Evaluation of the surfactant properties of ascorbyl palmitate sodium salt. *Langmuir*, **18**, 9219–9224.
120. Lo Nostro, P., Capuzzi, G., Pinelli, P. *et al.* (2000) Self-assembling and antioxidant activity of some vitamin C derivatives. *Colloids Surf. A*, **167**, 83–93.
121. Lo Nostro, P., Capuzzi, G. and Mulinacci, N. (2000) Self-assembly and antioxidant properties of octanoyl-6-ascorbic acid. *Langmuir*, **16**, 1744–1750.
122. Ambrosi, M., Lo Nostro, P., Fratoni, L. *et al.* (2004) Water of hydration in coagels. *Phys. Chem. Chem. Phys.*, **6**, 1401–1407.
123. Lo Nostro, P., Ninham, B. W., Ambrosi, M. *et al.* (2003) Hofmeister effect in coagels of ascorbic acid based surfactants. *Langmuir*, **19**, 9583–9591.
124. Takahashi, M., Komuro, F., Niki, E. and Tanaka, K. (1992) Action of fatty acid esters of L-ascorbic acid as antioxidants in phosphatidylcholine liposomal membranes. *Bull. Chem. Soc. Jpn*, **65**, 679–684.
125. Liu, Z. L., Han, Z. X., Yu, K. C. and Zhang, Y. L. (1992) Kinetic ESR study on the reaction of vitamin E radical with vitamin C and its lipophilic derivatives in cetyltrimethylammonium bromide micelles. *J. Phys. Org. Chem.*, **5**, 33–38.

126. Mukai, K., Fukuda, K., Ishizu, K. and Kitamura, Y. (1987) Stopped-flow investigation of the reaction between vitamin E radical and vitamin C in solution. *Biochem. Biophys. Res. Commun.*, **146** (1), 134–139.
127. Bisby, R. and Parker, A. W. (1995) Reactions of excited triplet duroquinone with alpha-tocopherol and ascorbate: a nanosecond laser flash photolysis and time-resolved resonance Raman investigation. *J. Am. Chem. Soc.*, **117**, 5664–5664.
128. Capuzzi, G., Lo Nostro, P., Kulkarni, K. and Fernandez, J. E. (1996) Mixtures of stearoyl-6-*O*-ascorbic acid and alpha-tocopherol: a monolayer study at the gas/water interface. *Langmuir*, **12**, 3957–3963.
129. Capuzzi, G., Lo Nostro, P., Kulkarni, K. *et al.* (1996) Interactions of 6-*O*-stearoylascorbic acid and vitamin K1 in mixed Langmuir films at the gas/water interface. *Langmuir*, **12**, 5413–5418.
130. Capuzzi, G., Kulkarni, K., Fernandez, J. E. *et al.* (1997) Mixtures of ascorbyl-stearate and vitamin D3: a monolayer study at the gas/water interface. *J. Colloid Interface Sci.*, **186**, 271–279.
131. Mounanga, T. K., Gérardin, Ph., Poaty, B. *et al.* (2008) Synthesis and properties of antioxidant amphiphilic ascorbate salts. *Colloid Surf. A*, **318**, 134–140.
132. Han, Y.-K., Oh, S.-G., Shin, S.-I. *et al.* (2002) Stability of alkanoyl-6-*O*-ascorbates in various surfactant aggregates systems. *Colloids Surf. B: Biointerfaces*, **24**, 33–44.
133. Palma, S., Manzo, R. H., Allemandi, D. *et al.* (2002) Solubilization of hydrophobic drugs in octanoyl-6-*O*-ascorbic acid micellar dispersions. *J. Pharm. Sci.*, **91** (8), 1810–1816.
134. Ambrosi, M., Fratini, E., Alfredsson, V. *et al.* (2006) Nanotubes from a vitamin C-based bolaamphiphile. *J. Am. Chem. Soc*, **128**, 7209–7214.
135. Lo Nostro, P., Ramsch, R., Fratini, E. *et al.* (2007) Organogels from a vitamina C-based surfactant. *J. Phys. Chem. B*, **111**, 11714–11721.
136. Palma, S., Manzo, R., Lo Nostro, P. and Allemandi, D. (2007) Nanostructures from alkyl derivatives (ASCn): properties and potential platform for drug delivery. *Int. J. Pharm.*, **345**, 26–34.
137. Palma, S., Manzo, R., Lo Nostro, P. *et al.* (2003) Vehiculization of anthralin into n-alkyl ascorbic acid derivatives coagels. *Acta Farm. Bonaer.*, **22** (4), 305–312.

Part 3

New Ways of Making Renewable Building Blocks

6

Ethylene from Renewable Resources

Anna Lundgren[1] **and Thomas Hjertberg**[2]

[1]*Stiftelsen Chalmers Industriteknik, Gothenburg, Sweden*
[2]*Borealis AB, Stenungsund, Sweden*

6.1 Introduction

This chapter aims to provide information concerning ethylene from renewable resources. The intention is to give the reader an insight into the present development and the possibilities when using renewable resources.

Ethylene is the most important intermediate in the chemical industry. The production volume was about 120 metric tonnes/year in 2007 and is expected to increase to approximately 180 metric tonnes/year by 2020 [1]. The main outlet for ethylene, roughly 60%, is used for polyethylene, followed by ethylene oxide, vinyl chloride and styrene. Ethylene oxide is a key material in the production of surfactants and detergents. It is mainly converted to ethylene glycol which ends up in, for example, polyethylene terephthalate and glycol ether solvents. Vinyl chloride and styrene are almost exclusively used to produce polyvinyl chloride and polystyrene, respectively. Ethylene is an intermediate for more than 50% of the polymer production volume.

Today the feedstock for ethylene is different fossil fuel streams such as ethane, propane, butane and naphtha. They are obtained as products from natural gas processing and petroleum refining. However, the total production of chemicals based on fossil fuel streams only consumes a small fraction compared to energy and transportation. On a global basis approximately 3% of the oil and gas market is used for chemical production. This corresponds to approximately 1% of the global energy consumption. Over the years

Surfactants from Renewable Resources Edited by Mikael Kjellin and Ingegärd Johansson

the chemical industry has optimized its processes, which are now quite effective. Of the incoming feedstock about three-quarters of the mass and about half of the energy is retained in these products.

The increased environmental awareness, in particular with respect to global warming, has led to an interest in renewable feedstocks. The main emphasis has been directed towards renewable fuels, primarily for the transportation sector. Development of sustainable polymers has been on the agenda for at least the last two decades. The dominating driving force has, however, not been the feedstock. A search on the Internet using 'sustainable polymer' gives an enormous amount of hits and in most cases this expression is associated with biodegradable polymers. A lot of research effort has been directed towards this area and some products have been developed to a commercial level. Some examples are starch-based plastics (e.g. MaterBi from Novamont), polyhydroxybutyrate (PHB) (originally developed by ICI) and polylactic acid (PLA). Of these the latter is produced on a relatively large scale (140 000 tonnes/year) by Nature Works in the USA. In fact, this constitutes almost 50% of the total global production capacity of bioplastics, estimated at 300 000 tonnes/year, in 2008 by European Bioplastics [2]. Although the production volume of these polymers is expected to increase it is very small compared to the total production of synthetic polymers; the market share is about 0.1% of the global plastics production [2].

The debate on global warming during the last two to three years has also spurred an increasing interest in renewable feedstocks for polymers. To have a significant impact on, for example, CO_2 release the conversion from fossil to renewable feedstock should thus be directed towards large-volume polymers such as bulk polymers like polyethylene, polyvinyl cloride (PVC), polypropylene and polystyrene. In this perspective it is very important to consider renewable feedstock for ethylene. The intensive debate around plastic bags and the resulting political decisions on restriction of fossil feed stock for plastic bags or demands on biodegradability has added an additional argument to look for renewable feedstock for ethylene. In June 2007 the petrochemical company Braskem, Brazil, announced that they had a pilot plant to convert ethanol from sugar cane to ethylene and that they had produced the first polyethylene based on renewable resources, that is *green polyethylene* [3]. A commercial plant capable of producing 200 000 tonnes/year of green ethylene is under construction and is expected to start in 2011 [4]. Shortly after Braskem's announcement Dow published that they were running a study together with Crystalsev, a Brazilian ethanol producer, aiming at a production of 350 000 tonnes/year, also expected to start in 2011. A Chinese company, Songyuan Jiàn Biochemical, is also aiming at an ethanol-based ethylene plant having a production capacity of 300 000 tonnes/year. The year for start-up is unknown [5]. Approximately 1.3 million tonnes of sugar (from sugar cane) is needed for an annual ethylene production of 350 000 tonnes.

It is obvious that we will see changes in the petrochemical industry in the coming 20–30 years. There are a number of different factors that will influence this development. First of all there are the normal considerations such as the cost of raw material and its availability, process development, capital requirement, and so on. In addition there will be a number of other factors that may be more difficult to predict, such as political decisions concerning a CO_2 tax and directives concerning renewable feedstock and public opinion, which may influence the behaviour of the consumer.

This chapter discusses both the technical–economical side of production of ethylene from renewable resources and the driving forces to do this, having the environmental arguments and the public/political consequences in mind. Polyethylene will be the main outlet for ethylene from renewable resources but other end-products will also benefit from this development. Solvay Indupa has announced that they will start to produce 60 000 tonnes/year of green ethylene in Brazil, aiming at the production of PVC. Start-up is scheduled for late 2010 [6]. Production of the hydrophilic part of green surfactants, that is from green polyethylene oxide, will certainly be another area as soon as price levels of green ethylene can compete with existing market prices for ethylene from petrochemical sources.

6.2 Why Produce Ethylene from Renewable Resources?

There are several reasons why the use of renewable feedstock to produce ethylene should be considered. The reason may also vary with the position in the value chain from feedstock to the final product or even the consumer. The following list is perhaps most valid with respect to a producer of polyethylene.

6.2.1 Price and Availability of Fossil Feedstock

The price of oil shows large variations. After the oil crises in the 1970s and 1980s there was a stable period during the 1990s when the oil price was around \$20–25/barrel, excluding short-term variations. After 2000 the price started to increase slowly and by 2004 it accelerated. During autumn 2008 the price even reached \$140/barrel. After that the demand decreased due to the economic crisis and the price fell to roughly \$40/barrel in the beginning of 2009. Looking ahead it is generally assumed that the demand for petroleum products will increase and many argue that the production capacity of oil will reach a maximum sometime in the next decades, the so-called 'peak oil scenario'. The general belief is that this will lead to an increasing oil price, and levels of \$100–200/barrel are often mentioned as realistic. In many cases an oil price of \$70–80/barrel is often given as a level where renewable feedstock starts to be competitive. The cost level alone should therefore be a strong reason for a petrochemical company to consider an alternative feedstock.

In some markets availability of fossil feedstock like ethane from natural gas starts to become a limiting factor for the industry. In the USA some petroleum refineries (crackers) have been closed down due to expected shortage of feedstock. In other areas the reverse situation is at hand. For instance, in the Middle East the potential supply from flare gas is enormous. If not used for chemical production these volatiles are just flared and therefore the price is much lower than in Europe, USA or Japan. It is thus not surprising that a large part of the new investments in the petrochemical industry are in countries like Saudi Arabia, Kuwait and The Emirates. It is expected that this new capacity will mainly be directed towards the expanding economies in the East such as China and India. From the CO_2 perspective it is also better that the flare gas is used to produce polymer instead of being directly burnt without producing any benefits.

In contrast to its derivatives, ethylene is more difficult to ship and in most cases it is converted to derivates, including polyethylene, in facilities close to the cracker or next to a pipeline. This means that the surplus in the Middle East will not be exported as ethylene. In markets with a limited supply of suitable petroleum or gas fractions other alternative feedstocks such as biomass should become increasingly important from an availability perspective.

6.2.2 The Environmental Image of Plastics

The image of plastics within the general public is low in many countries. This has recently become very clear in the debate around plastics bags that is in progress in many countries. Although many people have a very outspoken opinion that plastics are negative in relation to the environment, most cannot give a distinct answer supporting their opinion. Arguments such as 'it litters' and 'plastics are made of oil – an ending resource' are often given. The first argument may partly be explained due to the use of polymers in disposable products and packaging, both very visible to the general public and often seen in the nature of litter. Almost 40% of the polymer production is used for packaging, which most people encounter on a daily basis.

The second is related to our fear of using an ending resource. The general population is, however, not aware that transportation and energy consumption is the main consumer of fossil fuel and they do not know that only 3–4% of fossil fuel is used for polymer production. Polymer products are simply more visible than energy. The fact that only a small fraction of fossil fuel is used to make polymers surprises many. This is not, however, an excuse for the polymer industry to avoid the issue of using alternative feedstock. Instead, they could gain a lot with respect to their environmental image by working on ways to exchange the fossil feedstock (black carbon) to renewable (green carbon). When this possibility is presented the majority, and also those clearly negative towards plastics, respond very positively. The issue of environmental image has been a natural part for a long time for many companies producing consumer products. The polymer producers are not directly exposed to the opinion of the private consumers as they sell to industrial end users. However, the environmental profile of a chemical company will become more and more important. A shift towards renewable feedstock should give a positive contribution in this respect.

6.2.3 To Be Prepared for Political Decisions

The debate around global warming has made most people aware of the problem and that actions must be taken. This also influences politicians to take legal decisions that are supposed to be positive for the climate. Compared to earlier national actions these measures must be global. One example is the EU directive concerning renewable fuel [7]. Within the EU countries the proportion of renewable fuel should have been 2% in 2005 and this fraction should increase to 5.75% by 2010. Another example is the EU directive concerning the proportion of electricity consumption from renewable resources [8]. A similar decision for polymers, demanding a certain fraction of the feedstock to be renewable, would most probably be very easy for the politicians to take.

There are already decisions influencing the economy of trading with polymers. For packaging materials a fee must be paid to secure the recycling of the products and in some countries exemptions will be made for biobased materials. Another fee is the CO_2 tax, which is expected to be introduced in the coming climate agreement. A biobased polymer, that is a polymer made from renewable feedstock, would thus not be burdened by these kinds of fees, which will make it more competitive.

6.2.4 Demands from Customers

It is obvious that environmental arguments have become important in many sectors. One clear example is hygiene products such as diapers where large companies like Procter & Gamble and SCA Hygiene Products are evaluating alternative materials based on renewable resources. Companies like these, both plastic converters and users, are now starting to put pressure on polymer-producing companies, in particular polyethylene producers, to deliver sustainable solutions where renewable feedstock is an important factor. It should thus be obvious that there is a large potential market waiting.

6.2.5 Recycling via Energy Recovery

Within the EU there are directives on recycling for markets like packaging, cars and electrical/electronic products. The general outline is that a certain fraction of the used product must be material recycled while another fraction can be energy recycled. This has led to increased material recycling, but in some areas it is difficult to do this in an economically sound way. Recycling by burning and recovery of the energy is then an option. In the case of synthetic polymers this will lead to the release of CO_2 and consequently energy recycling will be burdened by CO_2 fees. Materials based on renewable feedstock can be seen as biofuel and will not be considered for CO_2 fees. This gives an additional incentive to produce polymer from renewable feedstock.

6.2.6 Decreased Investments for Industrial End Users

The pressure on companies producing products from chemicals to use renewable feedstock will most certainly increase. From an economical point of view this may influence both material and production costs. Products made of ethylene from renewable resources, whether it is plastics or surfactants, will have the same properties as if they were produced from fossil feedstock. This means that there will be no need to invest in new production equipment.

6.3 Production of Ethylene from Renewable Feedstock

6.3.1 History

Before the growth of the oil industry, starting after the Second World War, several chemicals were produced from renewable resources such as agricultural crops. Now, with the debate around global warming these alternatives are revitalized (the focus in this chapter is on ethylene from renewable feedstock).

There are several ways to synthesize ethylene but only a few have been in commercial use. The most common have been:

- dehydration of ethanol;
- hydrogenation of acetylene;
- separation from coke gases;
- steam cracking of naphtha or ethane.

Dehydration of ethanol was the first experiment that yielded a gas rich in ethylene from ethanol [9]. By using sulfuric acid as catalyst Dutch chemists had demonstrated the formation of ethylene in 1795 [9]. Later on this was developed to commercial production, first using phosphoric acid and later aluminium oxide as catalyst. Before the Second World War this was how ethylene was produced by removal of water from ethanol on a commercial scale, although in much smaller units than the steam crackers of today. In markets with low availability of fossil fuel or with abundant low-cost biomass, such as India and South America, this technology was used well into the 1980s. However, the relative cost of petroleum made the ethanol-to-ethylene process noncompetitive and the production plants were closed.

Hydrogenation of acetylene and separation from coke gases have also been used earlier and with specific driving forces. In 1943, during the Second World War, Germany utilized the first route to produce 33 000 tonnes of ethylene and the second route to produce 18 000 tonnes of ethylene as they were cut off from petroleum supplies [10]. Separation from coke gases offered a possibility in countries with coal-fired power plants and was used, for example, in Great Britain until the beginning of the 1950s [11].

During the 1950s the petroleum industry experienced a rapid development. A new abundant and cheap feedstock, naphtha, became available for the chemical industry and all ethylene needed for polyethylene and other chemical products started to be made from fossil feedstock. Combined with the development of cracker technology this has led to the very cost-effective steam crackers operated today. A typical size of a cracker built today has an ethylene production capacity of up to 1 metric tonnes/year. Gradually ethane and propane obtained either by separation from natural gas or from flare gas in oilfields have been used as feedstock. In areas with large oilfields and low population the latter provides a cheap feedstock. This is an important reason why most of today's investments in cracker capacity are made in the Middle East.

6.3.2 Today

Today there are several reasons to consider the use of renewable feedstock for the production of ethylene as discussed above. Several options can then be considered, although the most obvious are:

1. Dehydration of ethanol obtained by fermentation of sugar, starch or cellulose.
2. Dehydration of methanol obtained from any biomass via gasification to synthesis gas.

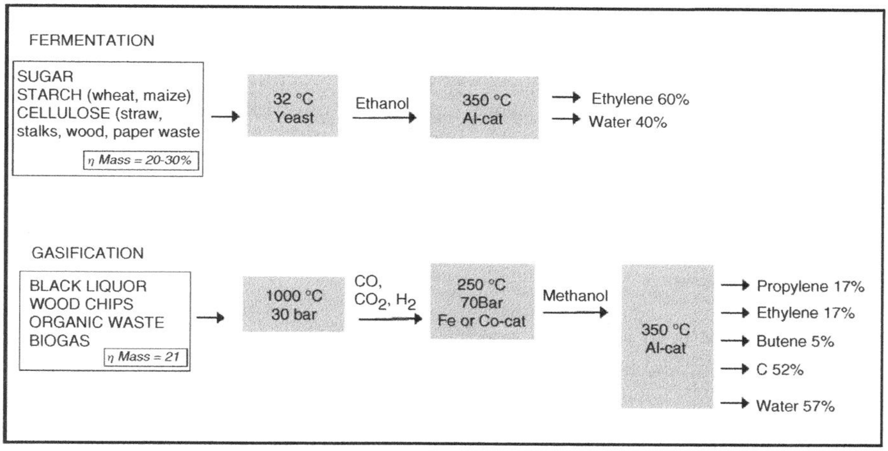

Figure 6.1 *Schematic process of fermentation and gasification routes for ethylene production from renewable feedstock.*

The choice will depend on a number of reasons such as the cost of feedstock and investment capital needed. A schematic process scheme is outlined in Figure 6.1.

6.3.2.1 Gasification

From the perspective of the petrochemical industry the second alternative could look more appealing as the dehydration process step has already been developed, although with natural gas as feedstock. Anything containing carbon and hydrogen could be used as feedstock, but for economic reasons different low-value waste streams are to be preferred. There are a number of projects around the world developing gasification of biomass with the primary aim to produce fuel, methanol or dimethyl ether (DME). One example is Choren, Germany, with the aim to use the synthesis gas formed in the gasification to produce synthetic diesel by a Fischer–Tropsch process [12]. Another is CHEMREC, Sweden, which is developing gasification of black liquor from the pulp industry to produce energy of higher quality, primarily electricity using the synthesis gas in a gas turbine or methanol/DME to be used as automotive fuel [13]. The high temperature needed for the gasification will contribute to a high investment cost for this route (see Figures 6.1 and 6.4 below). CHEMREC announced in September 2008 that they are building the world's first BioDME advanced biofuels plant. It will be ready in 2010 [13].

The technology to form methanol is well known and is used commercially. As already mentioned, the technology to convert methanol to olefins has already been developed and is available to apply on a commercial scale. There are at least two different options, Lurgi's methanol-to-propylene technology [14] and UOP/Hydro's methanol-to-olefin technology [15]. They are similar, but by applying different processing conditions the product mix is somewhat different. The UOP/Hydro process can provide a broad range of propylene-to-ethylene product ratios while Lurgi's process is optimized towards propylene. Note that there will always be a mix of both unsaturated and saturated hydrocarbons, which calls for separations using at least five distillation columns.

Looking at the yield based on the ingoing dry mass of the feedstock it will depend strongly on the carbon, hydrogen and oxygen balance in the feed. Using wood waste with a relatively high oxygen content from cellulose and hydrocellulose, the yield of methanol is approximately 27% while plastic waste containing only limited amounts of oxygen gives a much higher yield of methanol, approximately 70% [16]. The inevitable mass loss due to removal of water in the last step will give a final yield of ethylene–propylene mix of approximately 11 and 28%, respectively.

6.3.2.2 Fermentation

The other alternative is to remove water from ethanol obtained by fermentation of fermentable sugar. Fermentation is a well-known technology that has been used for a long time. The interest to replace fossil fuel (gasoline) in cars with renewable fuel has increased the interest in bioethanol tremendously. The main sources are sugar from sugar cane (Brazil) or starch from corn (USA) or wheat (Europe). The production has increased steeply in the last 10–15 years [17] (see Figure 6.2).

The development of the technology has led to lower costs. Increasing crop yields and development of cheaper enzymes have been the main factors. This has led to a situation where the production cost of ethanol has decreased in contrast to the situation of petroleum. The production cost of ethanol is, however, strongly dependent on the cost of the feedstock and the lowest possible cost is today obtained with sugar cane from Brazil. The price in Brazil is today approximately 3.5 Swedish krona (SEK/l)

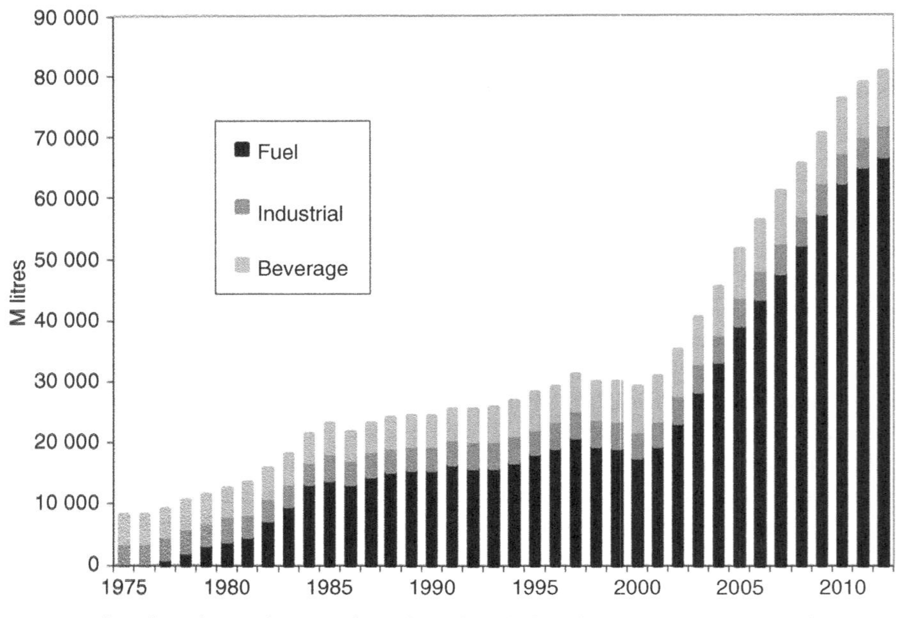

Figure 6.2 *Ethanol production by type. The industrial alcohol market is the smallest of the three.*
From C. Berg, World Fuel Ethanol – Analysis and Outlook. 2004, www.distill.com. Reproduced with permission from C. Berg, F.O. Licht.

freight-on-board (FOB) while the price in the USA and Europe is approximately 60–70% higher due to the lower yield per hectare and a somewhat more expensive process.

6.3.3 Production of Ethylene from Second Generation Ethanol

Conventional bioethanol production from sugar cane or starch-rich crops such as corn or wheat, so-called 'first generation bioethanol production', is well established and the industry is growing throughout the world. Bioethanol produced from lignocellulosic biomasses, such as agricultural residues, wood or energy crops, is a promising alternative as large-scale ethanol production based on crop yields is questioned for ethical and environmental reasons. This and the cost factor have been drivers for the development of what is called the 'second generation biofuels', which is mainly based on lignocelluloses, preferably using waste streams from forestry and farming.

Two aspects must be considered with lignocellulose:

1. Release of cellulose and its hydrolysis to fermentable sugar.
2. Fermentation of the pentoses from the hemicellulose fraction.

These aspects add costs to the process and although huge research resources have been devoted to development of this process technology no large-scale production plant is yet in operation. However, two illustrative examples from the Nordic countries can be mentioned. The Swedish company SEKAB is developing technology to use wood residues for ethanol production [18]. They are operating a pilot plant where they are evaluating weak acid and enzymatic hydrolysis. The yield of ethanol calculated on ingoing dry wood is 15–17%, indicating that the hemicellulose fraction is not utilized. In Denmark the company BioGasol has developed a process directed towards waste from farming, for example straw [19]. The pretreatment is a combination of steam explosion and wet oxidation. Thereafter enzymatic hydrolysis of the pretreated biomass material with combined hydrolysis and fermentation is used. The main product from the hydrolysis is glucose and xylose. The glucose is simultaneously fermented into ethanol by yeast and after the fermentation the process stream is separated into a solid and a liquid fraction. The liquid fraction is led to a xylose fermentation reactor where a specific thermophilic anaerobic bacterium is used for fermentation of xylose into ethanol or ethanol/hydrogen. They report an overall yield of ethanol of 24% and a cost level that approaches that in Brazil. BioGasol is currently expanding their technology and production plants are under construction in both Denmark and the US.

Another option that has appeared in the last few years is a combination of gasification and fermentation. The company Coskata has announced a process that gasifies biomass at 1000 °C and after purification and cooling the resulting synthesis gas is fed to a bioreactor where a proprietary microorganism converts H_2 and CO specifically to ethanol [5]. Coskata claims a high yield, approximately 30%, which can be explained by the fact that the lignin fraction is used as well. The production cost is claimed to be 0.18 euro/l, which is similar to the level in Brazil.

It is important to stress that the ethanol obtained in this process only represents a certain fraction of the energy in the starting feedstock. To come to a process that both gives a low production cost of ethanol and a low environmental impact it is essential that the fraction of energy utilized in different ways is as high as possible. In the fermentation process the lignin must be recovered and, for example, used for energy production and in the Coskata process the hot synthesis gas is used to generate steam to produce electricity. It is also important to use renewable energy in the processes to deliver ethanol, giving a large CO_2 reduction.

6.3.4 Ethanol-to-ethylene

As said above, the conversion of ethanol to ethylene is a rather old technology and there are companies delivering commercial processes, for example Chemtex Corp. [20]. The push to use renewable feedstock has regained interest in this technology, both industrially and from a technological point of view. There are two possible ways that can be used to dehydrate ethanol: intramolecular dehydration to ethylene and intermolecular dehydration to diethyl ether. With increasing temperature ethylene becomes the major product but at higher temperatures dehydrogenation of ethanol to acetaldehyde can occur.

The catalyst used in the earlier industrial process was γ-alumina (Al_2O_3) [21]. This catalyst requires a relatively high temperature for dehydration (450 °C) and results in a relatively low yield of ethanol, approximately 80%. In the 1990s some papers appeared where HZSM-5 zeolite was studied as catalyst for the dehydration of ethanol [22, 23]. At a conversion level of 98% of ethanol the yield of ethylene was reported to be 95% and a much lower temperature could be used, namely 300 °C compared to 450 °C. The strong acidity of HZSM-5 can, however, lead to coking deactivation.

In the development of the methanol-to-olefin process Union Carbide Corporation (UCC) synthesized SAPO-34 to be used as catalyst. SAPO-34 was considered to be the best catalyst, both with respect to activity and selectivity for the production of light olefins [24]. Ni substitution to this catalyst (NiAPSO-34) has been demonstrated to give high yields of ethylene from methanol [25–27]. In a recent paper [28] the four mentioned catalysts, HZSM-5, SAPO-34, NiASPO-34 and Al_2O_3, were compared with respect to selectivity, activity and stability. The optimal temperatures for the catalysts were 300, 350, 350 and 450 °C, respectively. With respect to conversion of ethanol and selectivity to ethylene, HZSM-5 and NiAPSO-34 showed the best performance followed by SAPO-34 and Al_2O_3 in that order. NiAPSO-34 and SAPO-34 showed much better stability and the final conclusion was that NiAPSO-34 is the best choice in the production of ethylene from ethanol.

The process technology of ethanol-to-ethylene has also been developed. Braskem filed a patent in 2007 where substantial improvements are claimed [29]. The most important is an improved process design to maximize the energy recovery from the outgoing stream of the dehydration reactor. Apart from decreased operational cost, this also leads to a reduction in the amount of equipment needed, leading to decreased capital cost.

6.3.5 Ethanol or Methanol to Ethylene

Looking at the announcements made concerning commercial production of ethylene from renewable resources, more or less all of them are based on the ethanol-to-ethylene route.

Those in Brazil, which will be the first to become operational, will use ethanol from sugar cane. This is what could be expected as this represents the cheapest feedstock available today. In China the ethanol is most likely to be produced from grains or cassava.

In the short term the choice of ethanol instead of methanol is obvious as bioethanol is available on a large scale today while there is hardly any production of biomethanol. Another advantage is that the composition of the crude ethylene flow from the reactor is much simpler than that obtained by dehydration of methanol. Mainly it contains water and ethylene with some nonreacted ethanol and minor traces of other substances. Consequently, there is no need for distillation equipment leading to a lower investment. Table 6.1 gives the composition of the crude ethylene stream from the reactor and the composition after water washing, compression and cooling as stated in Table II in the Braskem patent [29]. The ethylene to be sent for final purification is approximately 99.4%, with ethane (0.15%) and butene-1 (0.27%) as the most abundant by-products.

6.4 Commercialization of Bioethylene

The interest in bioethylene, i.e. green ethylene, has pushed process development and commercialization of the ethanol-to-ethylene process. The units that produced ethylene in this way until the 1980s were quite small, at most 50 000 tonnes of ethylene/year, compared to today's announced units, which are considerably larger, Braskem 200 000 tonnes/year and Dow-Crystalsev 350 000 tonnes/year, respectively.

Table 6.1 *Composition of ethylene for final purification (after washing with water, compression and cooling, sent for drying and cryogenic fractioning)*

Component	Molar fraction
H_2O	0.00063
Ethanol	0.00007
Acetic acid	0.00000
Acetaldehyde	0.00002
Ethyl acetate	0.00013
Alcohol C_4^+	0.00001
Methane	0.00058
Ethane	0.00149
Propene	0.00059
1-Butene	0.00270
Ethylene	0.99378
Total	1.00000
Other information	
Vapour fraction	1.0000
Temperature (°C)	16.00
Pressure (bar)	19.61
Molar flow (kgmol/h)	1104
Mass flow (kg/h)	30 956
Molecular weight of ethylene (g/mol)	28.04

Adapted from H. V. Barrocas and A. I. Lacerda, Process for production of ethylene from ethyl alcohol, WO 2007/134415 A2, 2007.

As will be discussed below, polyethylene produced from renewable resources leads to decreased environmental impact, in particular with respect to CO_2. However, will it become an economically viable process? Many question the option to compete with large-volume bulk chemicals whose production has been optimized during a long period of time. In a paper written by people at Dow the situation for production of ethylene from ethanol has been compared with the petrochemical route [30]. A number of different options have been evaluated, for example, with respect to starting feedstock and localization. In the calculation both variable costs (feedstock, maintenance) and capital cost were included for a production plant with an ethylene capacity of 1 metric tonnes/year. The result is given in Figure 6.3. The numbers are based on the price levels in the fall of 2007 and US conditions are assumed if not notified otherwise. The resulting production cost, which is given by the diagonal lines, includes the variable cost and return of investment.

It is possible to point out a number of conclusions from this figure. First of all it is understandable why much of the investment within the petrochemical industry is directed to the Middle East. The feedstock, ethane from flare gas, is cheap and a dedicated ethane cracker requires a relatively low investment. It is also seen that the coming investments in bioethylene in Brazil, with a production cost of approximately \$800/tonne, are very competitive with the fossil-based US alternatives, having a production cost of roughly \$1050/tonne. The variable cost is relatively low but compared to the Middle East ethane alternative the investment is higher due to the fermentation plant. Furthermore, it is important to point out that the investment for the ethanol-to-ethylene plant is very low as seen for ethanol-purchased alternatives. For these cases it is important to stress that these are valid for the conditions in the USA, i.e. the price level of ethanol in the USA. In Europe it is possible to get exemption from the import duty on ethanol imported from Brazil. This would lead to a variable cost of approximately \$700/tonne and thus a production cost of approximately \$900/tonne.

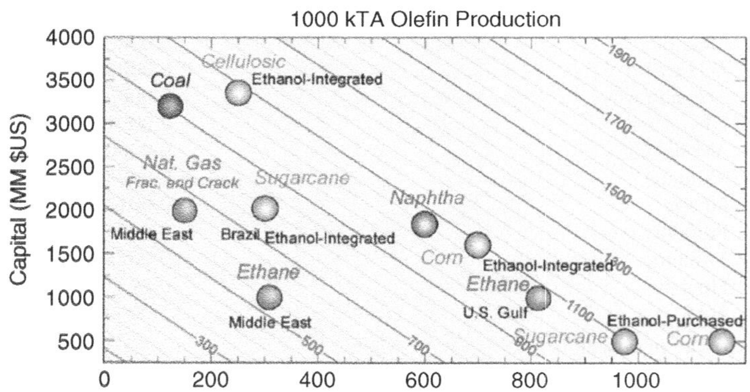

Figure 6.3 *Contour plot of production cost plus return of investment as a function of capital and variable costs (based on 1000 kilotonnes/year of olefin production). Diagonals = cost of production, MM \$US = million US dollars.*

From W. F. Banholtzer, K. J. Watson and M. E. Jones, How might biofuels impact the chemical industry? *ABI/INFORM Trade & Industry*, **104**, 7–14. Reproduced with permission from CEP (Chemical Engineering Progress), March 2008. Copyright 2008 American Institute of Chemical Engineers.

Figure 6.3 is also useful to discuss from a more strategic point of view. Should one go for a high-capital/low-variable-cost or a low-capital/high-variable-cost solution? The former, as, for example, represented by coal as feedstock followed by gasification, conversion to methanol and finally conversion to olefins, represents one extreme solution. It requires a large investment but due to the low cost of coal the production cost will to a large part depend on the depreciation, which is known. On the other hand, the large investment will lock more capital that could have been used for other purposes. A low-capital/high-variable-cost solution like that based on purchased ethanol is more risky as even relatively small changes in the variable cost, i.e. the price of ethanol, can change the profitability drastically.

In addition, there is one more factor that must be included. Within the EU, CO_2 trading has been introduced as one of the means to reduce CO_2 emissions. The petrochemical industry considers it likely that the renewal of the Kyoto protocol will result in a CO_2 tax. The extra cost on the final product will then depend on the fee per unit of emission and the amount of emission per unit of product. For the examples in Figure 6.3 above this would lead to changes in the competitive situation (see Figure 6.4 [30]). The emissions for coal are large, around 5–7 kg CO_2/kg olefin, and for a naphtha- or ethane-based cracker about 0.5–3 kg CO_2/kg olefin. Green ethylene would most probably be free from this kind of tax if biomass is used as the energy source.

6.5 Environmental Impact of Bioethylene

Life cycle assessment (LCA) is one method within the broader field of environmental system analysis. Environmental system analysis is concerned with the assessment of technical systems, in a broad sense, with regard to their impact on the environment and

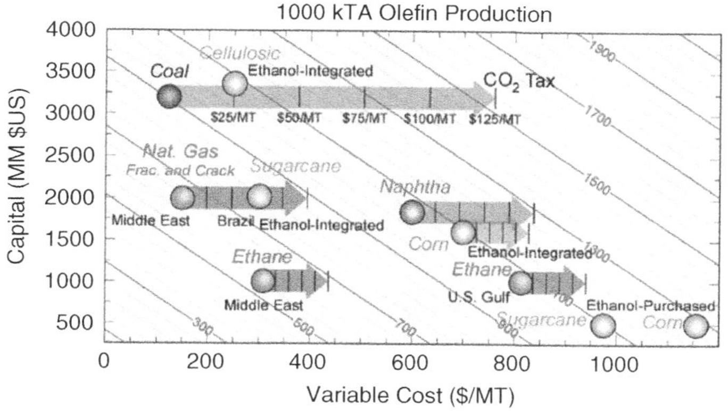

Figure 6.4 *Contour plot of production cost plus return of investment as a function of capital and variable costs including estimates of the potential impact of a CO_2 tax on the various technologies (based on 1000 kilotons/year of olefin production) Diagonals = cost of production, MM $US = million US dollars.*

From W. F. Banholtzer, K. J. Watson and M. E. Jones, How might biofuels impact the chemical industry? *ABI/INFORM Trade & Industry*, **104**, 7–14. Reproduced with permission from CEP (Chemical Engineering Progress), March 2008.

the practical use of such assessment [31]. Environmental assessment aims at informing decision makers, also in a broad sense, ranging from policy makers to producers and consumers, on the environmental consequences of their activities and decisions.

Chemical producers can start to be independent from an ending resource (fossil fuel) by implementing green ethylene and at the same time also stop contributing to the increasing amount of CO_2 emission to the atmosphere (provided that the biomass is from a durable forestry and agricultural system having balance in its carbon flows). However, there is one conflict between this and the effects an increased amount of biomass has on biodiversity, land use, water use and maybe also carbon flows. A change in land use may give an increased carbon footprint if the amount of climate gases increases. There is also a competitive situation between food, fuel and other use (such as plastics made from biomass) that needs to be taken into account.

The fermentation of biomass to ethanol and its subsequent use as a fuel has been investigated in numerous projects (e.g. Concawe/EUCAR 2007 and the BrewProject) [32, 33]. Different biomass feedstocks, e.g. wheat, sugar beet, wood and straw, have been taken into account in these studies. However, when comparing separate studies of the same raw material results vary between studies. Some possible reasons are variation in geographical scope, different alcohol contents (anhydrous or hydrous, if stated at all) under different methodological choices such as allocation and how the alternative use of the material for other purposes has been dealt with.

Apart from its use as a fuel, biomass gains increasing interest as feedstock for the chemical industry. The Brew project in which different biomaterials were assessed focus towards their potential in that sector [33]. A special role is played by the integration into the production of plastics, for which bio-based ethanol and subsequently polyethylene presents one alternative. As already stated, ethanol can be obtained from different sources. However, it seems that all data in the literature on greenhouse gas emissions and energy use for fermentation of Brazilian sugar cane and subsequent refinement to ethanol are based on the reports of one single researcher [34]. This is of course a limitation for the LCA.

As several of the processes 'ethylene from ethanol' do not yet exist, at least not on an industrial scale, process simulation and other data for LCA need to be estimated to understand how 'ethylene from ethanol' compares to a conventional 'ethylene from fossil' route. For reasons of data availability today's LCAs will also probably have to be limited to certain impact categories, i.e. emissions of greenhouse gases and resource use (in terms of raw materials, land use, water use and energy use). For a complete LCA everything from extraction of resources to ultimate disposal must be included and well defined.

6.6 Certificate of Green Carbon Content

Carbon dating is a radiometric dating method that uses the naturally occurring radioisotope carbon-14 (^{14}C) to determine the age of carbonaceous materials up to about 60 000 years. The technique was developed in 1949 [35]. Although fossil fuels have their origin in ancient biomass, they are not considered biomass because they contain carbon that has been 'out' of the carbon cycle for a very long time. This means that the C14 content is zero compared to biomass that is by far younger.

Carbon dating is used in ASTM D6866-08, which is the standardized test method from the American Society for Testing and Materials used to quantify the biobased content of a given product using radiocarbon techniques [36]. The test directly discriminates between carbon resulting from contemporary carbon input (biomass) and that derived from fossil-based input. This standard will be most useful both from the consumer and legislation perspective for determining the carbon origin of chemicals made from ethylene from renewable resources.

6.7 Concluding Remarks

The recent development shows that production of ethylene from biomass is possible, from a technical point of view and, perhaps most importantly, also from an economical perspective. Although the first investments were made in Brazil due to the availability of large volumes of bioethanol at low cost, it is most likely that the ongoing development of production of the second generation of bioethanol will drive the expansion to other parts of the world.

With the large investments needed it is understandable that the chemical industry is hesitating for a rapid investment in bio-based ethylene. It is our belief that the environmental discussion relating to climate change will lead to an increased pressure from public opinion favouring renewable solutions. This will also influence the chemical industry to make a gradual shift in this direction. In fact, companies that take the first steps might be able to gain from this opinion via premium prices.

References

1. (2007) Petrochemical Market Dynamics: Olefins, www.nextant.com (accessed 15 May 2009)
2. European Bioplastics (2008) Bioplastics, Frequently Asked Questions, p. 4, http://www.european-bioplastics.org/index.php?id=191.
3. (2007) Braskem has the First Certified Green Polyethylene in the World, Press release, www.braskem.com (accessed 5 May 2009).
4. (2007) Braskem Confirms Investment to Produce 200 kton/year of Green Polyethylene, 30/10/2007, Press release, http://www.braskem.com.br/site/portal_braskem/en/sala_de_imprensa/sala_de_imprensa_detalhes_6970.aspx (accessed 5 May 2009).
5. www.coskata.com (accessed 15 May 2009).
6. (2007) Solvay News, Polyvinyl Chloride (PVC) Derived from Sugar Cane and Salt, www.solvay.com (accessed 15 May 2009).
7. (2003) European Directive 2003/30/EC on the Promotion of the Use of Biofuels or Other Renewable Fuels for Transport.
8. (2001) European Directive 2001/77/EC on the Promotion of Electricity Produced from Renewable Energy Sources in the Internal Electricity Market.
9. *Crell's Annalen*, **1795** (2), 195, 310, 430.
10. Howat (1946) *Coke Smokel. Fuel Age*, **8**, 177.
11. Stanley, H. M. (1969) History of ethylene and its industrial utilisation, in *Ethylene and Its Industrial Derivatives* (ed. S. A. Miller), Ernest Benn Ltd, London.
12. Choren (2009) The Key Element in the Technology: The Carbo-V Process, www.choren.com (accessed 12 May 2009).

13. (2008) CHEMREC to Build World's First BioDME Advanced Biofuels Plant, Press release, www.chemrec.se (accessed 12 May 2009).
14. Gas to Chemicals, http://www.lurgi.com/website/fileadmin/user_upload/1_PDF/1_Broshures_Flyer/englisch/B_0305e_Gas-to-Chemicals.pdf (accessed 12 May 2009).
15. (2004) UOP/HYDRO MTO Process Methanol to Olefins Conversion, www.uop.com (accessed 12 May 2009).
16. Nouri, S. and Kaggerud, K. (2006) Waste-to-Plastics: Process Alternatives, Department of Energy and Environment, Environmental Systems Analysis Division, p. 71.
17. Berg, C. (2004) World Fuel Ethanol – Analysis and Outlook, www.distill.com (accessed 15 May 2009).
18. Ethanol Pilot Plant – Introduction, www.sekab.com (accessed 15 May 2009).
19. Concept and Technology, www.biogasol.com (accessed 5 May 2009).
20. Bio-Ethylene/Ethylene Oxide/Ethylene Glycol, www.chemtex.com/templates/ethylene.html (accessed 15 May 2009).
21. Winter, O. and Ming, E. (1976) Make ethylene from ethanol. *Hydrocarb. Process.*, **55** (11), 125–133.
22. Phillips, C. B. and Datta, R. (1997) *Ind. Engng Chem. Res.*, **36**, 4466–4475.
23. Nguyen, T. M. and Mao, R. L. V. (1990) *Appl. Catal.*, **58**, 119–129.
24. Wilson, S. and Barger, P. (1999) *Microporous Mater.*, **29**, 117–126.
25. Inoue, M., Dhupatemiya, P. and Inui, T. (1999) *Microporous Mesoporous Mater.*, **28**, 19–24.
26. Kang, M. (1999) *J. Molec. Catal. A: Chem.*, **159**, 205–212.
27. Kang, M. (2000) *J. Molec. Catal. A: Chem.*, **160**, 437–444.
28. Zhang, X., Wang, R., Yang, X. and Zhang, F. *et al*. (2008) Comparison of four catalysts in the catalytic dehydration of ethanol to ethylene. *Microporous Mesoporous Mater.*, **116** (1–3), 210–215.
29. Barrocas, H. V. and Lacerda, A. I. (2007) Process for production of ethylene from ethyl alcohol, WO 2007/134415 A2.
30. Banholtzer, W. F., Watson, K. J. and Jones, M. E. (2008) How might biofuels impact the chemical industry?, *ABI/INFORM Trade & Industry*, **104** (3), 7–14.
31. Baumann, H. and Tillman, A.-M. (2004) *The Hitch Hiker's Guide to LCA. An Orientation in Life Cycle Assessment Methodology and Application*, Studentlitteratur, Lund.
32. (2007) JRC/CONCAWE/EUCAR, http://ies.jrc.ec.europa.eu/WTW.
33. Patel, M. K., *Medium and long-term opportunities and risks of the biotechnological production of bulk chemicals from renewable resources*, http://www.chem.uu.nl/brew/.
34. Macedo, I. d. C. (2004) Assessment of Greenhouse Gas Emissions in the Production and Use of Fuel Ethanol in Brazil, Government of the State Sao Paulo, Brazil.
35. Arnold, J. R. and Libby, W. F. (1949) Age determinations by radiocarbon content: checks with samples of known age. *Science*, **110** (2869), 678–680.
36. ASTM Standard D6866-08 (2008) *Standard Test Methods for determining the Biobased Content in Solid, Liquid, and Gaseous Samples Using Radiocarbon Analysis*, ASTM International, www.astm.org.

7

Fermentation-Based Building Blocks for Renewable Resource-Based Surfactants

Kris Arvid Berglund, Ulrika Rova and David B. Hodge
Division Chemical Engineering, Luleå University of Technology, Sweden

7.1 Introduction

In this chapter we will discuss the current and potential use of fermentation-based building blocks for surfactants. Surfactants are generally synthesized by the chemical bonding of two moieties, usually a hydrophilic head and a hydrophobic tail. The features that are important for a head group include both its production cost and chemical properties relating to the synthesis of the surfactant and its desired application such as charge and solubility. Functional groups such as hydroxyl, carboxyl and carbonyl can be used for reactively linking the head group to the tail. Organic acids are an excellent choice as raw materials for surfactants because they generally possess an anionic charge as well as functional groups. A number of carboxylic acid cellular metabolites are currently used as building blocks for head groups in numerous applications. An example is lactic acid, which has both acid and alcohol functionalities and thus can participate in a wide range of reactions including condensation, esterification, reduction and substitution reactions.

Microbes are capable of synthesizing complete surfactants with a wide range of structures and functionalities [1]. However, a more promising route for the economic production of surfactants through bioprocessing is to focus on only the organic acid metabolites for use as the hydrophilic moiety. The diverse fields of existing applications for common metabolites are listed in Table 7.1 to give an overview of the range of structures and potential for functional versatility.

Surfactants from Renewable Resources Edited by Mikael Kjellin and Ingegärd Johansson

Table 7.1 Production and applications of carboxylic acid produced on a commercial scale

Carboxylic acid	World production (tonnes/year)	Production method	Current application	References
Citric acid	1 600 000	Biochemically by aerobic cultivation of *Aspergillus niger*	Food and beverage acidulant, pharmaceuticals	[2]
Lactic acid	120 000	Biochemically by anaerobic cultivation of *Bacillus* and *Lactobacillus* spp.	Polylactic acid used in the production of biodegradable polymers	[3]
Acetic acid	70 000 000	Petrochemical by carbonylation of methanol	Vinyl acetate used for the production of polymers, solvents	[4]
	190 000	Biochemically using *Acetobacter* (aerobic) and *Clostridium* (anaerobic) spp.	Food purposes (i.e. vinegar)	
Succinic acid	15 000	Petrochemical by hydration of maleic anhydride	Food acidulant and anti-microbial agent, fuel oxygenate, detergents and surfactants, ion chelator, pharmaceuticals	[5]
Fumaric acid	90 000	Petrochemical by hydration of maleic anhydride	Food and beverage additives, polymer production	[6]
Malic acid	25 000	Petrochemical by hydration of maleic anhydride Enzymatic conversion of fumaric acid to L-malic acid	Food and beverage additives, pharmaceuticals	[4]

Besides their functional versatility as outlined above, much of the growing interest in carboxylic acids as building blocks is driven by factors relating to both their appeal as green chemicals and positive process economics. Many carboxylic acids are capable of being produced industrially through biocatalytic processes as a result of their placement in cellular metabolic pathways (Figure 7.1). Important implications of these chemicals being cellular metabolites are low toxicity, since these are present in living cells, and quick biodegradability due to easy reintegration back into common metabolic pathways. Compared to chemical synthesis, bioconversion processes are ultimately catalysed by an enzyme or a cascade of enzymes resulting in high product specificity at mild process conditions such as low temperature. The raw materials for biocatalytic conversion are carbohydrates rather than hydrocarbons, with the consequence that further processing may be in an aqueous phase, which requires the development of new catalysts and reaction schemes for further processing of these biologically derived building blocks. Another important consequence of using plant carbohydrates such as sucrose, starch and potentially lignocellulosic biomass is that these are ultimately derived from sequestered

	Monoesters	Diesters
R_1	Alkyl Alcanolamide Ethoxylate alkyl	Alkyl
R_2	H	Alkyl

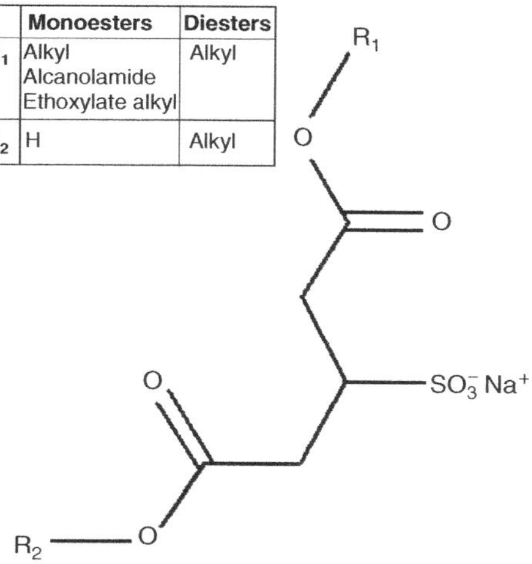

Figure 7.1 *General structure of the sulfosuccinate class of surfactants.*

CO_2, making these products both renewable and carbon neutral in contrast to petroleum-based feedstocks. As a result of these distinct features, fermentation-based building blocks can be considered as green chemicals. As such, these compounds have appeal in consumer applications such as food and cosmetics.

Fermentation-based building blocks are becoming increasingly more competitive as a result of both changes in the feedstock economics and improved technologies for biological production. The recent development of tools for the engineering of entire metabolic pathways of microorganisms at the genetic level have resulted in improvements in yields and productivities for biotechnological processes and increased the number of potential raw materials available for bioconversion to include lignocellulosic biomass sugars [7]. These advances, in combination with rising costs for petroleum and natural gas and increasing awareness of environmental concerns, make bio-based alternatives to petrochemical building blocks more attractive. Several bioprocesses have recently surpassed traditional chemical synthesis processes from hydrocarbon feedstocks in terms of economic competitiveness. Lactic acid is such an example, where production through bioprocessing has seen rapid growth and replacement of the synthetic processes that once dominated the market [8].

7.2 Existing and Potential Classes of Surfactants from Biologically Derived Metabolites

7.2.1 Carboxylic Acid Esters

One class of surfactants commonly used for applications in the food industry consists of carboxylic acids esterified with a single tail group such as a monoglyceride or, alternatively, for lactylates esterification of the alcohol group on lactic acid or its polymers with a fatty acid. These anionic surfactants are produced commercially using lactic, acetic and citric acid as well as the cyclic cognates of fumaric and succinic acid, such as maleic anhydride [9, 10]. As food additives, biologically produced natural fatty acids, monoglycerides, carboxylic acids and many compounds synthesized from these starting molecules have been approved for food and consumer applications by government regulatory agencies such as the Food and Drug Administration (FDA) due to being well characterized and able to be readily integrated into existing metabolic pathways.

Natural fatty acid monoglycerides alone are among the most widely used food additives [11] and linking these monoglycerides with a carboxylic acid through esterification yields an important class of compounds with applications primarily as food emulsifiers [9]. For example, citric acid esters of monoglycerides are used as emulsion stabilizers to prevent fat separation and are an alternative to soya-lecithin. Lactic acid esters of monoglycerides, such as glyceryl lactostearate and glyceryl lactopalmate, are commonly used to improve the whipping properties of dairy and nondairy creams due to their strong foaming behaviour. Succinic acid esters of monoglycerides, currently produced using succinic anhydride, are used as stabilizers for protein–starch mixtures and emulsifiers for shortening. Acetic acid esters of monoglycerides are not as effective as emulsifiers but have applications as plasticizers and as a foam stabilizer in food.

Another class of carboxylic acid-derived surfactant is the sodium or calcium salts of fatty acid esters. Commercial examples include metal salts of stearic acid esterified with a dimer of lactic acid (sodium stearoyl lactylate) or maleic anhydride (sodium stearoyl fumarate), which are used as emulsifiers in bread making and in bread preservation due to their properties of preventing starch crystallization and dispersing fats [12, 13].

Besides the examples listed above, a wide range of other compounds with diverse properties can be envisaged using different tail groups other than natural fatty acids and monoglycerides. As an example, alkoxylated alcohols have been esterified with carboxylic acids such as citric acid with potential applications for high-viscosity surfactants in cosmetics [14]. In addition to esters, amides of carboxylic acids linked with lipophilic groups are also known from the patent literature [15].

7.2.2 Amino Acids

Amino acids as the polar head groups for surfactants have been produced using aspartic acid, glutamic acid, arginine, alanine, glycine, leucine, praline, serine and protein hydrolyzates. Amino acids have the ability to be linked to tail chains through the α-amino, α-carboxylic or side-chain groups [16]. Three general classes are those with a single tail, those with a glycerolipid-like tail structure and dimeric or 'gemini' surfactants [17]. Currently, surfactants from amino acids are not widely produced due to low yields and productivities and significant costs in downstream processing.

7.2.3 Sulfonates

Sulfonates make up the largest share of the market for anionic surfactants. Sulfosuccinates (Figure 7.1) are composed of a sulfosuccinic acid metal salt, such as sodium, linked with a mono- or diester to tail groups (R_1 and R_2) such as alkyl groups for diesters or alkyl, alcanolamide and ethoxylated alkyl groups for monoesters [18].

Sulfosuccinates often have improved properties for solubilization, wetting, foaming and dispersing relative to other anionic surfactants, and have applications in dry cleaning solvents, printing, textile dying, paints and coatings, emulsion polymerization, cosmetics and shampoos [19]. Diesters are used as wetting agents and dispersants and are not used in personal care products.

In current production processes, maleic anhydride is reacted with any variety of hydroxyl containing lipophilic compounds in a ring opening followed by sulfonation of the double bond on the maleic ester [20]. The reaction conditions can be controlled such that the mono- or diester is the preferred product, and it is possible to tailor the properties of the monoesters by changing the chain length or saturation of the alkyl group or by using an ethoxylated fatty alcohol or alkanolamides of fatty acids. For the diesters it is possible to change the properties by using either linear, branched or cyclic alkyl groups [20].

While sulfosuccinates are currently not manufactured with fermentation-based carboxylic acids, it is reasonable that bio-based succinic, fumaric or maleic acid could be used as the four-carbon head group, replacing maleic anhydride. For these feedstocks to be successful, effective processes and catalysts for aqueous esterification [21] and sulfonation [22] of these carboxylic acids need to be better characterized.

7.3 Fermentation-Based Building Blocks with Large Existing Markets

Citric, acetic and lactic acid are the top three industrially significant carboxylic acids (Table 7.1). The industrial production of these three chemicals can be characterized in terms of the historical developments relating to process technology and process economics. Both lactic and acetic acid have been used in food processing since ancient times. Industrial production of both of these chemicals in the early twentieth century was through fermentation processes. With the development of the petrochemical industry and the associated technology for industrial catalysis and separations, the synthetic route from hydrocarbons became economically superior and largely replaced the bio-route. As mentioned previously, lactic acid production by fermentation has regained dominance in the industry, while economics are still strongly in favour of the petro-route for acetic acid due to difficulties in recovery and purification from aqueous solutions. Citric acid, on the other hand, has always been a profitable biological metabolite either by extraction from plants or fermentation.

7.3.1 Citric Acid

Citric acid is an industrially important six-carbon tricarboxylic acid with global production in 2007 of over 1.6 million tonnes [2] with an expected increase in demand and consumption of up to 3.5–4.0% annually [23]. Current markets for citric acid include food and beverage, pharmaceuticals, detergents, metal cleaning, textile dyes and cosmetics [24]. The majority of citric acid is used in the food and beverage industry as an acidulant and flavour additive [25] and it is also used as an additive to detergents due to its properties as a chelating agent and for pH regulation (Table 7.1).

Citric acid is currently produced commercially from sucrose from sugar cane and sugar beet molasses or glucose from hydrolysed starch by the filamentous fungi *Aspergillus niger* and *Candida* yeast strains. Many other low-cost substrates show potential, such as polymeric xylan and arabinoxylan from oat spelt [26] or glycerol [23], which is a growing by-product of biodiesel production.

The majority of citric acid is currently produced by *A. niger* in submerged fermentation, although other technologies such as solid-state fermentation or continuous processes are economically promising as well. Submerged culture production with *A. niger* was developed with the finding that growth limiting concentrations of nitrogen and phosphorus as well as trace metal levels can trigger acid production [27]. Submerged fermentation is performed in a two-stage process utilizing first a growth phase followed by a production phase, with the entire batch process requiring between 3 and 14 days [28]. The media composition and reactor conditions for each phase are optimized for their respective function. Critical variables in the growth and production phase are sugar type and concentration, pH, temperature, aeration and mineral composition. In particular, manganese, a cofactor necessary for isocitric acid dehydrogenase activity, must be limited to encourage citric acid accumulation and high sugar concentrations are necessary to ensure that the glycolytic enzymes are active for feeding into the tricarboxylic acid (TCA) cycle. Cell growth suppression is necessary during the production phase of the process in order to maximize the flow of carbon to citric acid rather than into cell mass.

The primary technology for citric acid recovery from fermentation broths is through precipitation with $CaCO_3$, although other technologies include ion exchange, carbon adsorption, membrane filtration, chromatography, and liquid extraction [29]. Unlike other TCA cycle carboxylic acids, citric acid is excreted in the acid form rather than precipitated as citrate salt [30], which has the implication that the cell mass can first be filtered, followed by crystallization with lime with subsequent separation by filtration. The filter cake is next acidified with sulfuric acid, which precipitates the calcium as gypsum and the aqueous citric acid can next be decolourized and recrystallized by evaporation [29].

7.3.2 Acetic Acid

Acetic acid is a two-carbon monocarboxylic acid with many applications in food and as a building block for a wide range of industrial chemicals (Table 7.1). While much of the industrial acetic acid production is used directly for applications such as solvents, its primary use by volume is as a feedstock for the production of other chemicals such as vinyl acetate, which is used in plastics, adhesives and paints. Other products include cellulose acetate, acetic anhydride and acetate esters. The synthetic production route has been the dominant process since 1950 [31] using either hydrocarbon-derived ethylene or methanol as the feedstock. The 10% currently produced by microbial conversion is directed towards food applications.

Acetic acid fermentation has been known since antiquity and both aerobic and anaerobic biological routes exist for acetic acid production [31] with *Acetobacter* species being the most significant aerobic strains and *Clostridium* for anaerobic strains. Ethanol and sucrose are the primary feedstocks although there is interest in utilizing sugars from lignocellulosic biomass as well [32]. Submerged bacterial cultivation is the most important process for biological acetic acid production. Important variables include controlling the pH above the pK_a value of acetic acid and sufficient aeration for the aerobic process. For the ethanol conversion process, substrate concentrations greater than 100 g/l can be inhibitory to bacteria [31], and fed-batch techniques have been employed to circumvent substrate inhibition. Recovery of acetic acid from fermentation broths can be done through liquid–liquid extraction (ethyl acetate or diethyl ether) followed by azeotropic distillation, although anion exchange chromatography is also a promising technology. These recovery issues comprise the processing bottleneck that currently limit economic biological acetic acid production on a large scale.

7.3.3 Lactic Acid

Lactic acid is a three-carbon alpha-hydroxy acid that exists in two enantiomers, whose carboxyl and hydroxyl groups can undergo a number of reactions that make it a versatile building block. The global annual demand for lactic acid is estimated to be at least 130 000 tonnes with approximately 85% of this demand directed towards food and food-related applications [8]. Out of the food applications, more than 50% is used for producing emulsifiers such as lactic acid esters of fatty acids and monoglycerides discussed previously [3]. However, the new applications, such as green solvents and polylactic acid (PLA) plastics, have driven a sharp increase in production in recent years [33].

The current commercial synthetic route is based on reacting hydrogen cyanide with acetaldehyde followed by hydrolysis of the lactonitrile, producing a racemic mixture of lactic acid. However, the vast majority of lactic acid is currently produced industrially via the biological route [4]. One advantage of the fermentation route is that either the D- or L- or specific ratios of the lactic acid enantiomers can be produced depending on the microbial species used. All commercial fermentation processes use one of several *Lactobacillus* species as well as an *A. niger* and utilize a wide range of sugar substrates such as sucrose from beet and cane sugar, whey and glucose from hydrolysed starch [3]. Future feedstocks may also include lignocellulosic biomass sugars with proficient xylose fermentation demonstrated with, for example, the yeast *Pichia stipitis* [34] and metabolically engineered *Escherichia coli* strains [7, 35] and hydrolyzates of wheat straw and softwood by *Lactobacillus* [36] and thermophilic *Bacillus* species [37].

Lactic acid is neutralized with lime as it is produced up to a concentration of 100 g/l since low pH inhibits the fermentation. After sterilization, the broth is acidified with sulfuric acid to 1.8 pH to precipitate gypsum and yield free lactic acid. The precipitated salts and biomass are next removed by filtration and the liquor is treated with activated carbon. The free lactic acid is concentrated with ion exchange and evaporated. To achieve high-purity lactic acid for use as a feedstock for other processes, the lactic acid is esterified with an alcohol, purified with distillation and hydrolysed with water [3].

7.4 New Fermentation-Based Building Blocks

Two recent international studies focusing on potential chemicals that can be produced from renewable biomass and serve as economic drivers for a biorefinery have both identified three C_4-dicarboxylic acids as top candidates: succinic, fumaric and malic acids

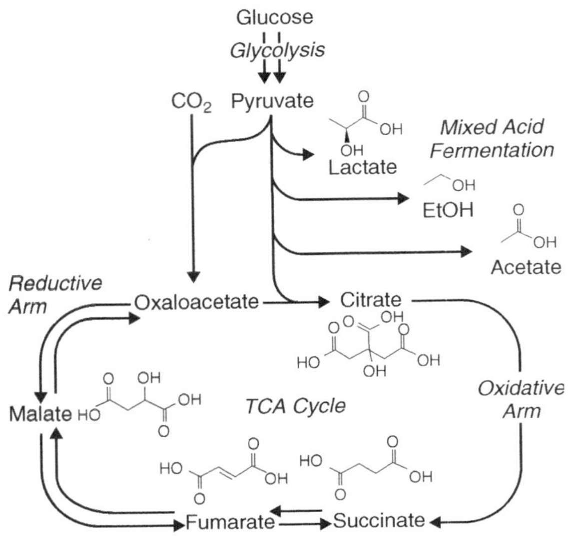

Figure 7.2 Placement of carboxylic acid metabolites within the central catabolic pathways for sugar metabolism.

[4, 38]. Currently, they are all produced chemically from petrochemical resources, using maleic anhydride as an intermediate. All three acids are metabolites synthesized in the TCA cycle (Figure 7.2) in many microbes and in all plants and animals. Increased interest for production of novel biodegradable polymers has intensified research activities focusing on bio-based production of these acids as a starting material for polymerization reactions. Manufacturing costs for the carboxylic acids are affected by productivity and yield, raw material cost and utilization and recovery methods. Therefore, exploration of microorganisms with well-defined metabolic pathways enabling a flexible use of different feedstocks in combination with optimization of the recovery process would allow cost competitive biochemical production of the acids from renewable resources. In addition to the production of biodegradable building blocks, a bio-based production pathway would give extra environmental effects, as, in addition to CO_2 fixation in sugar raw material, it involves CO_2 fixation during metabolism by one of two reductive carboxylation steps.

7.4.1 Succinic Acid

Succinic acid produced from maleic anhydride is primarily used as a surfactant/detergent extender/foaming agent [5]. In addition to its current use, succinic acid has the potential to produce a wide range of products and derivatives: diesel fuel oxygenates; biodegradable de-icing chemicals; glycol-free engine coolants; and polybutylene succinate (PBS), a biodegradable polymer that can replace polyethylene and polypropylene and solvents that can replace chlorinated and other volatile organic compound (VOC)-emitting solvents.

However, most of the potential applications for succinic acid are currently covered by maleic anhydride and therefore development of new succinic acid derivates has been limited. The increased interest for production of novel biodegradable polymers has intensified research activities focusing on bio-based production of succinic acid. To obtain a fermentation process that can be commercialized, two approaches have mainly been pursued, (a) the isolation of natural succinic acid producing organisms and (b) the metabolic engineering of *E. coli*. *A. niger* can also produce succinic acid, but due to complex fermentation, separation and purification processes, fermentation using filamentous fungi has not at the present time been subjected to further development. Most attention has been towards bacterial fermentation using *Anaerobiospirillum succiniciproducens*, *Actinobacillus succinogenes*, *Mannheimia succiniciproducens* and *E. coli*, which all produce succinic acid by phosphoenolpyruvate (PEP) carboxylation combined with both the reductive and oxidative arms of the TCA cycle [39]. *A. succinogenes* and *M. succiniciproducens* produce succinic acid as the major fermentation product, while *E. coli* produces succinic acid as one of several products during mixed-acid fermentation. *A. succinogenes* is a facultative anaerobe that has many benefits as an industrial producing strain; for example, it can utilize a wide range of substrates (glucose, cellobiose, lactose, xylose, arabinose and fructose) with a succinate yield of up to 1.33 mol per mol of glucose [40]. It can also tolerate elevated concentrations of succinate salts. By-product formation (acetic, formic, propionic and pyruvic acid) is a major obstacle since it has a negative impact on subsequent separation and purification processes. If *A. succinogenes* should find industrial use, it is important to develop genetic tools and methods to engineer the strain to produce succinate as the sole fermentation product.

Both *M. succiniciproducens* and *A. succiniciproducens* are interesting candidates for cost-effective succinic acid production from renewable resources, as fermentation of wood hydrolysate has been demonstrated [41, 42]. Industrial use of *A. succiniciproducens* might be complicated since it is a strictly anaerobic organism. Although *M. succiniciproducens* fermentation also generates a mixture of acids, a number of genetic and proteomic studies has resulted in a number of available genetic engineering tools and methods [43]. In spite of the problems with the formation of mixed acids, the most easily genetically manipulated succinic acid producer is *E. coli*. A number of standard genetic tools and methods facilitate introduction of heterlogous genes or inactivation of pathways competing with succinic acid production. The flux of carbon can then be directed towards succinic acid, avoiding the production of mixed acids, resulting in an increased yield from 0.2 to 1.714 mol succinate/mol of glucose [44]. The growth requirements and sugar uptake and utilization system have been thoroughly studied, which will also facilitate further industrial exploitation. A number of engineered strains have been developed and characterized. For example, the strains originating from the US Department of Energy's alternative feedstock program (AFP), AFP111 and AFP184, produce succinic acid as the main product [45, 46]. AFP184 produces succinic acid to final concentrations of up to 77 g/l with productivities in the range of 1.5–3 g/l h from glucose, fructose and xylose using a low-cost industrially relevant medium [47]. AFP184 can also efficiently ferment wood hydrolysates (unpublished work).

7.4.2 Fumaric Acid

Fumaric acid along with its isomer maleic acid is the *trans* version of an unsaturated four-carbon dicarboxylic acid (Figure 7.2). Current industrial production is based on the conversion of petrochemical feedstocks, but part of the production pathway can be replaced with enzymatic conversion using *Pseudomonas alcaligenes* [48] or thermostable maleate isomerase from *Bacillus* spp. [49]. The enzymatic conversion allows isomerization of maleic acid into fumaric acid without the unfavourable by-product formation in the chemical conversion process. Fumaric production by fungal fermentation was conducted at commercial quantities in the United States during the 1940s, but was later replaced with chemical synthesis when petroleum became a cheaper feedstock [30]. Extensive screening for efficient production strains has identified a number of *Rhizopus* strains that produce up to 1.3 mol of fumaric acid per mol of glucose under both aerobic and anaerobic conditions [30]. Fumaric acid is mainly produced by filamentous fungi as an intermediate of the TCA cycle by two different pathways. The first pathway, which is a reductive carboxylation of pyruvate during aerobic conditions, involves fixation of CO_2 via carboxylation of pyruvate resulting in oxaloacetate formation. During the growth phase, the TCA intermediates formed are withdrawn for further increase of biomass, but growth-limiting conditions stop the growth phase and instead intermediates such as fumaric acid accumulate as pyruvate carboxylation continues. The most critical parameter for overproduction of fumaric acid is the ratio of glucose to nitrogen, where a low nitrogen concentration is preferred, although phosphorus limitation can also be employed [50]. The second pathway is the oxidative arm of the TCA cycle. If eukaryotes are used, fumaric acid can also be produced through a cytosolic pathway by the isoenzymes of pyruvate carboxylase, nicotinamide adenine dinucleotide (NAD)-malate dehydrogenase and fumarase, respectively. Feedstocks such as xylose have been tested, but were not as successful as glucose with respect to yield. Xylose fermentation using *R. arrhizus* resulted in a volumetric productivity of 0.087 g/l h [51]. As in the case with succinic acid, fumarate production is inhibited when the acid reaches concentrations of 30–40 g/l. Another difficulty associated with the use of *Rhizopus* spp. as the fermenting organism is the complex morphology typically associated with filamentous fungi. Typically growth and organic acid production are dependent on a specific morphology, which in turn is dependent on the chemical and physical culturing conditions [52]. In the case of *R. oryzae*, fermentation at low initial pH favoured formation of small cell pellets, which facilitated oxygen transfer and high fumaric acid yields [53]. Development of growth conditions that favour a morphology that allows a high oxygen transfer is one challenge needed to be solved before biochemical conversion can replace the petrochemical route.

Continuous neutralization of pH is needed during fumaric acid production. The production cost would therefore be reduced if an acid-resistant microorganism could be identified and used. $CaCO_3$ has been the preferred neutralizing agent as it, besides buffering capacity, provides a source for CO_2. It also precipitates the fermentation product as calcium fumarate which has a positive effect on reducing product inhibition. However, the low solubility of calcium fumarate causes a viscosity problem that reduces the oxygen transfer, resulting in limited growth. During downstream processing, the use of $CaCO_3$ requires additional heating to dissolve the precipitated salt. A solution to circumvent these process problems is to use Na_2CO_3, since sodium fumarate has a higher solubility than the calcium fumarate [54]. An efficient recovery system for fumaric acid needs to be developed. Approaching this problem, Cao *et al.* [55] were able to obtain a productivity of 4.25 g/l h by integrating an adsorption column coupled to a rotary biofilm contactor with the fermentation. As a result, simultaneous and continuous removal of fumaric acid decreased product inhibition and an increased fermentation rate could be achieved.

7.4.3 Malic Acid

In addition to the chemical synthesis of a racemic mixture of malic acid from maleic anhydride, an alternative production process is via enzymatic conversion where fumarase expressed by immobilized cells, typically *Brevibacterium* strains, is used to catalyse hydration of fossil-based fumaric acid, yielding L-malic acid [56]. Potential new uses of malic acid, preferentially as a building block for biodegradable polymers, have resulted in increased requirements for a cost-efficient fermentation process, also because the biopathway result in the L-isomer, which is preferred for polymer production. A number of microorganisms producing malic acid have been identified. One of the most promising microorganisms identified is *Aspergillus flavus*, a filamentous fungi that, in response to stress, i.e. nitrogen limitation, overproduces malic acid up to a molar yield of 1.28, corresponding to 63% of the maximum theoretical yield on glucose [57]. As in fumaric acid production, malic acid accumulates by flux through the reductive arm of the TCA cycle, where additional CO_2 is provided by the neutralizing agent $CaCO_3$. However, the concurrent production of by-products, in this case succinic and fumaric acid, lowers the yield. The additional potential production of alfatoxin from *A. flavus* limits its industrial use. In order to decrease by-product formation and instead increase the flux towards malic acid, metabolic engineering of both *Saccharomyces cerevisiae* and *E. coli* has been conducted [58], but the quantities produced were too low for industrial utilization to be practical.

7.4.4 Other Carboxylic Acids as Building Blocks

Other carboxylic acids that could potentially be produced from renewable biomass and used as building blocks are, for example, muconic and aconitic acid. Muconic acid is a six-carbon dicarboxylic acid that has a potential use as raw material for the production of new resins, pharmaceuticals and agricultural products and with the potential to be converted to adipic acid, an essential chemical in the production of nylon-6,6 [59]. Several microorganisms producing *cis,cis*-muconic acid from aromatic compounds have been identified, e.g. *Pseudomonas*, *Mycobacterium*, *Sphingobacterium* and *Acinetobacter* spp. Fermentative production using *E. coli* with a modified pathway for aromatic amino acid synthesis [60] resulted in a molar yield of 0.22 from glucose after 48 hours of culturing under fed-batch fermentor conditions. Aconitic acid, a metabolite in the TCA cycle, can be produced by *A. terreus or A. itaconicus* under aerobic conditions.

7.5 Conclusion

Besides the existing success stories of citric and lactic acid, organic acid metabolites need to become more economically competitive with respect to existing petrochemically derived cognates to become successful as building blocks for surfactants. Improvements are required to overcome these obstacles in the biological processes with respect to yield, productivity and product recovery. Yield and productivity are primarily related to the microorganism used and the cultivation conditions. The separation and recovery of dilute concentrations of organic acids from aqueous fermentation broths is one of the biggest challenges towards the positive economics of a bioprocess. This is dependent on the chemical properties of individual metabolites and how they are produced. The primary separation methods for these are based on solubility using precipitation, chromatography, electrodialysis and liquid–liquid extraction.

Considering the 'new' top-ranked building blocks, succinic acid has the greatest potential to succeed in the commercial production plant. Already today the bio-based production route is cost competitive to the petro-based route [39]. A large number of microorganisms have been identified and, most importantly, this includes organisms with genomes that have been completely sequenced, enabling genome-based metabolic engineering for an increased succinate production. Succinic acid-producing organisms also have the advantage of being very flexible with respect to the feedstock sugar, and a number of studies have been conducted where successful fermentation using industrial hydrolyzates as the feedstock have been demonstrated. Potential hydrophobic tail groups such as natural fatty acids and glycerides from such sources as vegetable oil (e.g. palm oil) and tall oil give rise to the possibility of producing completely bio-based surfactants from renewable resources. Overall, this represents an encouraging development for the efficient use of renewable biomass.

References

1. Desai, J. D. and Banat, I. M. (1997) Microbial production of surfactants and their commercial potential. *Microbiol. Molec. Biol. Rev*., **61**, 47–64.
2. Berovic, M. and Legisa, M. (2007) Citric acid production. *Biotechnol. Annu. Rev*., **13**, 303–343.
3. Datta, R. and Henry, M. (2006) Lactic acid: recent advances in products, processes and technologies – a review. *J. Chem. Technol. Biotechnol*., **81**, 1119–1129.
4. Patel, M. K., Dornburg, V., Hermann, B. *et al.* (2006) BREW – Medium and Long-term Opportunities and Risks of the Biotechnological Production of Bulk Chemicals from Renewable Resources, The European Commission's GROWTH Programme, Department of Science, Technology and Society (STS)/Copernicus Institute.
5. Zeikus, J. G., Jain, M. K. and Elankovan, P. (1999) Biotechnology of succinic acid production and markets for derived industrial products. *Appl. Microbiol. Biotechnol*., **51**, 545–552.
6. Engel, C. A. R., Straathof, A. J. J., Zijlmans, T. W. *et al.* (2008) Fumaric acid production by fermentation. *Appl. Microbiol. Biotechnol*., **78**, 379–389.
7. Shanmugam, K. and Ingram, L. O. (2008) Engineering biocatalysts for production of commodity chemicals. *J. Molec. Microbiol. Biotechnol*., **15**, 8–15.

8. John, R. (2007) Fermentative production of lactic acid from biomass: an overview on process developments and future perspectives. *Appl. Microbiol. Biotechnol.*, **74**, 524–534.
9. Gaupp, R. and Adams, W. (2004) Acid esters of mono- and diglycerides, in *Emulsifiers in Food Technology* (ed. R. J. Whitehurst), Blackwell Publlishing Ltd, Oxford.
10. van Haften, J. L. (1979) Fat-based food emulsifiers. *J. Am. Oil Chem. Soc.*, **56**, 831–835.
11. Moonen, H. and Bas, H. (2004) Mono- and diglycerides, in *Emulsifiers in Food Technology* (ed. R. J. Whitehurst), Blackwell Publishing Ltd, Oxford.
12. Boutte, T. and Skogerson, L. L. (2004) Stearoyl-2-lactylates and oleoyl lactylates, in *Emulsifiers in Food Technology* (ed. R. J. Whitehurst), Blackwell Publishing Ltd, Oxford.
13. Thomas, P. D. (1967) Cereal products having incorporated therein alkali salts of monoalkyl fumarates and maleates, US Patent 3,343,964.
14. Mueller, M. (2005) Oil-containing surfactant gels comprising a mixture of ethoxylated citric acid esters, US Patent 7,407,926.
15. Slone, C. S., Lassila, K. R. and Meier, I. K. (2001) Citric acid tri-alkylamide surfactants, US Patent 6,399,088.
16. Zerkowski, J. A., Solaiman, D. K. Y., Ashby, R. D. and Foglia, T. A. (2006) Head group-modified sophorolipids: Synthesis of new cationic, zwitterionic, and anionic surfactants. *J. Surfactants Deterg.*, **9**, 57–62.
17. Infante, M. R., Pérez, L., Pinazo, A. *et al.* (2004) Amino acid-based surfactants. *C. R. Chim.*, **7**, 583–592.
18. Rosen, M. J. (2004) *Surfactants and Interfacial Phenomena*, John Wiley & Sons, Ltd., New York.
19. Edser, C. (2006) What does the future hold for anionic surfactants? *Focus on Surfactants*, **2006** (12), 1–2.
20. Tyagi, V. K. (2006) Sulfosuccinates as mild surfactants. *J. Oleo Sci.*, **55**, 429–439.
21. Budarin, V., Luque, R., Macquarrie, D. J. and Clark, J. H. (2007) Towards a bio-based industry: benign catalytic esterifications of succinic acid in the presence of water. *Chem. Eur. J.*, **13**, 6914–6919.
22. Gilbert, E. E. (1962) The reactions of sulfur trioxide, and its adducts, with organic compounds. *Chem. Rev.*, **62**, 549.
23. Anastassiadis, S., Morgunov, I. G., Kamzolova, S. V. and Finogenova, T. V. (2008) Citric acid production patent review. *Recent Pat. Biotechnol.*, **2**, 107–123.
24. Magnuson, J. K. and Lasure, L. L. (2004) Organic acid production by filamentous fungi, in *Advances in Fungal Biotechnology for Industry, Agriculture, and Medicine* (eds J. S. Tkacz and L. Lange), Springer, New York.
25. Grewal, H. S. and Kalra, K. L. (1995) Fungal production of citric acid. *Biotechnol. Adv.*, **13**, 209–234.
26. Kirimura, K., Watanabe, T., Sunagawa, T. and Usami, S. (1999) Citric acid production from xylan and xylan hydrolysate by semi-solid culture of *Aspergillus niger*. *Biosci. Biotechnol. Biochem.*, **63**, 226–228.
27. Gadd, G. (1999) Fungal production of citric and oxalic acid: importance in metal speciation, physiology and biogeochemical processes. *Adv. Microb. Physiol.*, **41**, 47–92.
28. Verhoff, F. H. (1986) Citric acid, in *Ullmann's Encyclopedia of Industrial Chemistry* (ed. W. Gerhartz), Wiley-VCH Verlag GmbH, Weinheim.
29. Gluszcz, P. and Ledakowicz, S. (1999) Downstream processing in citric acid production, in *Citric Acid Biotechnology* (eds B. Kristiansen, J. Linden and M. Mattey), Taylor & Francis Ltd, London.
30. Goldberg, I. (2006) Organic acids: old metabolites, new themes. *J. Chem. Technol. Biotechnol.*, **81**, 1601–1611.

31. Ghose, T. K. and Bhadra, A. (1985) Acetic acid, in *Comprehensive Biotechnology*, vol. 3 (ed. M. Moo-Young), Pergamon Press, New York.
32. Borden, J. R., Lee, Y. Y. and Yoon, H. H. (2000) Simultaneous saccharification and fermentation of cellulosic biomass to acetic acid. *Appl. Biochem. Biotechnol.*, **84–86**, 963–970.
33. Wee, Y. J., Kim, J. N. and Ryu, H. W. (2006) Biotechnological production of lactic acid and its recent applications. *Food Technol. Biotechnol.*, **44**, 163–172.
34. Ilmen, M., Koivuranta, K., Ruohonen, L. *et al.* (2007) Efficient production of L-lactic acid from xylose by *Pichia stipitis*. *Appl. Environ. Microbiol.*, **73**, 117–123.
35. Dien, B. S., Nichols, N. N. and Bothast, R. J. (2002) Fermentation of sugar mixtures using *Escherichia coli* catabolite repression mutants engineered for production of L-lactic acid. *J. Ind. Microbiol. Biotechnol.*, **29**, 221–227.
36. Iyer, P. V., Thomas, S. and Lee, Y. Y. (2000) High-yield fermentation of pentoses into lactic acid. *Appl. Biochem. Biotechnol.*, **84–86**, 665–677.
37. Neureiter, M., Danner, H., Madzingaidzo, L. *et al.* (2004) Lignocellulose feedstocks for the production of lactic acid. *Chem. Biochem. Engng Q.*, **18**, 55–63.
38. Werpy, T. and Petersen, G. (2004) Top Value Added Chemicals from Biomass: Volume I – Results of Screening for Potential Candidates from Sugars and Synthesis Gas, DOE/GO.
39. Song, H. and Lee, S. Y. (2006) Production of succinic acid by bacterial fermentation. *Enzyme Microb. Technol.*, **39**, 352–361.
40. Guettler, M. V., Jain, M. K. and Rumler, D. (1996) Method for making succinic acid, bacterial variants for use in the process, and methods for obtaining variants, US Patent 5,573,931.
41. Kim, D. Y., Yim, S. C., Lee, P. C. *et al.* (2004) Batch and continuous fermentation of succinic acid from wood hydrolysate by *Mannheimia succiniciproducens* MBEL55E. *Enzyme Microb. Technol.*, **35**, 648–653.
42. Lee, P. C., Lee, S. Y., Hong, S. H. *et al.* (2003) Biological conversion of wood hydrolysate to succinic acid by *Anaerobiospirillum succiniciproducens*. *Biotechnol. Lett.*, **25**, 111–114.
43. Hong, S. H., Kim, J. S., Lee, S. Y. *et al.* (2004) The genome sequence of the capnophilic rumen bacterium *Mannheimia succiniciproducens*. *Nat. Biotechnol.*, **22**, 1275–1281.
44. Vemuri, G. N., Eiteman, M. A. and Altman, E. (2002) Effects of growth mode and pyruvate carboxylase on succinic acid production by metabolically engineered strains of *Escherichia coli*. *Appl. Environ. Microbiol.*, **68**, 1715–1727.
45. Chatterjee, R., Millard, C. S., Champion, K. and Donnelly, M. I. (2001) Mutation of the ptsC gene results in increased production of succinate in fermentation of glucose by *Escherichia coli*. *Appl. Environ. Microbiol.*, **67**, 148–154.
46. Donnelly, M. I., Sanville-Millard, C. Y. and Nghiem, N. P. (2004) Method to produce succinic acid from raw hydrolysates, US Patent 6,743,610.
47. Andersson, C., Hodge, D., Berglund, K. A. and Rova, U. (2007) Effect of different carbon sources on the production of succinic acid using metabolically engineered *Escherichia coli*. *Biotechnol. Prog.*, **23**, 381–388.
48. Nakajima-Kambe, T., Nozue, T., Mukouyama, M. and Nakahara, T. (1997) Bioconversion of maleic acid to fumaric acid by *Pseudomonas alcaligenes* strain XD-1. *J. Ferment. Bioengng*, **84**, 165–168.
49. Goto, M., Nara, T., Tokumaru, I. *et al.* (1998) Method of producing fumaric acid, US Patent 5,783,428.
50. Riscaldati, E., Moresi, M., Federici, F. and Petruccioli, M. (2000) Direct ammonium fumarate production by *Rhizopus arrhizus* under phosphorus limitation. *Biotechnol. Lett.*, **22**, 1043–1047.
51. Kautola, H. and Linko, Y. Y. (1989) Fumaric-acid production from xylose by immobilized *Rhizopus arrhizus* cells. *Appl. Microbiol. Biotechnol.*, **31**, 448–452.

52. Gibbs, P. (2000) Growth of filamentous fungi in submerged culture: problems and possible solutions. *Crit. Rev. Biotechnol.*, **20**, 17–48.
53. Zhou, Y., Du, J. and Tsao, G. (2000) Mycelial pellet formation by *Rhizopus oryzae* ATCC 20344. *Appl. Biochem. Biotechnol.*, **84–86**, 779–789.
54. Gangl, I., Weigand, W. A. and Keller, F. A. (1990) Economic comparison of calcium fumarate and sodium fumarate production by *Rhizopus arrhizus*. *Appl. Biochem. Biotechnol.*, **24–25**, 663–677.
55. Cao, N., Du, J., Gong, C. S. and Tsao, G. T. (1996) Simultaneous production and recovery of fumaric acid from immobilized Rhizopus oryzae with a rotary biofilm contacter and an adsorption column. *Appl. Environ. Microbiol.*, **62**, 2926–2931.
56. Peleg, Y., Stieglitz, B. and Goldberg, I. (1988) Malic acid accumulation by *Aspergillus flavus*. *Appl. Microbiol. Biotechnol.*, **28**, 69–75.
57. Battat, E., Peleg, Y., Bercovitz, A. *et al.* (1991) Optimization of L-malic acid production by *Aspergillus flavus* in a stirred fermenter. *Biotechnol. Bioengng*, **37**, 1108–1116.
58. Moon, S. Y., Hong, S. H., Kim, T. Y. and Lee, S. Y. (2008) Metabolic engineering of *Escherichia coli* for the production of malic acid. *Biochem. Engng J.*, **40**, 312–320.
59. Wu, C. M., Lee, T. H., Lee, S. N. *et al.* (2004) Microbial synthesis of *cis,cis*-muconic acid by *Sphingobacterium* sp. GCG generated from effluent of a styrene monomer (SM) production plant. *Enzyme Microb. Technol.*, **35**, 598–604.
60. Frost, J. W. and Draths, K. M. (1997) Bacterial cell tranformants for production of *cis,cis*-muconic acid and catechol, US Patent 5,616,496.

Part 4

Biosurfactants

8

Synthesis of Surfactants Using Enzymes

Patrick Adlercreutz and Rajni Hatti-Kaul

Department of Biotechnology, Lund University, Sweden

8.1 Introduction

Environmental and health concerns about the effects of the conventional surfactants have increased the demand for surfactants from natural raw materials that possess good biodegradability and low toxicity characteristics, along with desired functional performance. Biological catalysts in the form of living microorganisms and enzymes constitute important tools for production of such surfactants. Microbial biosurfactants are structurally diverse and complex and are produced in a biosynthetic process catalysed by a cascade of enzymes. Major classes of biosurfactants include glycolipids (e.g. rhamnolipids, sophorolipids, etc.), lipopeptides and lipoproteins (e.g. surfactin, polymyxins, gramicidins, etc.), phospholipids and fatty acids, and other complex combinations of biopolymers (e.g. emulsan, liposan, etc.) [1, 2]. There are examples of some biosurfactants already being applied on a large scale, e.g. for enhanced oil recovery and environmental bioremediation.

The enzymatic synthesis of surfactants, on the other hand, is in essence a chemical reaction in which an enzyme (in a form isolated from its source or even as whole cells) replaces a conventional chemical catalyst. In contrast to the microbial surfactants mentioned above, the surfactants obtained by the use of single enzymes are simpler in structure but can be designed to have the desired physicochemical features. The scope of this chapter is limited to production of surfactants using enzymes and will not include microbial surfactants.

Surfactants from Renewable Resources Edited by Mikael Kjellin and Ingegärd Johansson

8.2 Enzymes as Catalysts for Synthesis of Surfactants

Enzymes have attracted considerable attention as catalysts for synthesis of chemicals [3, 4], and they also provide a unique approach for synthesis of biosurfactants [5, 6]. Enzymes are naturally adapted to processing renewable feedstocks and the enzyme features that are of prime interest for synthetic chemistry are their ability to catalyse a wide range of reactions, their unique selectivity and specificity. The selectivity is recognized at three levels: chemo-, regio- and enantioselectivity. This implies that reactions that are not easily conducted by classical organic chemistry or processes that require several chemical steps are facilitated by biocatalysis. Moreover, the high selectivity of enzyme catalysis results in the production of fewer by-products, resulting in higher atom economy, easier product purification and hence improved overall process yields. Also, the use of hazardous chemicals required for protection/de-protection of functional groups is often avoided. Yet another advantage of biocatalysis is energy efficiency due to the ability of the enzymes to catalyse reactions under mild conditions, e.g. at lower temperature and pressure, as compared to the chemical processes. A major limitation, however, with respect to their application is that enzymes are fragile molecules by nature and are sensitive to harsh environmental conditions.

Chemical synthesis invariably involves an organic medium for solubilising reactants and products and to drive the reaction equilibrium towards product formation. High temperatures are required for reactions utilizing, for example, oils and fats. Developments in nonaqueous enzymology for the past 25 years or so have clearly shown that enzymes are catalytically active in a low water environment and may even display altered selectivities in different organic solvents [7, 8]. Hydrolytic enzymes are able to perform synthesis when the water is removed from the environment. The amount of water required for optimal activity varies for different enzymes [9]. The solvent influences the rate of enzymatic reactions by solvating the substrate with varying efficiency and by direct interaction with the enzyme [9]. Often hydrophobic solvents are less inactivating to enzymes than hydrophilic ones, but several exceptions have been found [10, 11]. For use in organic media, the enzymes need to be immobilized to a solid support or cross-linked in order to avoid any protein aggregation [12, 13].

Although much of the research in the area has involved the use of organic solvents to increase the solubility of the reactants and products, the trend is towards development of solvent-free processes since they are safer, more environmentally benign, provide higher space–time yields and are hence more attractive for industrial applications [14].

Modern biotechnology has furthermore provided tools for accessing and developing novel biocatalysts that have desired features for applications of interest. Enzymes from extremophiles – the organisms that live under extreme conditions of temperature, pH, salinity or pressure, such as hot springs, alkaline and salt lakes and deserts, deep sea vents, and so on – have structural–functional characteristics that are adapted to the environmental conditions from where they are isolated [15]. For example, the enzymes produced by (hyper)thermophiles are normally thermostable, and are also likely to have higher stability in organic medium as compared to their counterparts from mesophilic microorganisms [16]. New biocatalysts may even be accessed directly from environmental DNA samples without the need for isolation of microorganisms. Evolution of enzymes is also possible *in vitro* by altering the amino acid composition of the enzymes by mutation in a rational or random mode. This approach has provided enzymes with enhanced thermostability, higher stability in organic solvents, altered specificity, and so on [17].

Biocatalysis has been used for synthesis of surfactants with different hydrophilic groups that are linked to the hydrophobic fatty chain via ester, amide or glycosidic bonds. Among the most crucial parameters to be considered for enzymatic synthesis are the different solubility characteristics of the hydrophobic and hydrophilic reactants. The use of hydrophilic solvents for increasing the solubility of hydrophilic components increases the toxicity of the medium and affects enzyme activity. An alternative would be to modify the hydrophilic moiety in order to increase its solubility in the hydrophobic component. Solubility may also be improved by increasing the temperature, for which the use of thermostable enzymes would be beneficial. Various groups of surfactants for which enzymatic synthesis has been applied are described below.

8.3 Enzymatic Synthesis of Polar Lipids Useful as Surfactants

A wide variety of polar lipids occurs naturally in large amounts. Some of these are important surfactants in the food industry and for other applications. Mono- and diacylglycerols constitute typical examples, which are chemically related to the carbohydrate esters discussed below. Technical production of these normally uses a neutral fat or oil (triacylglycerol) as the starting material. Enzymatic production methods can be advantageous when high selectivity is desired or when sensitive fatty acids are involved. It is thus possible to make 2-monoacylglycerols by selective removal of the fatty acids in the 1- and 3-positions of triacylglycerols by hydrolysis or alcoholysis using a regiospecific lipase as catalyst in a suitable reaction medium [18]. Mild reaction conditions, such as those commonly used for enzymatic reactions, are needed to avoid isomerization to the more stable 1-monoacylglycerol. Glycerolysis of triacylglycerols is attractive because in principle all the starting materials can be converted to mono- and diacylglycerols. Chemical glycerolysis is carried out on a large scale but the reaction conditions are rather harsh, and enzymatic glycerolysis is therefore being explored in order to make mono- and diacylglycerols containing unsaturated (health-promoting) fatty acids for food applications [19].

Glycerophospholipids are important components of biological membranes and are thus widely spread in nature. Partially purified products are used for a variety of applications, with soya lecithin as a typical example. Enzymes can be used to modify glycerophospholipids in various ways and in the surfactant area removal of one of the fatty acids to make lysophospholipids is the most important example. Sometimes this reaction is carried out only to make it easier to remove the phospholipids fraction from the neutral fat, such as in the processing of vegetable oils. This enzymatic de-gumming is an important industrial process [20]. In other applications, lysophospholipids are produced in order to improve the emulsifying properties of the lipids. One such example is in the preparation of mayonnaise, with improved emulsion stability [21]. In this application, phospholipase A2 is used selectively to remove the fatty acid in the sn-2 position.

8.4 Carbohydrate Esters

Lipases have been used to synthesize a wide range of surfactant esters from fatty acids and carbohydrates. The fatty acid can be used either in the free form or as an ester. When free fatty acids are used as an acyl donor, it is beneficial to remove the water formed in the reaction continuously to achieve a favourable equilibrium position and a high yield. Lipases normally express good catalytic activity even in media with quite low water content, which makes it possible to remove water quite efficiently without negative effects on the enzymatic reaction. Another way to get high yields is to use activated fatty acid esters as acyl donors, with vinyl or trichloroethyl esters as typical examples [22, 23]. The carbohydrate component often contains several hydroxyl groups of which normally a primary one forms the ester bond with the fatty acid. Esterification of glucose with octanoic acid using *Candida antarctica* lipase as catalyst and acetonitrile as solvent has been demonstrated [24]. With moderate amounts of octanoic acid, only the primary hydroxyl group of glucose was acylated, but with a large excess of the acid diesters were formed as well. The synthesis of carbohydrate esters is complicated by the fact that the fatty acid and the carbohydrates are poorly miscible and it is difficult to find a solvent that dissolves considerable concentrations of both. Acetonitrile used in the synthesis mentioned above worked reasonably well with the relatively short acid – octanoic acid – and with moderate amounts of glucose. However, to make efficient synthesis of esters involving long fatty acids, other methods are needed. Below, different approaches to this problem are reviewed.

8.4.1 How to Make Lipase-Catalysed Carbohydrate Ester Synthesis Efficient

In order to make it possible to find a solvent that efficiently dissolves both a fatty acid and a carbohydrate, it has been found suitable to reversibly modify the carbohydrate to make it more hydrophobic and thereby more similar to the fatty acid. In a typical case some of the secondary hydroxyl groups of glucose can be derivatized with a hydrophobic reagent, leaving the primary hydroxyl free to react with a fatty acid in the lipase-catalysed reaction. One method is to use acetals (Figure 8.1) which are miscible with fatty acids, making solvent-free ester synthesis possible [25]. The acetal groups can be removed easily by acid catalysis after the enzymatic step. An alternative method is to make use of a reversible complex formation between the carbohydrate and a hydrophobic boronic acid [26]. Complex formation with phenylboronic acid was thus used to make fructose soluble in hexane and thereby promote its esterification using lipase as catalyst. It is noteworthy that lipases are able to accept acetals and boronic acid derivatives, which are both rather bulky substrates. Among the lipases used are those from *Rhizomucor miehei*, *Chromobacterium viscosum*, *Pseudomonas* and *Candida antarctica* [25–27].

Still another method to make the carbohydrate more hydrophobic is to convert it to an alkyl glycoside. After enzymatic acylation, the resulting alkyl glycoside ester can be a useful surfactant. This approach has been successful even when relatively short alcohols (such as ethanol) are used for making the glycoside [28]. It is possible to achieve high yields in solvent-free reactions, and one can thus start from a monosaccharide and in one alkylation and one acylation step produce an alkyl glycoside ester (Figure 8.2). The

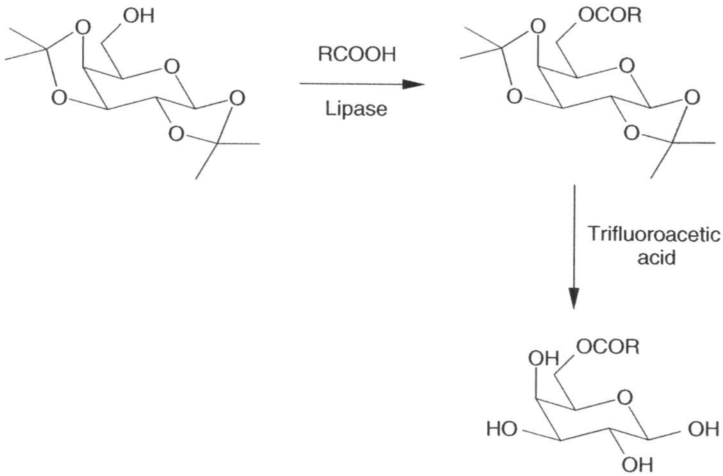

Figure 8.1 Lipase-catalysed acylation of D-galactose protected via acetal formation to make it more hydrophobic. After the enzymatic step, the protecting groups are removed by acid catalysis.

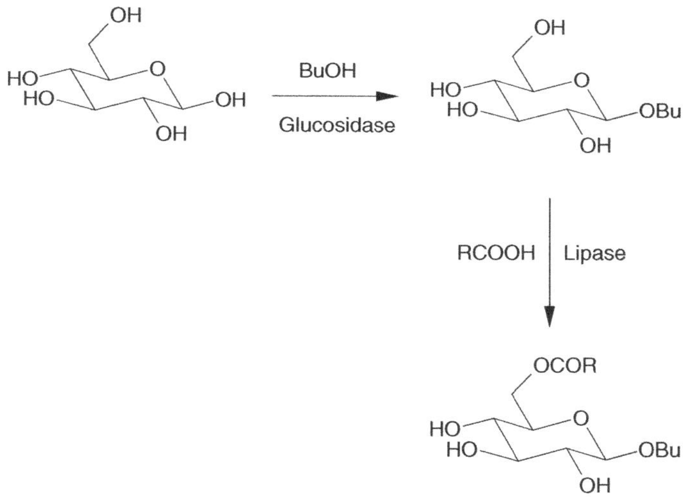

Figure 8.2 Synthesis of an alkyl glycoside ester by a two-step enzymatic route.

alkylation step is most favourable with short alcohols (see below), with butanol as a good example [29]. Enzymatic esterification of butyl glucoside with unsaturated fatty acids has generated products of interest for dermo-cosmetic applications [30].

Most studies of enzymatic sugar ester synthesis have focused on the esterification of monosaccharides, since the problems with poor substrate miscibility increase significantly with increasing size of the carbohydrate. However, by careful choice of reaction conditions it has been shown possible to acylate several di- and trisaccharides efficiently. Solvent mixtures of 2-methyl-2-butanol and dimethylsulfoxide have thus been used in combination with vinyl esters (C8–C18) to make successful acylation of maltose, maltotriose and leucrose possible [22].

Ionic liquids are promising nonaqueous solvents for the dissolution of carbohydrates and they have been used in several studies on enzymatic sugar ester synthesis. *Candida antarctica* lipase B was the best enzyme for synthesis of glucose esters in a two-phase system containing an ionic liquid and *tert*-butanol [31]. Recently, the use of supersaturated sugar solutions [32] and ultrasonic treatment [33] has been reported to make sugar ester synthesis in ionic liquids even more efficient.

Efficient enzymatic conversion can be achieved even though most of the reactants are present as solids, provided that there is a liquid phase in which the reaction can occur. This approach has been successfully used for carbohydrate ester synthesis with synthesis of glucose esters of fatty acids between C12 and C18 as typical examples [34]. It is important that the substrates dissolve during the reaction, and often the products precipitate as they are formed, which can be an advantage due to a favourable effect on the equilibrium position. *Candida antarctica* lipase B is an efficient catalyst in this system and solvents used (in moderate amounts) include ethyl methyl ketone, acetone or dioxane. In order to increase the ester yield, water formed in the reaction can be removed by azeotropic distillation and the solvent (e.g. ethyl metyl ketone) can after condensation be dried by pervaporation, giving a practically useful complete process [35].

8.4.2 Sugar Ester Synthesis Catalysed by Proteases

Although lipases constitute the dominating group of enzymes used for sugar ester synthesis, there are several reports on protease-catalysed syntheses of this kind. In an early study, 6-*O*-butyryl-D-glucose was prepared using subtilisin (a protease from *Bacillus subtilis*) as catalyst and trichloroethyl butyrate as acyl donor in anhydrous dimethylformamide [23]. It is noteworthy that oligosaccharides as long as maltoheptaose were also acylated under these conditions, although at lower rates. Another *Bacillus* protease has been used to synthesize sucrose laurate from sucrose and vinyl laurate in dimethylformamide [36].

8.4.3 Carbohydrate Esters Prepared by Enzymatic Synthesis

Lipases can accept a very wide variety of nucleophiles for the deacylation of acyl enzyme intermediates, and using cleverly chosen reaction conditions it has thus been possible to acylate a wide range of carbohydrates successfully (Table 8.1). The main products are those obtained by esterification of the primary hydroxyl group. In carbohydrates having more than one primary hydroxyl group, the enzyme can selectively acylate one of them in some cases (with maltose), while in other cases product mixtures are obtained (with fructose). In some cases (exemplified in Table 8.1 with sucrose) acylation in various positions can be achieved by the proper choice of enzyme. Most carbohydrate esters have been prepared using monosaccharides as substrates. Due to poor substrate miscibility, synthesis becomes increasingly difficult, with substrates having longer hydrophobic and especially longer hydrophilic groups. However, there are a few successful examples with di-, tri- and longer saccharides. Another, related, group of polar compounds that can be

Table 8.1 *Enzymatically synthesized carbohydrate esters*

Carbohydrate	Acyl donor	Main products	Enzyme	References
D-Glucose	C8 acid	6-*O*-Octanoyl-D-glucose	*C. antarctica* lipase	[24]
D-Glucose	C12–C18 acids	6-*O*-Acyl-D-glucose	*C. antarctica* lipase	[34]
D-Glucose	C4 trichloroethyl ester	6-*O*-Butanoyl-D-glucose	Subtilisin	[23]
D-Galactose	C12–C18 acids	6-*O*-Acyl-D-galactose	Various lipases	[25]
D-Fructose	C12–C18 acids	Fructose monoesters	*R. miehei* lipase	[26]
D-Xylose	C12–C18 acids	5-*O*-Acyl-D-xylose	Various lipases	[25]
Sucrose	C12 vinyl ester	6-*O*-Dodecyl sucrose	*T. lanuginosus* lipase	[37]
Sucrose	C12 vinyl ester	2-*O*-Dodecyl sucrose	*Bacillus* protease	[36]
Sucrose	C12 vinyl ester	1′-*O*-Dodecyl sucrose	Subtilisin A	[36]
Maltose	C8–C18 vinyl esters	6′-*O*-Acyl-maltose	*H. lanuginosa* lipase	[22]
Maltotriose	C8–C18 vinyl esters	6″-*O*-Acyl-maltotriose	*H. lanuginosa* lipase	[22]
Maltoheptaose	C4 trichloroethyl ester	–	Subtilisin	[23]
Ethyl glucoside	C8–C18 acids	6-*O*-Monoesters	*C. antarctica* lipase	[28]
Methyl *β*-D-mannopyranoside	C8 acid	6-*O*-Monoester	Various lipases	[38]
Sorbitol	C18 : 1 acid	Mono- and diesters	*C. viscosum* lipase	[39]

enzymatically acylated to produce surfactants are the sugar alcohols, with sorbitol as a typical example. Product mixtures with varying amounts of mono-, di- and triesters have been obtained [36].

8.5 Fatty Amide Surfactants

Fatty acid derivatives with an amide bond possess properties useful for surfactants, such as enhancing foaming properties of cleaning and personal care products, stabilizing foams and enhancing detergency. The amide bond increases the hydrophilicity of the fatty acids and is furthermore chemically and physically very stable under alkaline conditions. Synthesis of amides could be done by proteases, but these enzymes are very specific for certain amino acids and are more sensitive to organic solvents. The ability of lipases to serve as catalysts for amide synthesis in organic solvents was first demonstrated by Inada *et al.* [40] and Zaks and Klibanov [41]. Lipases have since been used for synthesis of peptides, fatty amides and *N*-acylamino acids and the acylation of alkanolamines.

8.5.1 Sugar Amides

Glycamide surfactants are nonionic, biodegradable surfactants in which the hydrophilic moiety (an amino-sugar derivative) is linked to the fatty acid by an amide bond, e.g. glucamides and lactobionamides. A conventional method for preparing sugar fatty amide surfactants includes the Schotten–Baumann reaction between an amine and a fatty acid chloride, where the chloride salt produced with the amide needs to be removed. The regio- and enantioselectivity of enzymes provides a convenient means of acylation of sugars and sugar amines.

Chemoselective acylation of a secondary amine, *N*-methyl glucamine with fatty acid using commercially available lipases from *Rhizomucor meihei* (Lipozyme) and *Candida*

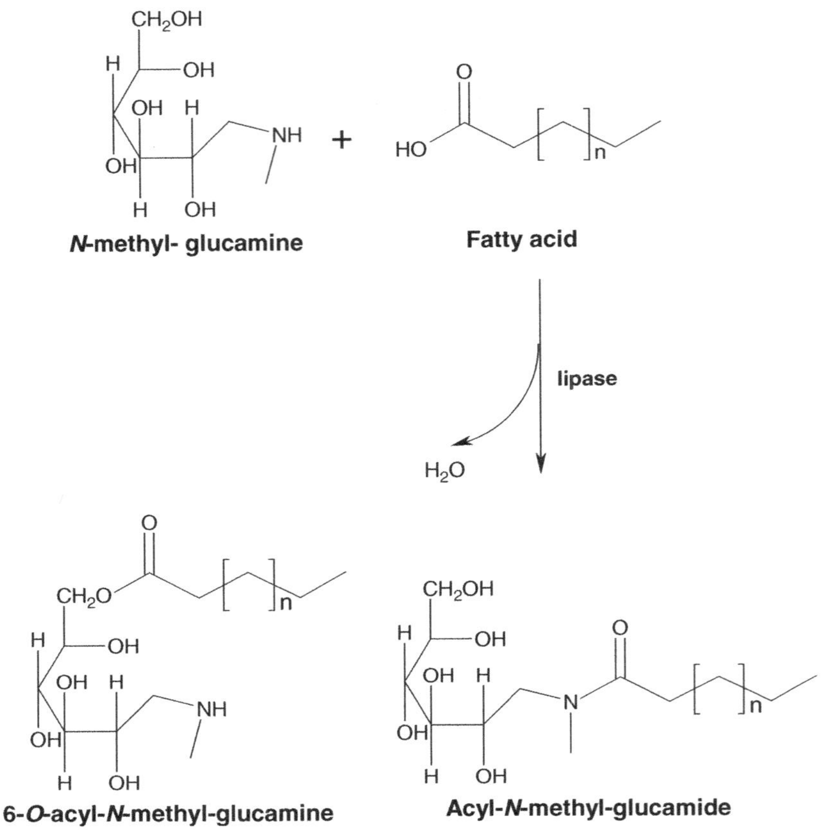

Figure 8.3 *Lipase-catalysed synthesis of a glucamide surfactant from N-methyl glucamine and a fatty acid. Solubilisation of the sugar amine is important for the reaction to occur. Amide formation is promoted when amine is in excess while ester is formed at a high acid/amine ratio.*

antarctica (Novozym®) has been reported [42, 43] (Figure 8.3). When hexane (nonpolar solvent in which *N*-methyl glucamine was not soluble) was used in the reaction medium, the sugar is solubilised by complexation with fatty acid, resulting in the formation of an ion pair between the acid and amine function that is stable in the solvent. As a result all the fatty acid is not available for the reaction and the conversion was limited to 50% [42]. The acid/amine ratio determines the amount of *N*-methyl glucamine solubilised in the organic phase as well as the chemoselectivity of the enzyme catalysis. By increasing the acid/amine ratio it was possible to improve the amine transformation rate and the ester synthesis at the expense of amide formation, while amidation was exclusive at a ratio lower than 1. For the reaction to occur, solubilisation of the sugar amine and ion pair formation was necessary. As the ion pair is stable in hexane, the amine function is blocked and cannot react with acyl-enzyme. When methyl oleate was used as the acyl donor no solubilisation of *N*-methyl glucamine occurred and there was very limited acylation due to the impossibility to form an ion pair.

The acid conversion was improved by using a polar protic solvent, 2-methyl-2-butanol, that solubilised the *N*-methyl-glucamine more efficiently. Immobilized *C. antarctica* lipase B was seen to be the most efficient as catalyst [43]. Complete acid conversion with an amide yield of 97 and 3% diacylated amide ester derivative was achieved in 50 hours at 90 °C at a fatty acid/amine ratio of 1. The amide ester was formed from the monoester that was formed as an intermediate. Even here, the chemoselectivity of the reaction seems to depend on the acid–base conditions in the medium that determines the ionization state of both the reactants and the enzyme [44]. The acyl-enzyme interacts with an alcohol or amine function of *N*-methyl glucamine, resulting in the formation of tetrahedral intermediates, which then yield amide and/or ester products. Under acidic conditions at an acid/*N*-methyl glucamine ratio higher than 1, the amine group is protonated and cannot react with the acyl enzyme. This enhances esterification of *N*-methyl glucamine.

Synthesis of glucamides by transacylation using fatty esters and triglycerides has also been tested [45]. The reaction was completed in a shorter time than for reverse hydrolysis. Initially both an amide and ester of *N*-methyl glucamine were produced and subsequently the ester was consumed to give rise to amide-ester.

8.5.2 Alkanolamides

Alkanolamides are industrially important fatty acid derivatives useful in a wide range of applications including personal care products and hard surface cleaners. They are characterized by their skin tolerance, good biodegradability and low toxicity [46]. Alkanolamides are produced by condensation of fatty acids or fatty acid esters or triglycerides with alkanolamines, such as monoethanolamine or diethanolamine using either high temperature (about 180 °C) or in the case of methyl ester a metal oxide catalyst at 100 °C [46]. Different production strategies provide the alkanolamides of different purities, superamides (produced from fatty acid methyl esters) being the most pure with about 90% purity.

Alkanolamide synthesis has also been achieved using lipase. Alkanolamines are susceptible to acylation both at the amine and hydroxyl group [47, 48]. However, the main product during a lipase-catalysed reaction is the amide, owing to spontaneous acyl migration from alcohol to the amine, the thermodynamically more favourable position (Figure 8.4).

Immobilized lipase B from *Candida antarctica* (Novozym®435) has been shown to catalyse the formation of alkanolamides from both monoethanolamine and diethanolamine when anhydrous reaction conditions are employed [49, 50]. Both free fatty acids as well as ethyl esters can be used as substrates; in the former case the fatty acid and ethanolamine form an ion pair. Ethanolamine is a stronger base than secondary amines and is totally complexed as the ion pair with fatty acid at equimolar concentrations. As described above for sugar amides, the reaction rate and product yield depends on the proportion of the reagents soluble in the reaction. In a hydrophobic solvent like n-hexane both ethanolamine and the ion pair have similar (low) solubilities, resulting in slow reactions. On the other hand, in polar solvents like acetonitrile and dioxane, ethanolamine is more soluble than the corresponding ion pair so that the transacylation procedure is faster than direct acylation. Selectivity of conversion to the amide was increased under conditions

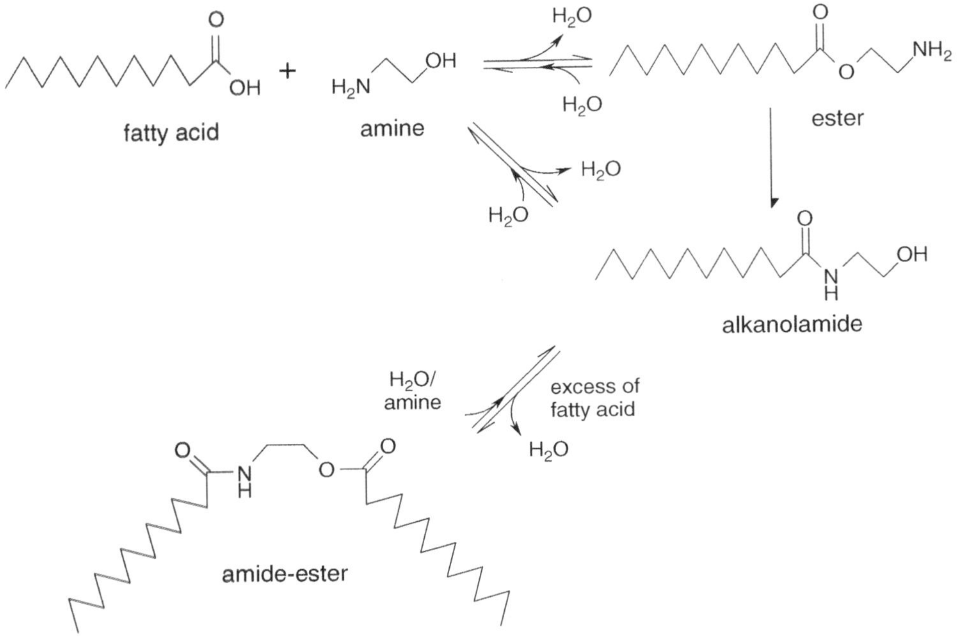

Figure 8.4 Alkanolamide synthesis by lipase-catalysed acylation of ethanolamine with a fatty acid. O-Acyl ester is formed as an intermediate product, which is immediately converted to the amide by acyl migration. When acid is in excess, the amide will react further with the acid, yielding the amide ester. The amide ester can be converted back into the amide through hydrolysis or aminolysis; in the latter case two moles of amide will be formed from one mole of amide ester.

when the product solubility was reduced and could be precipitated out from the reaction medium. Diacylated amide ester was formed as a minor product, but no monoacylated ester was detected due to the fast rate of spontaneous $O \rightarrow N$-acyl migration. Acyl migration is faster in shorter chain alkanolamines ($n < 3$) than in the longer ones [49].

In studies with acylation of diethanolamine, it was found that direct acylation of diethanolamine to amide is more selective in hexane than in dioxane [50]. As diethanolamine is highly viscous, an increase in temperature was important to reduce the viscosity and promote the reaction. *O*-acylation does not take place in hexane but is favoured in dioxane rather than *N*-acylation. Excess of the amine helped to increase the conversion to amide. The transacylation approach resulted in higher reaction rates and selectivities in both the solvents. Lipase-catalysed synthesis of secondary-amide surfactant (*N*-methyl lauroylethanolamide) in acetonitrile was also favoured in the presence of a molar excess of amine (*N*-methyl ethanolamine) [51].

More recently a solvent-free process has been developed for alkanolamide production, which is preferentially carried out at a temperature slightly above the melting point of the alkanolamide product (e.g. at 90 °C for alkanolamides from lauric acid) [14]. Formation of an ion pair salt at equimolar concentrations of ethanolamine and fatty acid results in extremely high viscosity of the medium, even at very high temperatures. This problem could be overcome by using only half of the amine initially; the excess of

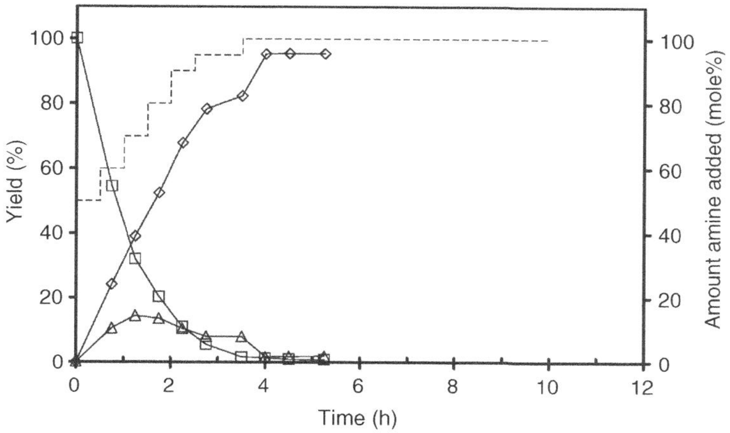

Figure 8.5 *Solvent-free synthesis of alkanolamide from lauric acid (0.5 mol) and monoethanolamine (0.5 mol) catalysed by Novozym®435 at 90 °C. Water is removed throughout the reaction by vacuum. Half the molar amount of ethanolamine is added at the start of the reaction and the remaining amount is added in steps over a period of 4 hours, indicated by the dotted line. Symbols: amide (◇), amide ester (Δ) and lauric acid (□).*

From P. Tufvesson, A. Annerling, R. Hatti-Kaul, and D. Adlercreutz, Solvent-free enzymatic synthesis of fatty alkanolamides, *Biotechnol. Bioeng.* Vol. 97, 2007, 447–453. Copyright 2007 Wiley Periodicals, Inc. Reprinted with permission of John Wiley & Sons, Ltd.

acid thus favoured the formation of amide ester. By adding the remaining alkanolamine stepwise in small portions and as quickly as possible, the formation of the amide ester was suppressed and also kept losses of amine to a minimum (Figure 8.5). Removal of water produced during the reaction by applying a vacuum was important for driving the reaction to complete conversion of the fatty acid. A product yield of 95% was obtained after 4 hours of reaction between lauric acid and monoethanolamine at a 1 : 1 ratio and an enzyme loading of 5% (w/w). The product was free from discoloration and odour and is probably of a similar quality as the superamides [14].

8.6 Amino Acid-Based Surfactants

Amino acid/peptide-lipid conjugates represent an interesting class of surfactants possessing good surface properties, excellent emulsifying properties, antimicrobial activity, low potential toxicity and high biodegradability. They are attractive for applications in foods, personal care products and pharmaceuticals [1, 5]. The large variety of amino acid/peptide structures combined with fatty acids of varying structure and carbon chain length can potentially provide surfactants with wide structural diversity and different physicochemical and biological properties [52, 53]. Depending on the free functional groups on the amino acids, nonionic, amphoteric, cationic as well as anionic surfactants can be obtained.

Different forms of amino acid surfactants have been synthesized enzymatically (Figure 8.6). Linear amino acid surfactants are made up of a natural amino acid linked

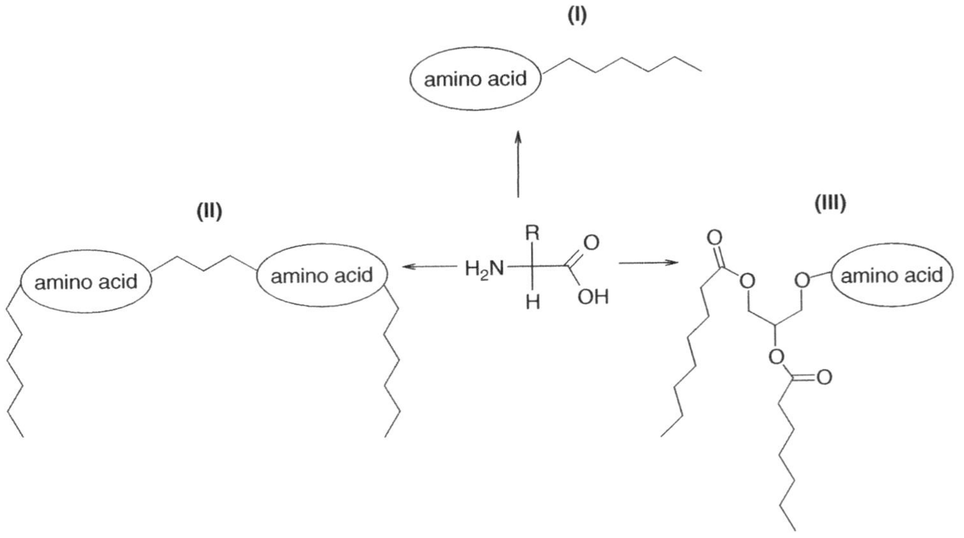

Figure 8.6 Enzymatic synthesis of amino acid-based surfactants: (I) linear, (II) dimeric or gemini and (III) diacylglycerol amino acid conjugates. The amino acid can be coupled to an alkyl chain via an ester or an amide linkage.

to an alkyl chain through the α-amino, α-carboxyl or a side-chain group. The amino group can be coupled with a fatty acid or alkyl halide chain yielding *N*-acyl and *N*-alkyl derivatives, respectively, while the carboxyl group can be coupled with alkyl amines and alcohols to give *N*-alkyl amides and esters, respectively. Proteases as well as lipases have been used as catalysts for the synthesis of these surfactants.

Synthesis of amino acid surfactants with amide bonds and ester bonds between *N*-carbobenzyoxy (Cbz)-arginine methyl ester and various long-chain alkyl amines and fatty alcohols has been achieved using papain from *Carica papaya* [54]. Papain deposited on polyamide was found to exhibit the highest activity. Crude preparation of papain was seen to be more stable and increased the performance of the reaction significantly. Arginine alkyl esters were synthesized in solvent-free systems with purity higher than 99%. The enzymatic synthesis provided higher overall yields and a cleaner product than chemical condensation methods, thus avoiding the need for product acid/base extractions for eliminating the by-products.

Synthesis of amino acid surfactants has also been achieved using certain lipases. Studies with immobilized lipases from *Candida antarctica* and *Rhizopus miehei* have shown that the enzymes could accept *N*-Cbz amino acids as acyl donors and catalyse the esterification with long-chain fatty alcohols with high yields [55, 56]. Removal of water produced during the reaction was essential to shift the equilibrium towards ester synthesis. Synthesis of *N*-acyl amino acids was also done by lipase-catalysed direct transacylation of amino acids with triglycerides or vegetable oils (e.g. soya bean, palm oil) [57–59].

The long-chain alcohols may be substituted with α,ω-diol analogues to generate alkanediyl-α,ω-bis(amino acid) derivatives (amino acid dimers linked by a hydrophobic spacer) with final yields depending on the ratio of the substrates used [56]. The diols may also be replaced by diamines for lipase-catalysed reaction with amino acids. Subsequent deprotection using conventional procedures results in the formation of bola-amphiphiles or further derivatisation can provide a range of gemini (dimeric) surfactants (Figure 8.6). Unlike conventional surfactants, bola-amphiphiles contain two polar heads connected with a hydrophobic linker while gemini surfactants consist of two molecules of monomeric surfactants linked through a flexible spacer. Bola-amphiphiles have the ability to form highly stable ultrathin lipid vesicles. Gemini surfactants display useful properties such as much lower Krafft points and higher solubility in water as compared to their monomeric counterparts. Ionic gemini surfactants have significantly lower critical micelle concentrations (CMCs) and are efficient in lowering surface tension. Papain immobilized on celite was used for the synthesis of arginine-based gemini cationic surfactants, bis(Args), in a two-step procedure: acylation of one amino group of the spacer α,ω-diaminoalkane by the carboxyl ester of the N^{α}-acyl-arginine followed by a papain-catalysed reaction between the product and another N^{α}-acyl-arginine alkyl ester [60]. This strategy has been applied for the synthesis of bis(Arg) analogues of 8, 10 and 12 carbon atoms using 1,3-diaminopropane and 1,3-diamino-2-hydroxypropane as hydrocarbon spacers. The overall yields after product purification were in the range of 51–65% of the pure product.

Diacylglycerol amino acid conjugates constitute a novel class of speciality biocompatible surfactants, which can be considered analogues of partial glycerides and phospholipids. They consist of two aliphatic chains and one polar head; i.e. the amino acid is linked together through a glycerol moiety (Figure 8.6). *N*-protected amino acid glyceryl esters can be prepared enzymatically by esterification of one of the primary hydroxyl groups of glycerol by the α-carboxyl group of the amino acid. Glycerol acts both as a reactant and solvent at elevated temperature. Both proteases and lipases have been effective for this reaction [53, 61]. In the case of aspartic and glutamic acids, selective esterification with the α-carboxyl group was obtained. None of the enzymes could differentiate between the two enantiotopic hydroxymethyl groups of glycerol, giving diastereoisomeric mixtures. In the second step, glyceryl esters of *N*-Cbz-protected amino acids served as excellent substrates for lipase-catalysed acylation, with fatty acid giving yields of 50–90% [55, 62]. All the reactions were performed in solvent-free media. Both Novozym®435 and Lipozyme showed a high regioselectivity towards the primary hydroxyl group of the amino acid glyceryl ester derivatives [55].

As in conventional glycerides, spontaneous acyl migration reactions were observed for the acyl and aminoacyl moieties of the glycerol backbone and both possible regioisomers were obtained [62]. Taking advantage of the acyl migration, mono- and diacylation of amino acid glyceryl ester can be achieved using selective lipases. Thus the 1(3)-acylated product may undergo 1(3) $\rightarrow$ 2-acyl migration and the resulting 1,2(2,3)-isomer may subsequently be acylated at the free primary hydroxyl group by the lipase. The yield of the diacylated product would depend on the rate of intramolecular acyl migration and the esterification of the newly generated free hydroxyl group.

8.7 Alkyl Glycosides

Carbohydrate esters are very useful surfactants, but the instability of the ester bond is a potential disadvantage, especially in applications at neutral or alkaline pH. Alkyl glycosides constitute an interesting group of surfactants, based on similar building blocks as the carbohydrate esters, but with improved stability. Chemical synthesis is useful for the production of mixtures of alkyl glycosides, but when pure isomers are needed, enzymatic synthesis is an attractive alternative [63]. Two major classes of enzymes that can be used for coupling the carbohydrate part to a hydrophobic alcohol are glycosyl transferases and glycosyl hydrolases. These enzymes have been used much less for synthesis than lipases and proteases and in addition they have properties making them more difficult to use in synthesis. Reactions based on glycosyl transferases are troublesome because these enzymes are scarce and require nucleotide phospho-sugar donors that are very expensive [64]. Although glycosyl transferases from genomes of several organisms and methods for the preparation of nucleotide-activated sugars using enzymatic or biological techniques are becoming available, the likelihood of their use for production of speciality biosurfactant products is rather low. Hence, glycosyl hydrolases have been the enzymes of choice for investigating the synthesis of alkyl glycosides. Rapid progress is being made in this field and thus the importance of these reactions can be expected to increase in the future.

A very wide variety of glycosyl hydrolases exists in nature and new ones are discovered at a high rate. The ones used for synthesis are primarily those working according to the 'retaining mechanism'. In this mechanism, the enzyme reacts with the glycosyl donor, which can be a mono-, di- or polysaccharide or a glycoside, to form a glycosyl enzyme intermediate. In the normal hydrolytic reaction, this intermediate reacts with water. However, it can also react with other nucleophiles, such as alcohols or carbohydrates. In alkyl glycoside synthesis the nucleophile is an alcohol and the reactions are usually carried out in a medium with quite a high alcohol concentration to favour this reaction [65]. The synthesis can be carried out as a condensation reaction (reversed hydrolysis) between a monosaccharide and an alcohol. This reaction is under thermodynamic control and in order to achieve a high yield, the equilibrium position must be favourable. Because the glycosyl hydrolases require a relatively high water content to be catalytically active [66], this is difficult to achieve. The reaction works well for the synthesis of products with short carbohydrate and alcohol parts, such as butyl glucoside, but with longer alcohol chains the product concentrations and yields are moderate [67].

An alternative route to alkyl glycosides is to use a transglycosylation reaction with an activated carbohydrate substrate (Figure 8.7). Nitrophenyl glycosides are often used as activated substrates in fundamental studies [68], and in practical applications di- or polysaccharides are commonly used. Glycosyl enzyme intermediates are formed as in the reversed hydrolysis case. However, the transglycosylation reaction is under kinetic control and therefore yields higher than the equilibrium yields can be obtained provided that the kinetics are favourable. It is thus important to find enzymes that normally have a high ratio of transglycosylation to hydrolysis activities. The β-glucosidase 3B from *Thermotoga neapolitana* is such an enzyme [68]. Protein engineering in order to increase

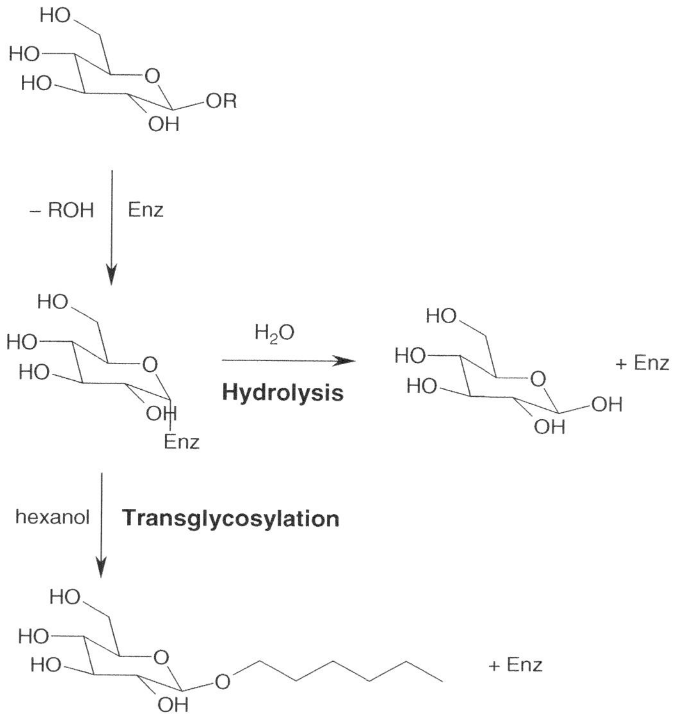

Figure 8.7 *Alkyl glycoside synthesis by a transglycosylation reaction catalysed by a glycosyl hydrolase (Enz). Hydrolysis is a competing side reaction.*

the transglycosylation-to-hydrolysis ratio is a promising methodology, as already shown for the *Pyrococcus furiosus* β-glucosidase [69].

Synthesis of alkyl glycosides with long carbohydrate chains is rather difficult. One approach is to start with a polysaccharide like starch and treat it with a glycosyl hydrolase in the presence of a high alcohol concentration. In this way, α-amylase yielded several alkyl glycosides from methanol, ethanol, propanol and butanol, and with benzyl alcohol glycosides having two, three or five glucose residues [70].

Another enzymatic approach to make surfactants with long carbohydrate groups is to start with one with a short carbohydrate group such as a glucoside and prolong the carbohydrate using an enzyme like dextransucrase or alternansucrase [71]. These enzymes use sucrose as the glycosyl donor and transfer one glucose residue at a time to an acceptor, which can be an alkyl glycoside. Using this approach, mixtures of alkyl glycosides have been produced with at least up to four consecutive glucose residues in their polar part [71]. A relatively recent approach is to mutate the glycosyl hydrolases to generate glycosynthases in which the hydrolytic activity of the enzyme is eliminated [64, 72]. This is achieved by replacement of the nucleophilic residue in the catalytic site by another residue that is unable to perform that function. When supplied with glycosyl fluoride substrates of the opposite anomeric configuration to that of the natural substrate, the enzyme is often able to transfer this activated glycosyl donor to a suitable acceptor.

Among the monosaccharides used for alkyl glycoside synthesis, glucose is the one most frequently used and the products are β-glucosides [65, 66]. The β-glucosidase from almond has been used to make β-fucosides and β-galactosides as well [73]. Some glycosidases accept both glucose and galactose as substrates and it has thus been possible to convert lactose directly to a mixture of glucose and galactose β-glycosides [74]. Production of alkyl-β-D-galactoside, using whole cells of *Bacillus halodurans* displaying β-D-galactosidase activity and lactose as the glycosyl donor, has been reported [75]. β-D-Fructofuranosides have been prepared from sucrose using invertase as catalyst [76].

While much of the work reported so far on alkyglycosides has dealt with hexose sugars, there are a few reports on the synthesis of pentose-based alkyglycosides that are effective next-generation surfactants. Starting from xylan, alkyl-β-xyloside and -β-D-xylobioside were produced by partial hydrolysis and simultaneous transglycosylation to various alcohols in a reaction catalysed by acetone powder of *Aureobasidium pullulans* cells with xylanase activity [77]. The yield of the xyloside product increased while that of the xylobioside decreased with time and with an increase in concentration of the cells. The product yields were significantly improved and even alkyl-β-xylotrioside was formed when the reaction was performed in supercritical fluids [78].

Most of the alkyl glycosides produced are β-glycosides because of the specificity of the enzymes used. However, the use of α-amylase made it possible to make α-glycosides of several alcohols although the yields were low [70].

The alcohols most frequently used for alkyl glycoside synthesis are primary straight-chain alcohols from methanol to 1-octanol and occasionally longer alcohols [67]. However, secondary alcohols have been used as well, e.g. with almond β-glucosidase [79]. More complex alcohols can also be used as substrates, but that is outside the scope of this review because the products are not used as surfactants.

8.8 Future Prospects

There are currently several examples of large-scale industrial processes using enzymes as catalysts for the production of fine chemicals [3, 80]. Use of biocatalysts is also attracting attention as there is a trend of shifting the resource base of chemical industry from fossil feedstocks to renewable raw materials. The potential of enzymatic synthesis of surfactants from renewables is obvious and it seems possible to produce a variety of surfactants using a limited group of enzymes. Among these enzymes, lipases have been used the most in organic syntheses and have also been employed on a large scale for modification of fats, e.g. in the production of *trans*-free fats by an interesterification reaction [81]. This is mainly due to the higher stability of lipases, making it possible to recycle the enzymes for several process runs and hence making them potentially economically competitive.

In general, however, enzymes may still be somewhat expensive to compete with the conventional chemical catalysts for the production of bulk chemicals. Research in industrial biotechnology has now provided tools for discovering as well as developing novel and robust catalysts. The enzyme industry has come of age and enzymes can indeed be produced at low cost provided there is demand. For example, enzymes are already widely used in bulk quantities as additives in detergents for improved functional performance.

Hence, with increasing interest in bio-based products and an increase in demand for environmentally benign processes and products, the development, availability and application of enzymes for the production of surfactants and other chemicals will be promoted.

Acknowledgements

The authors wish to thank the Swedish Foundation for Strategic Environmental Research (Mistra) for financial support. RHK would also like to thank Dr Cecilia Orellana Coca for help with some of the figures.

References

1. Desai, J. D. and Banat, I. M. (1997) Microbial production of surfactants and their commercial potential. *Microbiol. Molec. Biol. Rev.*, **61**, 47–64.
2. Gautam, K. K. and Tyagi, V. K. (2006) Microbial surfactants: a review. *J. Oleo. Sci.*, **55**, 155–166.
3. Straathof, A. J. J. and Adlercreutz, P. (eds) (2000) *Applied Biocatalysis*, Harwood Academic, Amsterdam.
4. Schmid, A., Dordick, J. S., Hauer, B. *et al.* (2001) Industrial biocatalysis today and tomorrow. *Nature*, **409**, 258–268.
5. Sarney, D. S. and Vulfson, E. N. (1995) Application of enzymes to the synthesis of surfactants. *Trends Biotechnol.*, **13**, 164–172.
6. Vulfson, E. (2003) Enzymatic synthesis of surfactants, in *Novel Surfactants. Preparation, Applications and Biodegradability*, Surfactant Science Series, vol. 114 (ed. K. Holmberg), Marcel Dekker, New York, pp. 257–278.
7. Klibanov, A. M. (2001) Improving enzymes by using them in organic solvents. *Nature*, **409**, 241–246.
8. Carrea, G. and Riva, S. (eds) (2008) *Organic Synthesis with Enzymes in Non-aqueous Media*, Wiley-VCH Verlag GmbH, Wenheim.
9. Adlercreutz, P. (2008) Fundamentals of biocatalysis in neat organic solvents, in *Organic Synthesis with Enzymes in Non-aqueous Media* (eds G. Carrea and S. Riva), Wiley-VCH Verlag GmbH, Weinheim, pp. 3–24.
10. Ghatorae, A. S., Guerra, M. J., Bell, G. and Halling, P. J. (1994) Immiscible organic solvent inactivation of urease, chymotrypsin, lipase, and ribonuclease: separation of dissolved solvent and interfacial effects. *Biotechnol. Bioeng.*, **44**, 1355–1361.
11. Ross, A. C., Bell, G. and Halling, P. J. (2000) Organic solvent functional group effect on enzyme inactivation by the interfacial mechanism. *J. Mol. Catal. B: Enzymes*, **8**, 183–192.
12. Persson, M., Mladenoska, I., Wehtje, E. and Adlercreutz, P. (2002) Preparation of lipases for use in organic solvents. *Enzyme Microb. Technol.*, **31**, 833–841.
13. Clair, N. L. S. and Navia, M. A. (1992) Cross-linked enzyme crystals as robust biocatalysts. *J. Am. Chem. Soc.*, **114**, 7314–7316.
14. Tufvesson, P., Annerling, A., Hatti-Kaul, R. and Adlercreutz, D. (2007) Solvent-free enzymatic synthesis of fatty alkanolamides. *Biotechnol. Bioeng.*, **97**, 447–453.
15. Madigan, M. T. and Marrs, B. L. (1997) Extremophiles. *Scientif. Am.*, **276**, 82–87.
16. Sellek, G. A. and Chaudhuri, J. B. (1999) Biocatalysis in organic media using enzymes from extremophiles. *Enzyme Microb. Technol.*, **25**, 471–482.

17. Reetz, M. T. and Jaeger, K.-E. (1999) Superior biocatalysts by directed evolution. *Top. Curr. Chem.*, **200**, 31–57.
18. Millqvist, A., Adlercreutz, P. and Mattiasson, B. (1994) Lipase-catalyzed alcholysis of triglycerides for the preparation of 2-monoglycerides. *Enzyme Microb. Technol.*, **16**, 1042–1047.
19. Damstrup, M. L., Jensen, T., Sparso, F. V. *et al.* (2006) Production of heat-sensitive monoacylglycerols by enzymatic glycerolysis in *tert*-pentanol: process optimization by response surface methodology. *J. Am. Oil Chem. Soc.*, **83**, 27–33.
20. Yang, B., Zhou, R., Yang, J.-G. *et al.* (2008) Insight into the enzymatic degumming process of soybean oil. *J. Am. Oil Chem. Soc.*, **85**, 421–425.
21. Kawai, S. (2004) Characterization of diacylglycerol oil mayonnaise emulsified using phospholipase A2-treated egg yolk. *J. Am. Oil Chem. Soc.*, **81**, 993–998.
22. Ferrer, M., Cruces, M. A., Plou, F. J. *et al.* (2000) A simple procedure for the regioselective synthesis of fatty acid esters of maltose, leucrose, maltotriose and n-dodecyl maltosides. *Tetrahedron*, **56**, 4053–4061.
23. Riva, S., Chopineau, J., Kieboom, A. P. G. and Klibanov, A. M. (1988) Protease-catalyzed regioselective esterification of sugars and related compounds in anhydrous dimethylformamide. *J. Am. Chem. Soc.*, **110**, 584–589.
24. Ljunger, G., Adlercreutz, P. and Mattiasson, B. (1994) Lipase catalyzed acylation of glucose. *Biotechnol. Lett.*, **16**, 1167–1172.
25. Fregapane, G., Sarney, D. B. and Vulfson, E. N. (1991) Enzymic solvent-free synthesis of acetal fatty acid esters. *Enzyme Microb. Technol.*, **13**, 796–800.
26. Scheckermann, C., Schlotterbeck, A., Schmidt, M. *et al.* (1995) Enzymatic monoacylation of fructose by two procedures. *Enzyme Microb. Technol.*, **17**, 157–162.
27. Ward, O. P., Fang, J. and Li, Z. (1997) Lipase-catalyzed synthesis of a sugar ester containing arachidonic acid. *Enzyme Microb. Technol.*, **20**, 52–56.
28. Bjorkling, F., Godtfredsen, S. E. and Kirk, O. (1989) A highly selective enzyme-catalyzed esterification of simple glucosides. *J. Chem. Soc. Chem. Commun.*, **14**, 934–935.
29. Monsan, P. F., Paul, F., Pelenc, V. and Boures, E. (1996) Enzymic production of α-butylglucoside and its fatty acid esters. *Ann. N.Y. Acad. Sci.*, **799** (Enzyme Engineering XIII), 633–641.
30. Bousquet, M.-P., Willemot, R.-M., Monsan, P. and Boures, E. (1999) Enzymatic synthesis of unsaturated fatty acid glucoside esters for dermo-cosmetic applications. *Biotechnol. Bioeng.*, **63**, 730–736.
31. Ganske, F. and Bornscheuer, U. T. (2005) Optimization of lipase-catalyzed glucose fatty acid ester synthesis in a two-phase system containing ionic liquids and *t*-BuOH. *J. Mol. Catal. B: Enzymes*, **36**, 40–42.
32. Lee, S. H., Dang, D. T., Ha, S. H. *et al.* (2007) Lipase-catalyzed synthesis of fatty acid sugar esters using extremely supersaturated sugar solution in ionic liquids. *Biotechnol. Bioeng.*, **99**, 1–8.
33. Lee, S. H., Nguyen, H. M., Koo, Y.-M. and Ha, S. H. (2008) Ultrasound-enhanced lipase activity in the synthesis of sugar ester using ionic liquids. *Process Biochem.*, **43**, 1009–1012.
34. Cao, L., Fischer, A., Bornscheuer, U. T. and Schmid, R. D. (1997) Lipase-catalyzed solid phase synthesis of sugar fatty acid esters. *Biocatal. Biotransform.*, **14**, 269–283.
35. Yan, Y. C., Bornscheuer, U. T., Stadler, G. *et al.* (2001) Regioselective lipase-catalyzed synthesis of glucose ester on a preparative scale. *Eur. J. Lipid Sci. Technol.*, **103**, 583–587.
36. Pedersen, N. R., Wimmer, R., Matthiesen, R. *et al.* (2003) Synthesis of sucrose laurate using a new alkaline protease. *Tetrahedron Asymmetry*, **14**, 667–673.
37. Ferrer, M., Plou, F. J., Fuentes, G. *et al.* (2002) Effect of the immobilization method of lipase from *Thermomyces lanuginosus* on sucrose acylation. *Biocatal. Biotransform.*, **20**, 63–71.

38. Cordova, A., Hult, K. and Iversen, T. (1997) Esterification of methyl glycoside mixtures by lipase catalysis. *Biotechnol. Lett.*, **19**, 15–18.
39. Janssen, A. E. M., Klabbers, C., Franssen, M. C. R. and Vantriet, K. (1991) Enzymatic-synthesis of carbohydrate esters in 2-pyrrolidone. *Enzyme Microb. Technol.*, **13**, 565–572.
40. Inada, Y., Nishimura, H., Takahashi, K. *et al.* (1984) Ester synthesis catalysed by polyethylene glycol-modified lipase in benzene. *Biochem. Biophys. Res. Commun.*, **122**, 845–850.
41. Zaks, A. and Klibanov, M. (1985) Enzyme-catalyzed processes in organic solvents. *Proc. Natl Acad. Sci. U.S.A.*, **82**, 3192–3196.
42. Maugard, T., Remaud-Simeon, M., Petre, D. and Monsan, P. (1997) Lipase-catalyzed chemoselective *N*-acylation of amino-sugar derivatives in hydrophobic solvent: acid-amine ion-pair effects. *Tetrahedron*, **53**, 7587–7594.
43. Maugard, T., Remaud-Simeon, M., Petre, D. and Monsan, P. (1997) Enzymatic synthesis of glycamide surfactants by amidification reaction. *Tetrahedron*, **53**, 5185–5194.
44. Maugard, T., Remaud-Simeon, M. and Monsan, P. (1998) Kinetic study of chemoselective acylation of amino-alditol by immobilized lipase in organic solvent: effect of substrate ionization. *Biochim. Biophys. Acta*, **1387**, 177–183.
45. Maugard, T., Remaud-Simeon, M., Petre, D. and Monsan, P. (1997) Lipase-catalysed synthesis of biosurfactants by transacylation of *N*-methyl-glucamine and fatty-acid methyl esters. *Tetrahedron*, **53**, 7629–7634.
46. Maag, H. (1984) Fatty acid derivatives: important surfactants for household and industrial purposes. *J. Am. Oil Chem. Soc.*, **61**, 259–266.
47. Kanerva, L. T., Kosonen, M., Vänttinen, E. *et al.* (1992) Studies on the chemo- and enantio-selectivity on the enzymatic monoacylations of monoalcohols. *Acta Chem. Scand.*, **46**, 1101–1105.
48. Furutani, T., Furui, M., Ooshima, H. and Kato, J. (1996) *N*-acylation of β-amino alcohol by acyl migration following enzyme-catalysed esterification. *Enzyme Microb. Technol.*, **19**, 578–584.
49. Fernández-Pérez, M. and Otero, C. (2001) Enzymatic synthesis of amide surfactants from ethanolamine. *Enzyme Microb. Technol.*, **28**, 527–536.
50. Fernández-Pérez, M. and Otero, C. (2003) Selective enzymatic synthesis of amide surfactants from diethanolamine. *Enzyme Microb. Technol.*, **33**, 650–660.
51. Sharma, J., Batovska, D., Kuwamori, Y. and Asano, Y. (2005) Enzymatic chemoselective synthesis of secondary-amide surfactant from *N*-methylethanol amine. *J. Biosci. Bioeng.*, **100**, 662–666.
52. Infante, M. R., Pérez, L., Pinazo, A. *et al.* (2004) Amino acid-based surfactants. *C. R. Chim.*, **7**, 583–592.
53. Morán, M. C., Pinazo, A., Pérez, L. *et al.* (2004b) 'Green' amino acid-based surfactants. *Green Chem.*, **6**, 233–240.
54. Clapés, P., Morán, C. and Infante, M. R. (1999) Enzymatic synthesis of arginine-based cationic surfactants. *Biotechnol. Bioeng.*, **63**, 333–343.
55. Valivety, R., Jauregi, P., Gill, I. and Vulfson, E. (1997) Chemo-enzymatic synthesis of amino acid-based surfactants. *J. Am. Oil Chem. Soc.*, **74**, 879–886.
56. Valivety, R., Gill, I. S. and Vulfson, E. N. (1998) Application of enzymes to the synthesis of amino acid-based bola and gemini surfactants. *J. Surfactants Deterg.*, **1**, 177–185.
57. Nagao, A. and Kito, M. (1989) Synthesis of *O*-acyl-L-homoserine by lipase. *J. Am. Oil Chem. Soc.*, **66**, 710–713.
58. Soo, E. L., Salleh, A. B., Basri, M. *et al.* (2003) Optimization of the enzyme-catalyzed synthesis of amino acid-based surfactants from palm oil fractions. *J. Biosci. Bioeng.*, **95**, 361–367.

59. Montet, D., Servat, F., Pina, M. *et al.* (1990) Enzymatic synthesis of *N*-ε-acyl-lysines. *J. Am. Oil Chem. Soc.*, **67**, 771–774.
60. Piera, E., Infante, M. R. and Clapés, P. (2000) Chemo-enzymatic synthesis of arginine-based Gemini surfactants. *Biotechnol. Bioeng.*, **70**, 323–331.
61. Morán, M. C., Infante, M. R. and Clapés, P. (2001) Synthesis of glycerol amino acid-based surfactants. Part 1. Enzymatic preparation of rac-1-*O*-(N^{α}-acetyl-L-aminoacyl)glycerol derivatives. *J. Chem. Soc. Perkin Trans.*, **1**, 2063–2070.
62. Morán, M. C., Pinazo, A., Pérez, L. *et al.* (2004a) Enzymatic synthesis and physicochemical characterization of glycerol arginine-based surfactants. *C. R. Chim.*, **7**, 169–176.
63. Van Rantwijk, F., Woudenberg-van Oosterom, M. and Sheldon, R. A. (1999) Glycosidase-catalysed synthesis of alkyl glycosides. *J. Mol. Catal. B: Enzymes*, **6**, 511–532.
64. Hancock, S. M., Vaughan, M. D. and Withers, S. G. (2006) Engineering of glycosidases and glycotransferases. *Curr. Opin. Chem. Biol.*, **10**, 509–519.
65. Vulfson, E. N., Patel, R., Beecher, J. E. *et al.* (1990) Glycosidases in organic solvents: I. Alkyl-ß-glucoside synthesis in a water-organic two-phase system. *Enzyme Microb. Technol.*, **12**, 950–954.
66. Ljunger, G., Adlercreutz, P. and Mattiasson, B. (1994a) Enzymic synthesis of octyl-β-glucoside in octanol at controlled water activity. *Enzyme Microb. Technol.*, **16**, 751–755.
67. Panintrarux, C., Adachi, S., Araki, Y. *et al.* (1995) Equilibrium yield of n-alkyl-ß-D-glucoside through condensation of glucose and n-alcohol by ß-glucosidase in a biphasic system. *Enzyme Microb. Technol.*, **17**, 32–40.
68. Turner, P., Svensson, D., Adlercreutz, P. and Karlsson, E. N. (2007) A novel variant of *Thermotoga neapolitana* β-glucosidase B is an efficient catalyst for the synthesis of alkyl glucosides by transglycosylation. *J. Biotechnol.*, **130**, 67–74.
69. Hansson, T. and Adlercreutz, P. (2001) Enhanced transglucosylation/hydrolysis ratio of mutants of *Pyrococcus furiosus* β-glucosidase: effects of donor concentration, water content, and temperature on activity and selectivity in hexanol. *Biotechnol. Bioeng.*, **75**, 656–665.
70. Larsson, J., Svensson, D. and Adlercreutz, P. (2005) alpha-Amylase-catalysed synthesis of alkyl glycosides. *J. Mol. Catal. B: Enzymes*, **37**, 84–87.
71. Richard, G., Morel, S., Willemot, R.-M. *et al.* (2003) Glucosylation of α-butyl- and α-octyl-D-glucopyranosides by dextransucrase and alternansucrase from *Leuconostoc mesenteroides*. *Carbohydr. Res.*, **338**, 855–864.
72. Perugino, G., Trincone, A., Rossi, M. and Moracci, M. (2004) Oligosaccharide synthesis by glycosynthases. *Trends Biotechnol.*, **22**, 31–37.
73. Kobayashi, T., Adachi, S. and Matsuno, R. (1999) Synthesis of alkyl fucosides through beta-glucosidase-catalyzed condensation of fucose and 1-alcohols. *Biotechnol. Lett.*, **21**, 105–109.
74. Hansson, T. and Adlercreutz, P. (2002) Enzymatic synthesis of hexyl glycosides from lactose at low water activity and high temperature using hyperthermostable beta-glycosidases. *Biocatal. Biotransform.*, **20**, 167–178.
75. Das-Bradoo, S., Svensson, I., Santos, J. *et al.* (2004) Synthesis of alkylgalactosides using whole cells of *Bacillus pseudofirmus* species as catalysts. *J. Biotechnol.*, **110**, 273–285.
76. Straathof, A. J. J., Vrijenhoef, J. P., Sprangers, E. *et al.* (1988) Enzymic formation of beta-D-fructofuranosides from sucrose – activity and selectivity of invertase in mixtures of water and alcohol. *J. Carbohydr. Chem.*, **7**, 223–238.
77. Matsumura, S., Sakiyama, K. and Toshima, K. (1997) One-pot synthesis of alkyl beta-D-xylobioside from xylan and alcohol by acetone powder of *Aureobasidium pullulans*. *Biotechnol. Lett.*, **19**, 1249–1253.
78. Nakamura, T., Toshima, K. and Matsumura, S. (2000) One-step synthesis of n-octyl beta-D-xylotrioside, xylobioside and xyloside from xylan and n-octanol using acetone powder of Aureobasidium pullulans in supercritical fluids. *Biotechnol. Letts*, **22**, 1183–1189.

79. Panintrarux, C., Adachi, S. and Matsuno, R. (1997) ß-Glucosidase-catalyzed condensation of glucose with 2-alcohols in buffer-saturated alcohols. *Biotechnol. Lett*., **19**, 899–902.
80. Straathof, A. J. J., Panke, S. and Schmid, A. (2002) The production of fine chemicals by biotransformations. *Curr. Opin. Biotechnol*., **13**, 548–556.
81. Holm, H. C. and Cowan, D. (2008) The evolution of enzymatic interestification in the oils and fats industry. *Eur. J. Lipid Sci. Technol*., **110**, 679–691.

9
Surfactants from Waste Biomass

Flor Yunuen García-Becerra, David Grant Allen and Edgar Joel Acosta
Department of Chemical Engineering & Applied Chemistry, University of Toronto, Canada

9.1 Introduction

As has been established in other chapters of this book, the reasons behind the search for bio-based alternatives to petroleum-based surfactants can be summarized in one word – sustainability. However, there are different ways to produce bio-based surfactants, all of which have different 'degrees' of sustainability. As an example, soaps – alkaline salts of fatty acids – can be considered bio-based surfactants, as they are derived from the saponification of triglycerides obtained from plants and animals. Methyl ester sulfonates are obtained from fatty acid methyl esters (FAMEs) which in turn are also obtained from triglycerides found in plants and animal tissues. Lecithin, lysolectihins and other phospholipids can also be extracted from plant and animal tissue. As will be explained in the next sections, bio-based surfactants can also be secreted by microbial cultures fed with specific substrates.

In all the above examples, there is no question about the bio-based origin of the surfactant, its biocompatibility and its ability to quickly biodegrade. However, all these attributes are necessary but not sufficient to answer the question of sustainability. Currently there are questions being raised about the sustainability of growing crops of palm oil in southeast Asia which are currently used as feedstock for soaps and other bio-based surfactants. Some of the issues being considered are reduction in biodiversity, deforestation, impact of fertilizers and pesticides, and reduction in the natural habitat for numerous species [1]. With respect to the production of biosurfactants by microorganisms (e.g. rhamnolipids and sophorolipids), valuable carbon sources such as glucose are often used. These carbon sources are expensive, and are also a food source for animals and humans.

Surfactants from Renewable Resources Edited by Mikael Kjellin and Ingegärd Johansson

In this chapter we look at the issue of feedstock sustainability from another angle. We will explore current alternatives of using waste biomass as a source of surface-active material. In the first part of the chapter, we look at the use of waste carbon sources (typically distillery waste) for the synthesis of biosurfactants by different microorganisms. In the second part of the chapter, we look at chemical treatment methods, including alkaline treatment, pyrolysis and simple extraction and separation, for producing surface-active material from waste biomass.

In this chapter we consider waste biomass to be any biologically derived waste produced out of food processing/agriculture operations, and excess activated sludge produced in the treatment of industrial and municipal wastewater. The reason for concentrating on these sources of waste biomass is because they are the largest, best characterized and poorly employed source of waste carbon. To give an idea of the magnitude of this source it is estimated that 30 g of dry activated sludge alone is produced per inhabitant in one day. This means that in the US alone, nearly 9000 tonnes per day of dry activated sludge are produced. This number has a similar order of magnitude to the total mass of surfactant used worldwide per day, which is estimated to be nearly 30 000 tonnes/day [2].

9.2 Surfactants Obtained from Biological Transformation of Waste Biomass

9.2.1 Current and Potential Market Value

The term biosurfactant is typically reserved for surface-active agents (surfactants) that are produced by microorganisms, as part of their metabolism [3–8]. However, it is necessary to clarify that there are other surfactants – such as lung surfactants – that are secreted by other species and could also be called biosurfactants. In this part of the chapter we will concentrate on the conventional definition of biosurfactants and the use of microbes to degrade waste biomass to secrete potentially useful biosurfactants.

The earlier work on biosurfactants was driven by the possibility of using microbes in oil reservoirs, and in aquifers and soils contaminated with a wide range of oily contaminants (also known as nonaqueous phase liquids or NAPLs). The idea was very simple – microorganisms exposed to these environments need to adapt their metabolism to use the available organic carbon as a food source. In order to adapt to such an environment the microorganism had to secrete some surface-active material to solubilize the oil in micelles, making it possible for the bacteria to consume that solubilized oil [9–11]. There are several reviews that describe the different strains of bacteria used to produce different surfactants, including growing conditions, kinetics, yields and separation methods [3, 5, 7, 12–15]. Other chapters in this book also review the specific case of the production of rhamnolipids, sophorolipids and saponins (the most common biosurfactants). The intention of this part of the chapter is not to review the whole field of biosurfactants and their applications, but to review the efforts on the use of waste biomass as the carbon source for the synthesis of these surfactants.

Before we discuss the use of waste biomass in biosurfactant production, it is important to discuss some basic concepts related to the current market situation for biosurfactants.

The market for biosurfactants can be divided into two: high value-added biosurfactants for medical use and commodity biosurfactants for general use in consumer products. The medical applications include the use of glycolipids for their antibacterial and antiviral effects, such as trehalose lipids (produced by mycobacteria and corynebacteria), rhamnolipids (produced by *Pseudomonas* sp.), sophorolipids (from yeasts) and, for their promising effects as anti-cancer agents, mannosylerythritol lipids (MELs, from *Candida antarctica*) [13]. Lipopeptides such as surfactin (from *Bacillus subtilis*), gramicidin (from *B. brevis*) and polymixins (from *B. polymyxa*) are also potent antibiotics [13], and in the case of polymyxins they have also been shown to serve as additives in lung surfactant therapy [16]. Figures 9.1 and 9.2 illustrate the structures of these glycolipid and lipopeptide surfactants. For medical applications, the price of the products (pharmaceuticals and some cosmetics) is sufficiently high that the cost of the carbon source

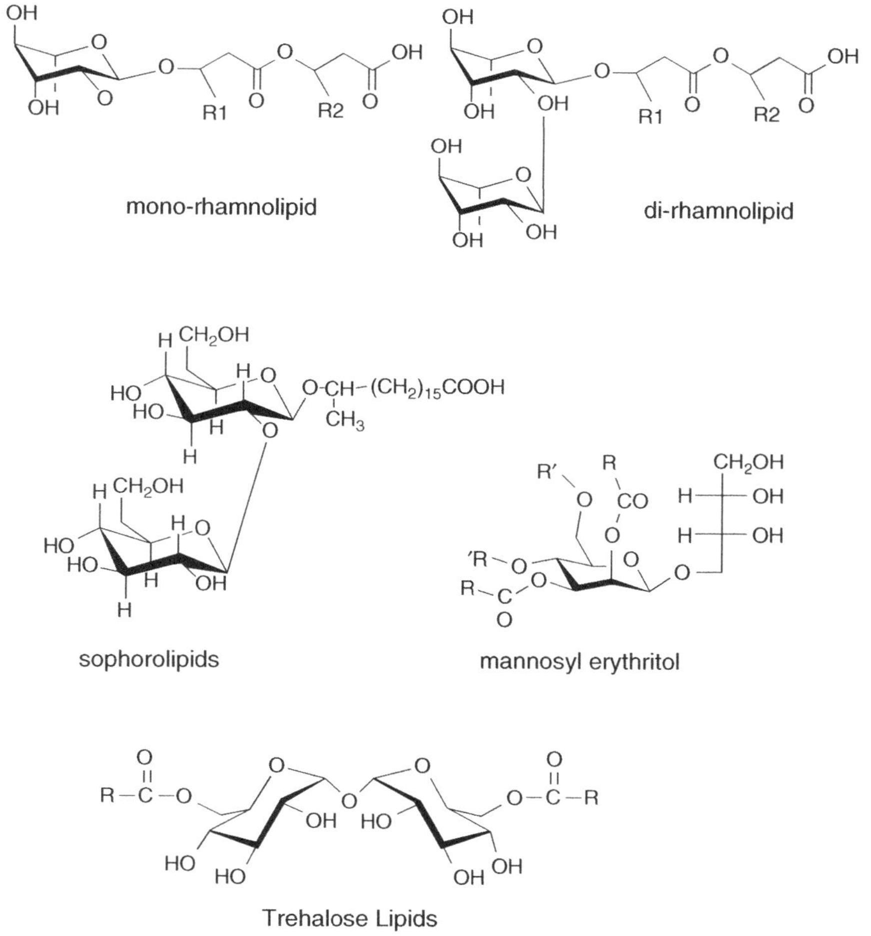

Figure 9.1 *Typical glycolipid biosurfactants.*
Adapted from http://www.lipidlibrary.co.uk/lipids.

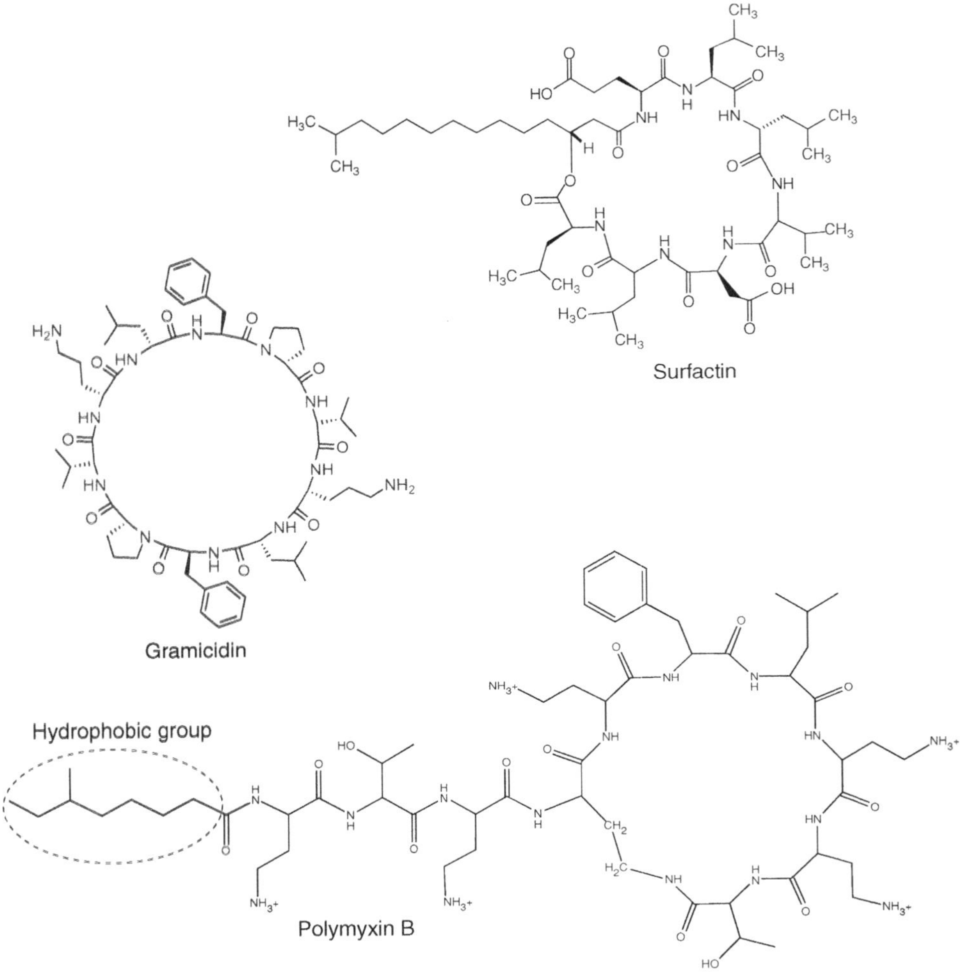

Figure 9.2 Examples of lipopeptide biosurfactants.
Adapted from http://www.lipidlibrary.co.uk and www.wikipedia.org.

to feed the microorganism is not of concern. Most of the current production of biosurfactants is used in these small-volume high value-added pharmaceutical and cosmetic products.

Currently the price of producing rhamnolipids, one of the less expensive biosurfactants to produce, has been estimated in the range of USD 5–20/kg in large fermenters (100–200 m^3) when using vegetable oils or ethanol as the carbon source [3, 17, 18]. This price is still higher than the price of USD 1–3/kg for bio-based surfactants synthesized through chemical processes. These numbers illustrate why bio-based surfactants can be used for biomedical and some cosmetic applications, but not as 'commodity' surfactants.

The efforts in reducing the production costs for biosurfactants have taken three basic routes: the use of less expensive carbon sources, the genetic modification of strains to improve surfactant yield and the improvement in the method of separation of the surfactant from the cell culture [19]. We will describe in more detail the efforts made in using waste biomass as the source of carbon and nutrients for the synthesis of biosurfactants.

9.2.2 Production Technology

Various inexpensive carbon sources have been evaluated for the production of biosurfactants [15]. Figure 9.3 shows a summary of the rhamnolipid yield and concentration obtained with different *Pseudomonas aeruginosa* strains using different carbon sources. The LB1 strain fed with vegetable oils like soya bean, olive and sunflower oil can only transform 15% of those oils into biosurfactants. To interpret this yield, it is important to consider that by subjecting the same oils to methyl ester transesterification followed by ethoxylation or sulfonation, more than 70% of the mass from these sources can be transformed into bio-based surfactants. These chemical processes produce a higher throughput and more concentrated surfactant solutions than the fermentation of these oils. Therefore, using highly valuable vegetable oils, or even used cooking oils as a carbon source for biosurfactant production, is not practical, given that all these sources can be transformed into FAMEs, which can be easily transformed into bio-based surfactants.

According to Figure 9.3, other carbon sources such as soapstocks (the water-soluble fraction obtained after the alkaline treatment of vegetable oil extracts), which contains a number of fatty acids, glycerides and phospholipids, produce yields of 25–40% rhamnolipids. This feedstock, unless used as soaps, cannot be easily transformed into methyl esters and bio-based surfactants. Therefore soapstock is still a potential alternative carbon source for biosurfactant production.

Recent work by George and Jayachandran [20] reveals that carbohydrate-rich waste such as orange peels can be fermented to produce yields of up to 35% biosurfactants. In this case, there is no chemical transformation that could turn the waste biomass in orange peels into some sort of bio-based surfactants. The same can be said about carrot waste, lime peelings and banana waste. The common denominator among all these waste sources is the presence of large molecular weight carbohydrates that can be easily degraded by the bacteria. In the case of wastes containing lower molecular weight carbohydrates [21, 22] the yield to biosurfactant production is slightly lower.

Figure 9.4 presents the biosurfactant yield and concentration produced with various waste carbon sources using different *Bacillus* species. In general, these studies show lower yields and surfactant concentrations than those presented in Figure 9.3. In the case of these *Bacillus* species, they tend to produce lipo-peptide surfactants, but in some cases they can also produce glycolipids [25]. Similar to the observations in Figure 9.3, sources rich in soluble carbohydrates, such as whey, cassava wastewater, simulated potato waste and molasses, produce poor yields (less than 10% in Figure 9.4) towards the production

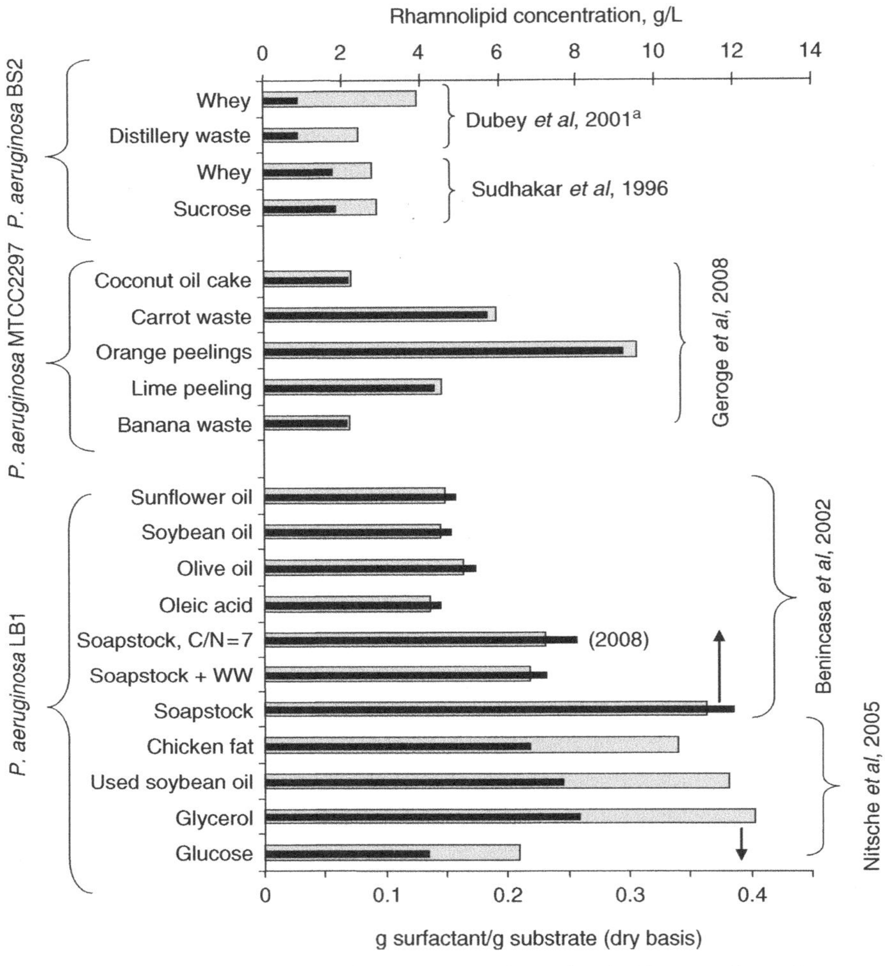

Figure 9.3 *Rhamnolipid carbon yield and concentration obtained with different carbon substrates. Note (a): the values of concentrations and yield were calculated based on the values of CMC and the critical micelle dilution (CMD) [20–24].*

of biosurfactants. On the other hand, a carbon source rich in carbohydrate fibre like peanut oil cake can yield up to 35% biosurfactants.

It is important to clarify that all the studies summarized in Figures 9.3 and 9.4 correspond to batch processes, and to highlight that the maximum biosurfactant concentration is close to 1% (10 g/l) for all the different combinations of carbon sources, bacteria species and strains. In most of these studies, the biosurfactant is separated by precipitation at a pH of ~2, followed by ultracentrifugation and other separation processes. Such separation processes are acceptable to produce low quantities of high value-added products, but are prohibitive for commodity-type surfactants. Furthermore, the low throughput of batch processes and the need for personnel to operate the process also increase the production costs of the biosurfactant.

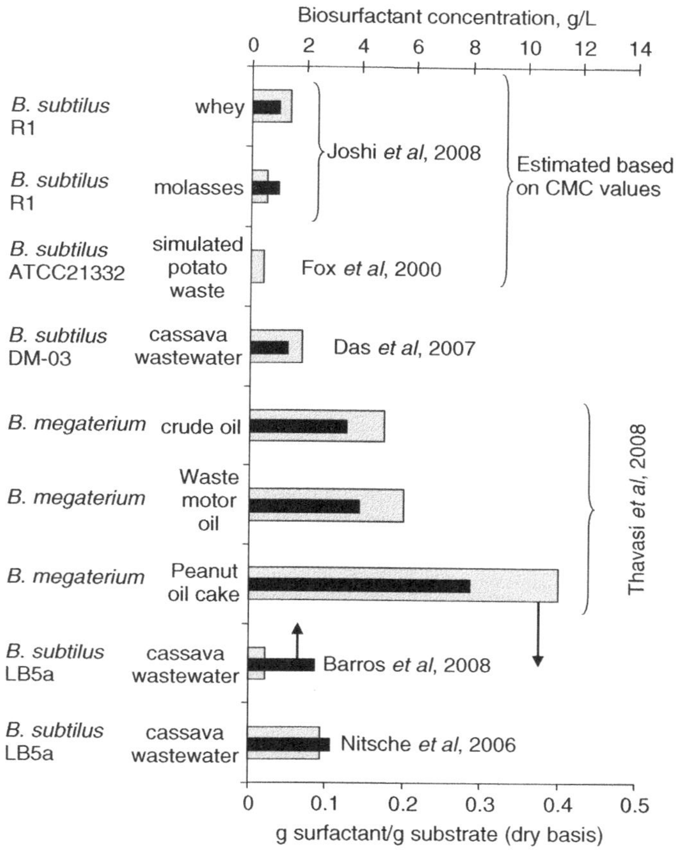

Figure 9.4 *Biosurfactant (lipopeptides in most cases) carbon yield and concentration obtained with Bacillus sp. fed with different carbon sources [19, 24–28].*

There have been certain advances in the design of continuous fermenters with surfactant concentrators based on foam fractionation. Figure 9.5 presents a schematic of this type of fermenter based on the designs given in References [26], [29] and [30]. The fermenter is continually fed with the carbon source and nutrients, and at the same time air is injected from the bottom of the reactor. The reactor is continuously agitated, and the excess biomass generated in the fermenter is withdrawn from the bottom (and a portion of it is extracted with the foam). The air injected from the bottom provides the necessary dissolved oxygen, and the liquid–air interface of the bubble helps to extract the CO_2 produced by the bacteria; as the air leaves the reactor, it produces a froth on the top of the reactor that carries the surfactant concentrated in the foam lamellae. This froth is broken at the top of the foam fractionation column, producing a liquid stream concentrated with the surfactant solution.

The 'separation' or enrichment factor obtained with foam fractionations is presented in Figure 9.6, as a function of surfactant concentration for two batch reactor-foam

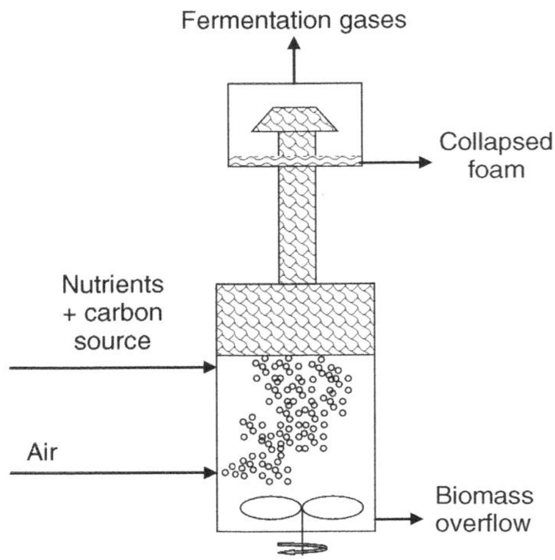

Figure 9.5 Schematic of a continuous/batch fermenter with a foam fraction unit.
Adapted from Barros et al. (2008), Chen et al. (2006), and Davis et al. (2001).

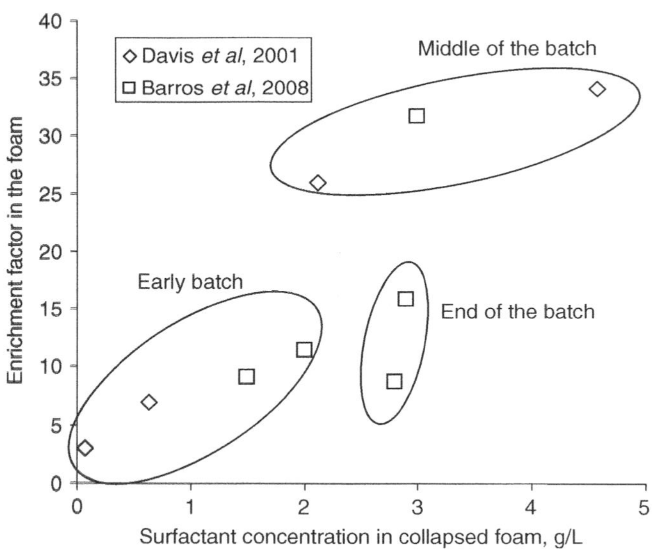

Figure 9.6 Performance of two foam fractionation units used in batch fermenters [26, 30].

fractionators. The data of Davis *et al.* [30] was obtained for surfactin production with *Bacillus subtillus* using glucose as a carbon source. The data of Barros *et al.* [26] was obtained for biosurfactant produced with cassava wastewater as a carbon source. In both cases, the concentration of the surfactant in the collapsed foam is close to 2–4 g/l (0.2–0.4%). The retention factor (mass of surfactant recovered in the foam/mass of surfactant left in the batch) varies from 5 to 15 during the first 24 hours of operation and reaches a maximum value of 25–35 at the peak of surfactant production at 24–48 hours, and then reduces to 10–15 after the peak surfactant production has been reached. It is important to highlight that even when the surfactant solution can be concentrated from a few hundreds of ppm in the batch to nearly 1% in the extract, this is still quite diluted for most surfactant applications. There is still much work to be done in terms of developing cost-effective methods of producing biosurfactants with the throughput and concentration needed to compete with conventional surfactants.

9.2.3 Characteristic Properties

The critical micelle concentration (CMC) for rhamnolipids produced with *Pseudomonas* sp. fed with synthetic media typically ranges from 10 to 50 mg/l [31]. In some cases, some rhamnolipids containing short alkyl chains (less than C10) can yield CMCs as high as 200 mg/l [17]. The surface tension after the CMC for rhamnolipids obtained from fermenting synthetic media (glucose in most cases) is typically 25–28 mN/m [17]. These systems also yield interfacial tensions less than 1 mN/m against hexadecane [17].

For rhamnolipids obtained from sunflower soapstock, a CMC of 120 mg/l was measured by Benincasa and Accorsini [32]. The surface tension after a CMC for those systems was just below 25 mN/m. The relatively high value of the CMC obtained in that study has been corroborated by other studies that have used waste carbon sources [33, 34]. Furthermore, using 100 mg/l of the rhamnolipid obtained by Benincasa and Accorsini [32] from sunflower soapstock, the interfacial tension against hexadecane was 1.3 mN/m, close to the values obtained for rhamnolipids produced from more expensive carbon sources. It is possible that the larger CMC of the surfactants produced with the waste source is due to the presence of other less surface-active material in the mixture that may interfere with the formation of micelles. However, the surface and interfacial tension obtained after the micelles are formed is quite similar to the values obtained with more pure carbon sources.

With respect to the emulsification properties of biosurfactants, one parameter that is often reported in the literature is the emulsification efficiency (E%). To measure this parameter, equal volumes of oil and water are vortexted for 1 minute and then left to settle for 24 hours (E24). After that time, the percentage of the total volume of the liquid occupied by the emulsion is reported (E24%) [35]. This emulsification index can be measured, in principle, against any oil but most of the studies use kerosene as the reference oil. In the work of Benincasa and Accorsini [32], the rhamnolipid produced from sunflower soapstock had an E24 index of 50 (or 50%) against kerosene. In the work of Mercadé *et al.* [35], who used wastewater from olive oil mills, they obtained E24 indices ranging from 15 to 75 with kerosene.

George and Jayachandran [20] showed that the E24 index for rhamnolipids produced using orange peels as carbon sources is highly dependent on the surfactant concentration and the stage of the fermentation process. In the early stages of the fermentation, this E24 index goes from 0 to about 45% in the first three days of incubation; then it reaches a peak value of 75% when the surfactant concentration reaches a maximum value (9 g/l) on day 7. This suggests that the emulsification potential of a given surfactant is not only a function of its composition but also of its concentration.

Similar observations with regards to the properties of rhamnolipids produced by *Pseudomonas* sp. can be made for lipopeptides like surfactin produced by *Bacillus* sp. For example, Nitschke and Pastore [36] produced a lipopeptide using *Bacillus subtillus* and cassava wastewater as the carbon source. This lipopeptide had an estimated CMC of 33 mg/l, and at concentrations above its CMC it produced surface tensions of 27 mN/m. The emulsification index (E24) for this surfactant was 70% for kerosene when the surfactant concentration reached 3 g/l. The interfacial tension for that lipopeptide solution against hexadecane was 0.96 mN/m. Using potato peals as the carbon source, Das and Mukherjee [19] produced a mixture of lipopeptides with CMCs of 120–140 mg/l , with surface tensions after their corresponding CMC of 30–35 mN/m, and emulsification indexes (E24) of 55–68. The authors discussed that, in the case of *Bacillus subtillus*, the ability of the specific strain to uptake nitrogen would affect the composition of the surfactant solution, where higher nitrogen uptake would lead to more surfactin production and desirable surface activity.

One important aspect that needs to be considered at the time of evaluating a surfactant solution is the ratio of surfactant concentration to the CMC. As shown in Figures 9.3 and 9.4, the biosurfactant concentration in batch fermenters can reach values of 1–10 g/l. Furthermore, the previous discussions regarding the CMC of the different biosurfactants typically range between 15 and 150 mg/l; therefore it is possible to obtain surfactant suspensions between 7 and 700 CMCs. Many properties of surfactant solutions are a function of the CMC; e.g. foaming, detergency and wetting tend to improve drastically as the surfactant concentration increases from below CMC to the CMC value, but then remain almost constant after the CMC. Therefore, solutions with low surfactant concentration, but still above their CMC, can be effective in various applications.

For enhanced oil recovery (EOR) and environmental remediation, an important property of any surfactant is its critical microemulsion concentration, or CμC, because this is the minimum surfactant concentration needed to achieve ultralow interfacial tensions ($<$0.1 mN/m) [37, 38]. Recently, it has been determined that for rhamnolipids with a CMC of 10 mg/l, the CμC is close to 100 mg/l [39], which is approximately 10 times lower than the CμC of anionic surfactants [38]. The study of the CμC for rhamnolipids and other biosurfactants obtained from waste sources has not been performed yet. The work of Nguyen *et al.* [39] on the formulation of microemulsions with rhamnolipids also suggests that they are relatively hydrophilic (i.e. tend to form micelles but not reverse micelles), and that it is best to use them in combination with other surfactants.

9.2.4 Current and Potential Areas of Application

Currently, there are some facilities that produce rhamnolipid surfactants; e.g. Rhamnolipid Inc. in United States (http://www.rhamnolipid.com/index.html) produces rhamnolipids on a relatively large scale. Their main target market is in EOR applications. A similar US-based company, Biosurfactant EOR, Inc. (http://www.bioeor.com/), also produces a range of rhamnolipids for the oil industry. The main principle in these two applications is to reduce the interfacial tension of the oil–water interface, therefore reducing the capillary forces that tend to keep the oil trapped in the reservoir. The advantage of the rhamnolipids is that they can reduce that interfacial tension to ultralow values using low surfactant concentrations (see the CμC discussion in the previous section). Furthermore, since the rhamnolipids are resistant to electrolytes and are not charged, it is unlikely that they will precipitate or adsorb in the reservoir, a common problem with anionic and some nonionic surfactants.

Another company, Paradigm Biomedical (http://www.paradigmbiomedical.com/), uses rhamnolipids in the treatment of psoriasis, in wound-healing applications. The low toxicity, biocompatibility and biodegradability of rhamnolipids is also the reason that a cosmeceutical company, Aurora Advanced Beauty Labs (http://www.aurorabeautylabs.com/), includes rhamnolipids in their cosmetic formulations. In the case of medical and cosmetic applications, both companies indicate that they use only highly purified forms of the biosurfactant, which suggests that biosurfactants produced from waste biomass are not suitable for these applications.

Besides EOR the other largest application of biosurfactants is in remediation of oil spills and in the removal of oil sludges from storage tanks. Rhamnolipid-based products like BIOREM and SLUDGER are now available in the market.

9.3 Surfactants Obtained from Chemical Transformation of Waste Biomass

9.3.1 Current and Potential Market Value

The chemical transformation of waste biomass into bio-based surfactants (surfactants produced using feedstocks of biological – renewable – origins) is an emerging area of research for which the potential market value has not yet been defined. The motivation behind using chemical transformations, as opposed to the biological ones (fermentations) described in the previous section, is to produce higher carbon yields from waste biomass, higher surfactant concentrations and an increase in throughput, since they can be designed to work on a continuous basis.

Currently, one of the technologies closer to being commercially available is the synthesis of methyl ester ethoxylates from FAMEs obtained by transesterification of used vegetable oils in the presence of salts of alkaline earths as catalysts [40]. One of the challenges in using waste biomass such as used vegetable oils and soapstock (not a waste, but a low-cost biomass by-product) as the carbon source to produce FAME intermediates is the relatively high free fatty acid content. To address these issues, acid catalysis technologies and single phase solvent technologies have been developed [41, 42]. Another issue with regards to the use of used vegetable oil as a source for waste biomass is its limited supply. Furthermore, the feedstock for FAME-based surfactants (FAME biodiesel) has a high demand on its own. In this chapter we will focus on unconventional but highly abundant feedstocks – waste biomass (compost) and activated sludge (wastewater biosolids).

The idea of obtaining surfactant-like material from compost has been explored by the group of Montoneri at the University of Torino [43–45]. This group based their studies on the fact that humic acids (HAs) and related compounds in alkaline solutions are capable of reducing the surface tension of water, forming micelles and solubilizing hydrophobic compounds [46]. Most of the applications for HA-like surfactants are currently concentrated on their potential use in remediation of soils contaminated with chlorinated hydrocarbons and polyaromatic hydrocarbons (PAHs) [47, 48]. One issue that will be discussed in more detail in the next section is that these surfactants or surfactant-like materials are not as surface active as biosurfactants or synthetic surfactants. However, it has been proposed that these surfactants might be a suitable alternative for environmental remediation and could possibly be used in the textile industry [45, 48, 49]

While the group of Montoneri used an alkaline treatment to convert compost into surfactant-like material, our group has concentrated on the use of alkaline solutions to produce surfactant-like material from wastewater activated sludge. This activated sludge (also known as wastewater biosolids) is the 'excess production' of the microbial biomass necessary to degrade the organic carbon dissolved or suspended in wastewater. Handling and disposal of wastewater biosolids represents approximately 50% of the total wastewater treatment costs [50]. Its environmental impact is also considerable as most of the sludge generated is directed to landfills, incineration or disposal in the sea [51]. Contrary to common belief, only a fraction of wastewater biosolids can be used as fertilizers, in part due to the costs of transportation and also to environmental regulations that limit the use of these biosolids because of the potential contamination with heavy metals [52].

It is estimated that nearly 30 g of dry mixed sludge (from aerobic and anaerobic treatment) are produced per day per inhabitant [53] in municipal treatment systems alone. This suggests that for a city of 1 million inhabitants, this constitutes approximately 30 tonnes of dry sludge produced every day. Wastewater biosolids can be an important source of carbon-based compounds [54]. The most important components of wastewater biomass are polysaccharides and proteins, either in pure form or in associations with other compounds, such as glycoproteins, lipopolysaccharides, lipoproteins, and so on. Other constituents include humic substances, deoxyribonucleic acid (DNA), lipids and uronic acids [54]. In the case of lipids, the content of lipids can vary from 2 to 30% [55].

The alkaline treatment is an effective analytical extraction technique for extracellular biopolymers from biomass, where NaOH is the most commonly used base. The increase in pH with alkaline addition induces the formation of anionic charges from carboxylic groups in proteins and polysaccharides, destabilizing the sludge flocs.

In the food industry, alkaline extraction is widely used. In the case of proteins, they have been historically extracted with strong alkali from waste livestock bone and hide for commercial applications as adhesives and gelling agents [56]. Proteins have also been recovered from waste by-products from beef bone rendering and tallow, and fishing and legume processing. Alkaline extraction (using up to 0.5 M NaOH) has also been used to recover 15–65% of polysaccharides from tubers [57] and nuts [58]. These extracted carbohydrates can be used as viscosity modifiers [57]. The alkaline extraction method is expected to release proteins, carbohydrates, lipids and humic material that can be further separated and used for different purposes. This idea has some elements of a biorefinery-type process. Certainly, saponifiable lipids and humic material are of interest for the production of bio-based surfactants.

Alkaline treatment of wastewater sludge is currently used as a method to stabilize it (inactivate pathogens), to dissolve part of the organic carbon as a pre-treatment before anaerobic digestion and as a method to reduce its heavy metal content [52, 59–62]. In the most aggressive of those treatments, pH 11–12 solutions are obtained after introducing NaOH into the activated sludge slurry [62]. Under those conditions, between 30 and 40% of the sludge is solubilized. The solubilized sludge is fed back into the wastewater treatment system to remove additional carbon. In our work, we used the solubilized organic carbon as a source of lipids, proteins, humic material and polysaccharides that can be further separated into products, including bio-based surfactants.

It is important to recall that the research into bio-based surfactants extracted by chemical methods from waste biomass is still in its early days. As such, the production technology that will be described in the next section is still quite close to laboratory protocols and not industrial processes. As will be described in more detail, the quality of these surfactants is inferior to the quality of the biosurfactants produced (higher CMC, higher surface tension after CMC). However, the potential for high throughput, abundance of the biomass and the current economics of wastewater biosolid handling make this an interesting alternative for some applications that will be described later in this chapter.

9.3.2 Production Technology

For the production of compost isolated humic acid-like material (cHAL), Quagliotto *et al.* [43] mixed food residues with green wastes from public parks and allowed this material to degrade for 15 days. They also evaluated solid wastes from municipal sources. The compost or solid waste was then treated for 24 hours at 65 °C under a nitrogen atmosphere with an aqueous solution of 0.1 M of sodium hydroxide and a solution of 0.1 M of tetrasodium pyrophosphate to maintain a pH of 10. These researchers mixed the solid waste with the alkaline solution to obtain a concentration of solids of approximately 20 g/l. At the end of the 24 hour incubation, the suspension was cooled to room temperature and the remaining solids were separated by centrifugation. The cHAL was recovered from the supernatant solution by reducing the pH to values lower than 1.5 (using sulfuric acid) in order to precipitate the cHAL components. The authors reported a 12% yield of cHAL (based on the weight of the original compost).

The authors conducted a series of chemical characterization studies involving H^1-NMR (nuclear magnetic resonance) and FTIR (Fourier transform and infrared) studies. The FTIR studies confirmed the existence of phenolic groups and carboxylic groups (similar to the case of lignins, but with more carboxylic residues). The NMR data confirmed that there is a significant fraction of hydrogen atoms linked to terminal CH_3 groups and linked to linear CH_2 groups. The relative proportion of carbon–nitrogen–oxygen can vary from one source to the other. One 'example' structure consistent with the compositional and spectrophotometrical analysis is presented in Figure 9.7 (adapted from the work of Quagliotto and co-workers). The hydrophilic groups in this surfactant-like molecule are the carboxylic groups and the phenol groups. The hydrophobic groups are residues in the oligomer/polymer chain (R1–R3) that seem to have between three and five carbons each with branched configurations. The asterisks in the structure indicate potential link sites to a larger polymer molecule. The overall molecular weight for these substances is estimated to be in the vicinity of 15–20 kDa [43].

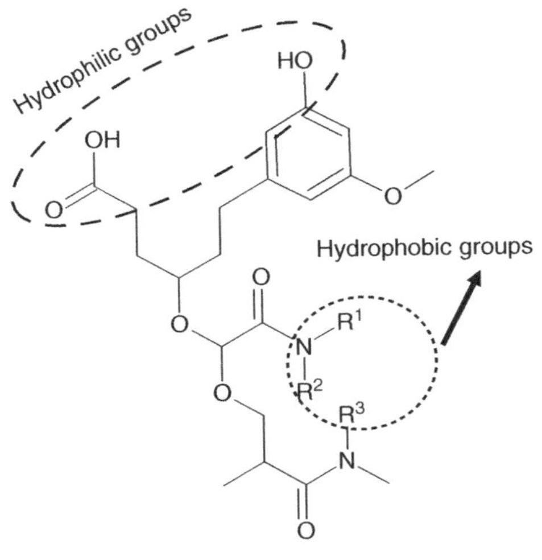

Figure 9.7 *Proposed molecular structure for surfactant-like humic material extracted from compost.* Adapted from Quagliotto *et al.* (2006).

The authors emphasize that the method of extraction and separation of cHAL was based on standard methods used for HA extraction for analytical purposes. They indicate that one of the research areas that require further exploration is the actual method of extraction, especially to make this process feasible [43, 44].

With respect to our process of extraction of surfactant-like material from wastewater biomass, we considered analytical methods (cation exchange resin (CER) extraction) and more conventional alkaline extractions currently used in the food industry [63].

The feedstock for our extraction process is return activated sludge (RAS) from the clarifier that follows the aeration section of the wastewater treatment process. This sludge contains about 4 g/l of biomass and can be concentrated by settling to about 6–8 g/l. Further mechanical dewatering is possible, but it has not been explored at this point. A concentrated sodium hydroxide solution (50% by weight) is then added to this suspension until the desired pH is reached. The suspension is agitated at room temperature for a certain period of time [63]. In our preliminary studies, we evaluated the extraction yield and surface tension of the supernatant liquid as a function of extraction pH for systems incubated for five weeks. We observed that for pH values larger than 11, a number of changes take place. For example, the fraction of solubilized carbon rapidly increases with increasing pH and the colour of the supernatant turns reddish (a common feature in alkaline extractions of food by-products). Furthermore, the surface tension decreases with increasing pH (after pH 11) and reaches values of 36 mJ/m^2 at pH values higher than 12. This level of surface tension approaches that of commercial surfactants. In order to optimize the extraction time we evaluated the extraction yield as a function of pH and extraction time [63]. A selected set of extraction yield (based on total organic carbon, or TOC) data is presented in Figure 9.8 as a function of extraction time for pH

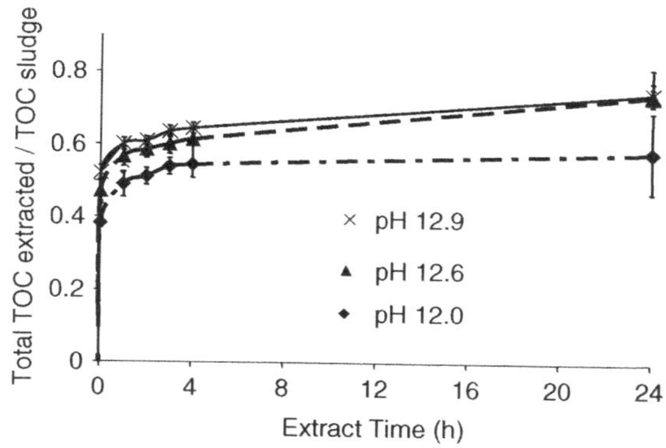

Figure 9.8 *Extraction yield and kinetics of the TOC solubilized in alkaline treatments at different pH values and at room temperature.*
Adapted from García-Becerra *et al.* (2009a).

values of 12, 12.6 and 12.9. According to these data, at pH 12.6 and after 30 minutes of contact time there is more than 55% yield to dissolved carbon compounds.

The chemical composition of the alkaline extract was determined on the basis of proteins, carbohydrates, saponifiable lipids and humic substances. The carbohydrates fraction was determined using the phenol-sulfuric acid method according to Masuko *et al.* [64] using D-glucose as standard. Proteins were measured using the bicinchoninic acid (BCA^{TM}) protein assay kit. The lipid (FAME) composition was analysed using the MIDI method of the Microbial Identification System, Microbial ID Inc., Newark, Delaware, using gas chromatography (GC). The fatty acids were identified according to their retention time using a standard mix of reference FAMEs (C13, C15, C17, C19 and C21; these number of carbons include the terminal methyl ester group). The humic substances were calculated as the difference of the total carbon and the carbon associated with the different fractions above. The molecular size of the extracted compounds was determined through size exclusion chromatography. It was found that the extracted biosolids contain nearly 35% proteins, 15% carbohydrates, close to 3% of saponifiable lipids and approximately 47% of humic-like substances. There are various molecular weight species ranging from 25 to 75 kDa. The extracted lipids are enriched in C14–C16 fatty acids (see Figure 9.9), which is consistent with the average composition of the lipid membranes of microorganisms. Increasing the pH to values near 13 results in the appearance of a fraction of lipids with more than 18 carbons.

9.3.3 Characteristic Properties

Quagliotto *et al.* [43] determined the surface tension, electrical conductivity and fenantrene solubility as a function of cHAL concentration. Selected surface tension data from that study is presented in Figure 9.10, but as a function of the equivalent cHAL carbon concentration. According to those studies, the CMC of cHAL is approximately 400 mg/l, and the surface tension after the CMC is approximately 36 mN/m. The value

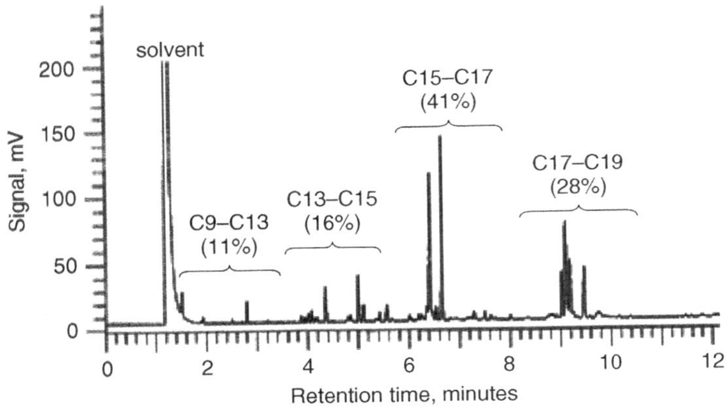

Figure 9.9 Gas chromatogram of the FAMEs obtained from the pH 12.6 extract. The percentage composition presented in parenthesis is the FAME composition on a weight basis. The remaining 4% of the FAMEs correspond to C19+ compounds. Note that the number of carbons (C#) include the terminal methyl ester group.

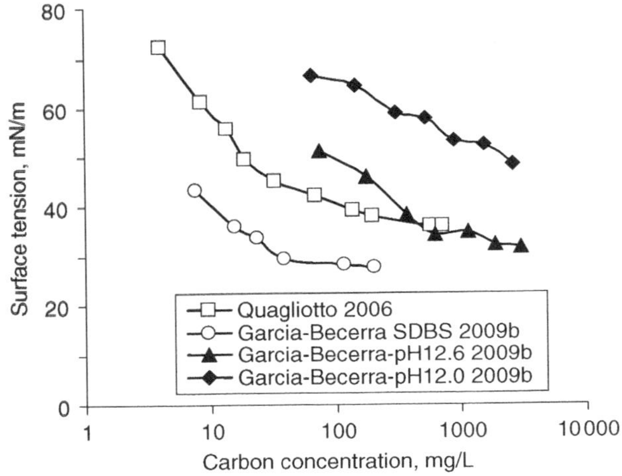

Figure 9.10 Surface tension–surfactant concentration curves for alkaline extracts from waste biomass. The data from García-Becerra was obtained in an electrolyte solution containing 1 g NaCl/100 ml. The sodium dodecyl benzene sulfonate data (SDBS) was obtained at a pH of 6 [43, 66].

of the CMC for cHAL is roughly one order of magnitude higher than the CMC of pure biosurfactants and the surface tension after the CMC for cHAL is approximately 10 mN/m higher than that of biosurfactants. These researchers did not specify the level of electrolyte used in that study, and therefore it is difficult to compare this value to the CMC of other anionic surfactants. The researchers suggest that this CMC is lower than the CMC of other common surfactants like sodium dodecyl sulfate. Other properties such as Kraft temperature, hardness tolerance, foaming ability, Draves wetting or characteristic curvature are yet to be evaluated for these extracts. Montoneri *et al.* [44] produce a similar cHAL extract, but from the green municipal waste before composting, finding that the alkaline treatment produced an extract with a CMC of nearly 1 g/l and surface tension after the CMC of 37 mN/m. These values of CMC and surface tension reveal that these cHAL extracts are more surface active than conventional HAs, which have a CMC in the order of 8 g/l and surface tension after the CMC of 48 mN/m [46].

Figure 9.10 also presents the surface tension of the activated sludge extracts obtained at pH values of 12 and 12.6 as a function of the carbon concentration (TOC) of the extract in the presence of 1 g NaCl/100 ml. It should be noted that those extracts have not been further processed, and therefore they contain a mixture of proteins, carbohydrates, lipids and humic material. According to Figure 9.10, the CMC of the pH 12.6 extract is close to 700 mg/l and the surface tension after the CMC is close to 34 mN/m. However, the extract at pH 12 is less surface active, suggesting that some of the most surface-active material is released at higher pHs. The pH 12.9 extract had a similar surface activity to the pH 12.6 extract.

While the values of surface tension quoted above were obtained using the alkaline solutions, the pH of the solutions could be reduced to 5 and the surface tension after the CMC would reduce from 37 to 35 mN/m; the CMC would also reduce by up to 50%. At pH 4 a fraction of proteins precipitate along with a fraction of the surface-active compounds in the extract, increasing the CMC and surface tension after the CMC.

To assess the activity of the extracts at the oil–water interface, we measured the interfacial tension of these extracts (at a concentration of two times the CMC) against hexadecane, in which case we obtained values close to 8 mN/m. This interfacial tension is substantially higher when compared to the 1 mN/m obtained with rhamnolipid solutions discussed in the first half of this chapter. Another point of comparison is that the interfacial tension obtained with a conventional sodium dodecyl benzene sulfonate (SDBS) surfactant against hexadecane was close to 2 mN/m. These studies suggest that these alkaline extracts are more hydrophilic than rhamnolipids and lipopeptides obtained by biological means.

The cytotoxicity of the extracts was also determined using MatTek keratocyte cultures (www.mattek.com) with a 6 hour exposure protocol [65]. It was found that when the keratocytes were exposed to a 0.1 wt% solution of the pH 12.6 extract whose pH was adjusted to pH 7 and in the presence of 1 g NaCl/100 ml, 50% of those keratocytes survived. In comparison, 50% of the keratocytes can only survive for 3.5 hours when exposed to a 1% sodium dodecyl sulfate (SDS) solution, suggesting that the alkaline extracts are relatively mild. Furthermore, no heavy metals have been found thus far in these extracts. The odour of the extracts is 'soapy' and their colour is yellow-red.

9.3.4 Current and Potential Areas of Application

Montoneri *et al.* [45] evaluated the use of their cHAL extracts as additives to solubilize the dye for textile dyeing and in the remediation of contaminated soils to solubilize PAHs. Figure 9.11 presents the change in colour (C) intensity of Nylon-6 microfibre tinted with

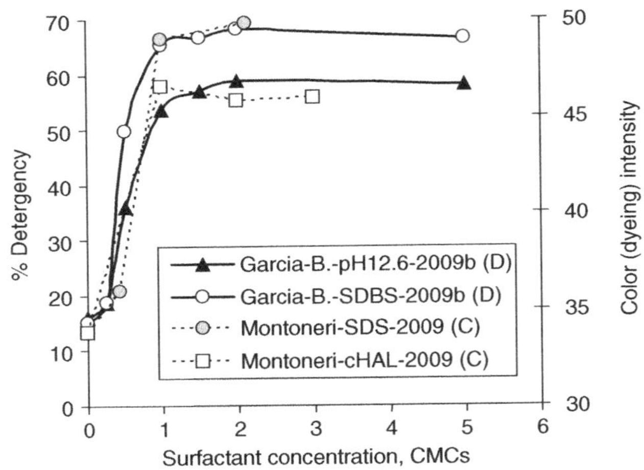

Figure 9.11 *Detergency (D) and colour dyeing (C) performance of alkaline extracts from waste biomass compared with conventional surfactants – SDBS and sodium dodecyl sulfate (SDS). The data from García-Becerra was obtained in an electrolyte solution containing 1 g NaCl/100 ml and pH readjusted to 7. The detergency experiments were carried out using 100% cotton swatches stained with dyed hexadecane [44, 66].*

a hydrophobic dye as a function of surfactant concentration expressed in terms of its CMC. When compared to a more conventional surfactant, SDS, the maximum change in colour intensity obtained with cHAL approaches that obtained with SDS. The difference could be due to the change in partition coefficient between the dye solubilized in micelles, the free dye in solution and the dye bound to the textile. The data certainly suggests a potential value for the extracts, but it is most likely that this surfactant might have to be combined with other surfactants to produce the desired effect. The same authors also observed potential value for these formulations in soil remediation technologies. The authors did not evaluate the relative hydrophobicity of their extract (e.g. through interfacial tension against hexadecane) and therefore it is difficult to assess the potential of these formulations in hard surface cleaning formulations.

We have also conducted preliminary evaluations of the potential value of the alkaline extracts from activated sludge. Figure 9.11 presents the percentage detergency (removal of hexadecane from cotton swatches) as a function of surfactant concentration expressed in terms of the CMC. When compared to a conventional detergent (SDBS), the alkaline extract at pH 12.6 (which was neutralized to pH 7 for the detergency studies) approaches, but does not match, the level of detergency obtained with the conventional detergent. The difference between the maximum detergency reached with the conventional detergent and the extract can be explained by the difference in the interfacial tension with hexadecane. Since the interfacial tension obtained with the extract is nearly four times higher than that obtained with SDBS, it suggests that the extract is not as hydrophobic as the SDBS, and therefore it is more difficult to solubilize or suspend the oil in the washing bath.

To understand the implications of these findings it is important to recall that the pH 12.6 extract is simply the supernatant solution obtained from the alkaline treatment, whose pH was adjusted to 7 and the electrolyte concentration to 1 g NaCl/100 ml. There are no further separation steps involved in obtaining the extract, the carbon yield is larger than 60% and the extraction time can be as low as 30 minutes. Furthermore, the carbon source is plentiful. These favourable process conditions suggest that if no further separation is involved, the economics of producing these extracts might be favourable, probably comparable to conventional surfactants.

It is unlikely that these extracts could be used in detergency, personal care products, dishwashing or other applications where there could be health concerns. However, there are numerous applications where human contact is less likely, such as car washing solutions, window washing solutions, pesticides and other agricultural applications, environmental remediation (soil washing), drilling fluids and others.

9.4 Summary and Outlook

In this chapter we have discussed the use of waste biomass to produce biosurfactant or bio-based surfactants, with particular emphasis on describing the carbon yield, process conditions and the properties of the surfactants obtained. The literature on biological routes to obtain biosurfactants is rich and continues to grow, particularly in the area of alternative feedstocks and in genetically modified strains that can produce larger carbon yields, higher surfactant production rates and higher surfactant concentrations. With regards to the use of alternative (waste biomass) feedstock, the use of waste or used vegetable oils is not particularly attractive, as this is also a feedstock for biodiesel which is currently in high demand. However, other feedstocks like soapstock, fruit or vegetable peelings or residues from vegetable oil processing are promising alternatives.

Bio-based surface-active material extracted from waste biomass by alkaline solutions also offers the potential of being mass-produced (plentiful carbon sources, high throughput process) with high carbon yields. The efforts in this area are recent and there are numerous questions to be answered. While there are opportunities to develop markets for these bio-based surfactants, this idea requires the collaboration of different sectors, including government organizations (wastewater treatment/composting facilities and regulatory bodies), surfactant manufacturers and users, and academia.

There are certainly the technical means to transform waste carbon sources into valuable surfactants. In our opinion, further cost reductions and more creative market developments should result in an increased use of biosurfactant or bio-based surfactants. The increased awareness of the consumers of the environmental impact of the products they use is creating the opportunity to explore these alternatives.

References

1. Brown, E. and Jacobson, M. F. (2005) Cruel Oil. Report published online by the Center for Science in the Public Interest, http://www.cspinet.org/palm/PalmOilReport.pdf.
2. Challener, C. (2003) Soaps and detergents – industry overview. *Chem. Mark. Rep.*, **263**, 4.
3. Singh, A., Van Hamme, J. D. and Ward, O. P. (2007) Surfactants in microbiology and biotechnology: Part 2. Application aspects. *Biotechnol. Adv.*, **25**, 99–121.

4. García-Junco, M., De Olmedo, E. and Ortega-Calvo, J. J. (2001) Bioavailability of solid and non-aqueous phase liquid (NAPL)-dissolved phenanthrene to the biosurfactant-producing bacterium *Pseudomonas aeruginosa* 19SJ. *Environ. Microbiol.*, **3**, 561–569.
5. Desai, J. D. and Banat, I. M. (1997) Microbial production of surfactants and their commercial potential. *Microbiol. Molec. Biol. Rev.*, **61**, 47–64.
6. Zhang, Y. and Miller, R. M. (1992) Enhanced octadecane dispersion and biodegradation by a *Pseudomonas rhamnolipid* surfactant (biosurfactant). *Appl. Environ. Microbiol.*, **58**, 3276–3282.
7. Fiechter, A. (1992) Biosurfactants: moving towards industrial application. *Trends Biotechnol.*, **10**, 208–217.
8. Guerra-Santos, L., Kappeli, O. and Fiechter, A. (1984) *Pseudomonas aeruginosa* biosurfactant production in continuous culture with glucose as carbon source. *Appl. Environ. Microbiol.*, **48**, 301–305.
9. Pareilleux, A. (1979) Hydrocarbon assimilation by *Candida lipolytica*: formation of a biosurfactant; effects on respiratory activity and growth. *Eur. J. Appl. Microbiol. Biotechnol.*, **8**, 91–101.
10. Cooper, D. G., Zajic, J. E. and Denis, C. (1981) Surface active properties of a biosurfactant from *Corynebacterium lepus*. *J. Am. Oil Chem. Soc.*, **58**, 77–80.
11. Parkinson, M. (1985) Bio-surfactants. *Biotechnol. Adv.*, **3** (1), 65–83.
12. Kitamoto, D., Isoda, H. and Nakahara, T. (2002) Functions and potential applications of glycolipid biosurfactants – from energy-saving materials to gene delivery carriers. *J. Biosci. Bioengng*, **94**, 187–201.
13. Cameotra, S. S. and Makkar, R. S. (2004) Recent applications of biosurfactants as biological and immunological molecules. *Curr. Opin. Microbiol.*, **7**, 262–266.
14. Rodrigues, L., Banat, I. M., Teixeira, J. and Oliveira, R. (2006) Biosurfactants: potential applications in medicine. *J. Antimicrob. Chemother.*, **57**, 609–618.
15. Mukherjee, S., Das, P. and Sen, R. (2006) Towards commercial production of microbial surfactants. *Trends Biotechnol.*, **24**, 509–515.
16. Stichtenoth, G., Jung, P., Walter, G. *et al.* (2006) Polymyxin B/pulmonary surfactant mixtures have increased resistance to inactivation by meconium and reduce growth of gram-negative bacteria *in vitro*. *Pediatr. Res.*, **59**, 407–411.
17. Lang, S. and Wullbrandt, D. (1999) Rhamnose lipids–biosynthesis, microbial production and application potential. *Appl. Microbiol. Biotechnol.*, **51**, 22–32.
18. Matsufuji, M., Nakata, K. and Yoshimoto, A. (1997) High production of rhamnolipids by *Pseudomonas aeruginosa* growing on ethanol. *Biotechnol. Lett.*, **19**, 1213–1215.
19. Das, K. and Mukherjee, A. K. (2007) Comparison of lipopeptide biosurfactants production by *Bacillus subtilis* strains in submerged and solid state fermentation systems using a cheap carbon source: some industrial applications of biosurfactants. *Process Biochem.*, **42**, 1191–1199.
20. George, S. and Jayachandran, K. (2009) Analysis of rhamnolipid biosurfactants produced through submerged fermentation using orange fruit peelings as sole carbon source. *Appl. Biochem. Biotechnol.* (in press). DOI: 10.1007/s12010-008-8337-6.
21. Dubey, K. and Juwarkar, A. (2001) Distillery and curd whey wastes as viable alternative sources for biosurfactant production. *World J. Microbiol. Biotechnol.*, **17** (1), 61–69.
22. Sudhakar Babu, P., Vaidya, A. N., Bal, A. S. *et al.* (1996) Kinetics of biosurfactant production by *Pseudomonas aeruginosa* strain BS2 from industrial wastes. *Biotechnol. Lett.*, **18**, 263–268.
23. Benincasa, M., Contiero, J., Manresa, M. A. and Moraes, I. O. (2002) Rhamnolipid production by *Pseudomonas aeruginosa* LBI growing on soapstock as the sole carbon source. *J. Food Engng*, **54**, 283–288.

24. Nitschke, M., Costa, S. G. V. A. O., Haddad, R., *et al.* (2005) Oil wastes as unconventional substrates for rhamnolipid biosurfactant production by *Pseudomonas aeruginosa* LBI. *Biotechnol. Program*, **21**, 1562–1566.
25. Thavasi, R., Jayalakshmi, S., Balasubramanian, T. and Banat, I. M. (2008) Production and characterization of a glycolipid biosurfactant from *Bacillus megaterium* using economically cheaper sources. *World J. Microbiol. Biotechnol.*, **24** (7), 917–925.
26. Barros, F. F. C., Ponezi, A. N. and Pastore, G. M. (2008) Production of biosurfactant by *Bacillus subtilis* LB5a on a pilot scale using cassava wastewater as substrate. *J. Ind. Microbiol. Biotechnol.*, **35**, 1071–1078.
27. Fox, S. L. and Bala, G. A. (2000) Production of surfactant from *Bacillus subtilis* ATCC 21332 using potato substrates. *Bioresour. Technol.*, **75**, 235–240.
28. Joshi, S., Bharucha, C., Jha, S. *et al.* (2008) Biosurfactant production using molasses and whey under thermophilic conditions. *Bioresour. Technol.*, **99**, 195–199.
29. Chen, C.-Y., Baker, S. C. and Darton, R. C. (2006) Continuous production of biosurfactant with foam fractionation. *J. Chem. Technol. Biotechnol.*, **81**, 1915–1922.
30. Davis, D. A., Lynch, H. C. and Varley, J. (2001) The application of foaming for the recovery of surfactin from B. subtilis ATCC 21332 cultures. *Enzyme Microbiol. Technol.*, **28**, 346–354.
31. Syldatk, C., Lang, S. and Wagner, F. (1985) Chemical and physical characterization of four interfacial–active rhamnolipids from *Pseudomonas* sp. DSM 2874 grown on n-alcanes. *Z. Naturforsch.*, **40**, 51–60.
32. Benincasa, M. and Accorsini, F. R. (2008) *Pseudomonas aeruginosa* LBI production as an integrated process using the wastes from sunflower-oil refining as a substrate. *Bioresour. Technol.*, **99**, 3843–3849.
33. Abalos, A., Pinazo, A., Infante, R. *et al.* (2001) Physicochemical and antimicrobial properties of new rhamnolipids produced by *Pseudomonas* AT10 from soy bean oil refinery wastes. *Langmuir*, **17**, 1367–1371.
34. Haba, E., Pinazo, A., Ja'uregui, O. *et al.* (2003) Physicochemical characterization and antimicrobial properties of rhamnolipids produced by *Pseudomonas aeruginosa* 47T2 NCBIM 40044. *Biotechnol. Bioengng*, **81**, 316–322.
35. Mercadé, M. E., Manresa, M. A., Robert, M. *et al.* (1993) Olive oil mill effluent (OOME). New substrate for biosurfactant production. *Bioresour. Technol.*, **43** (1), 1–6.
36. Nitschke, M. and Pastore, G. M. (2006) Production and properties of a surfactant obtained from *Bacillus subtilis* grown on cassava wastewater. *Bioresour. Technol.*, **97**, 336–341.
37. Aveyard, R., Binks, B. P., Clark, S. and Fletcher, P. D. I. (1990) Effects of temperature on the partitioning and adsorption of C12E5 in heptane–water mixtures. *J. Chem. Soc., Faraday Trans.*, **86**, 3111–3115.
38. Acosta, E. J., Harwell, J. H. and Sabatini, D. A. (2004) Self-assembly in linker-modified microemulsions. *J. Colloid Interface Sci.*, **274**, 652–664.
39. Nguyen, T. T., Youssef, N. H., McInerney, M. J. and Sabatini, D. A. (2008) Rhamnolipid biosurfactant mixtures for environmental remediation. *Water Res.*, **42**, 1735–1743.
40. Smith, G. A., O'Neil, J., Sneed, G. and Whewell, Ch. J. (2006) Alkaline earth-based alkoxylation catalysts, Patent No. WO/2006/025898.
41. Boocock, D. G. B. (2001) Single-phase process for production of fatty acid methyl esters from mixtures of triglycerides and fatty acids, Canadian Patent No. 2,381,394, February 22.
42. Gerpen, J. V. (2005) Biodiesel processing and production. *Fuel Process. Technol.*, **86**, 1097–1107.
43. Quagliotto, P., Montoneri, E., Tambone, F. *et al.* (2006) Chemicals from wastes: compost-derived humic acid-like matter as surfactant. *Environ. Sci. Technol.*, **40**, 1686–1692.
44. Montoneri, E., Boffa, V., Savarino, P. *et al.* (2009) Use of biosurfactants from urban wastes compost in textile dyeing and soil remediation. *Waste Manage.*, **29**, 383–389.

45. Montoneri, E., Boffa, V., Quagliotto, P. L. *et al.* (2008) Humic acid-like matter isolated from green urban wastes. Part I: structure and surfactant properties. *Bioresources*, **3**, 123–141.
46. Guetzloff, T. F. and Rice, J. A. (1994) Does humic acid form a micelle? *Sci. Total Environ.*, **152**, 31–35.
47. Conte, P., Agretto, A., Spaccini, R. and Piccolo, A. (2005) Soil remediation: humic acids as natural surfactants in the washings of highly contaminated soils. *Environ. Pollut.*, **135**, 515–522.
48. Quadri, G., Chen, X., Jawitz, J. W. *et al.* (2008) Biobased surfactant-like molecules from organic wastes: the effect of waste composition and composting process on surfactant properties and on the ability to solubilize tetrachloroethene (PCE). *Environ. Sci. Technol.*, **42** (7), 2618–2623.
49. Savarino, P., Montoneri, E., Biasizzo, M. *et al.* (2007) Upgrading biomass wastes in chemical technology. Humic acid-like matter isolated from compost as chemical auxiliary for textile dyeing. *J. Chem. Technol. Biotechnol.*, **82**, 939–948.
50. Kroiss, H. (2004) What is the potential for utilizing the resources in sludge? *Water Sci. Technol.*, **49**, 1–10.
51. Hospido, A., Moreira, Ma. T., Martín, M. *et al.* (2005) Environmental evaluation of different treatment processes for sludge from urban wastewater treatments. *Int. J. Life Cycle Assess.*, **10**, 336–345.
52. Dewil, R., Baeyens, J. and Neyens, E. (2006) Reducing the heavy metal content of sewage sludge by advanced sludge treatment methods. *Environ. Engng Sci.*, **23**, 994–999.
53. Von Sperling, M. (2007) *Activated Sludge and Aerobic Biofilm Reactors: Biological Wastewater Treatment*, vol. 5, IWA Publishing, London, p. 340.
54. Garnier, C., Görner, T., Lartiges, B.Sy. *et al.* (2005) Characterization of activated sludge exopolymers from various origins. *Water Res.*, **39**, 3044–3054.
55. Boocock, D. G. B., Konar, S. K., Leung, A. and Ly, L. D. (1992) Fuels and chemicals from sewage sludge. 1. The solvent extraction and composition of a lipid from a raw sewage sludge. *Fuel*, **71**, 1283–1289.
56. Haag, A. P. (2006) Mechanical properties of bacterial exopolymeric adhesives and their commercial development, in *Biological Adhesives* (eds A. M. Smith and J. A. Callow), Springer, Berlin, pp. 1–19.
57. Daiuto, É., Cereda, M., Sarmento, S. and Vilpoux, O. (2005) Effects of extraction methods on yam (*Dioscorea alata*) starch characteristics. *Starch*, **57**, 153–160.
58. Somboonpanyakul, P., Wang, Q., Cui, W. *et al.* (2006) Malva nut gum. (I): extraction and physicochemical characterization. *Carbohydr. Polym.*, **64**, 247–253.
59. Vlyssides, A. G. and Karlis, P. K. (2004) Thermal-alkaline solubilization of waste activated sludge as a pre-treatment stage for anaerobic digestion. *Bioresour. Technol.*, **91**, 201–206.
60. Neyens, E., Baeyens, J. and Creemers, C. (2003) Alkaline thermal sludge hydrolysis. *J. Hazard. Mater.*, **B97**, 295–314.
61. Lin, J.-G., Chang, Ch-N. and Chang, Sh. (1997) Enhancement of anaerobic digestion of waste activated sludge by alkaline solubilization. *Bioresour. Technol.*, **62**, 85–90.
62. Mizoguchi, T., Watanabe, K., Kobori, T. *et al.* (2008) Solubilization and reduction of activated sludge from petroleum refinery using high speed mixer and alkaline treatment. *J. Jpn Petrol. Inst.*, **51**, 245–251.
63. García-Becerra, F. Y., Acosta, E. and Allen, D. G. (2009) Alkaline Extraction of wastewater activated sludge biosolids. *Bioresour. Technol.* (submitted).

64. Masuko, T., Minami, A., Iwasaki, N. *et al.* (2005) Carbohydrate analysis by a phenol-sulfuric acid method in microplate format. *Anal. Biochem.*, **339**, 69–72.
65. Yuan, J., Ansari, M., Samaan, M. and Acosta, E. J. (2008) Linker-based lecithin microemulsions for transdermal delivery of lidocaine. *Int. J. Pharm.*, **349**, 130–143.
66. García-Becerra, F. Y., Allen, D. G. and Acosta, E. (2009) Surfactant-like properties of alkaline extracts from wastewater biosolids. *J. Surfactants Deterg.* (in press).

10
Lecithin and Other Phospholipids

Willem van Nieuwenhuyzen
Lecipro Consulting, Limmen, The Netherlands

10.1 Introduction

Phospholipids are important biochemical intermediates in the growth and functioning of cells in plants and animals. Almost all body cells contain phospholipids. Vegetable lecithins are derived commercially from oil-bearing seeds such as soya beans, sunflower kernels and rapeseed. Vegetable lecithins contain predominantly phosphatidylcholine (PC), phosphatidylethanolamine (PE), phosphatidylinositol (PI), phosphatidic acid (PA) and minor amounts of lysophosphatidylcholine (LPC) and other glycerol phospholipids of complex fatty acid composition. These lecithins with their surface-active properties are used as emulsifiers in a vast range of foods, feed, pharmaceutical and technical applications. The molecular structure of the main phospholipids is given in Figure 10.1. Methods for isolation of lecithins from vegetable and animal tissues, modification and the purification of specific phospholipids have been extensively investigated. In biochemistry and medicine, the name lecithin is exclusively given to the 3-sn PC. This chapter focuses on the commonly used definition of lecithin as a mixture of phospholipids with adherent glycolipids and oil.

10.2 Sources and Production

Soya beans have been the primary source of vegetable lecithin for many years, making it a leading food emulsifiers group. Although the production has grown steadily to a current estimated 180 000–200 000 million tonnes per year, only a 15–20% share of the potential available lecithin gums of the crushed seeds is processed worldwide. The main sectors of use of these natural-derived emulsifiers are the food and feed industries. Vegetable

Surfactants from Renewable Resources Edited by Mikael Kjellin and Ingegärd Johansson

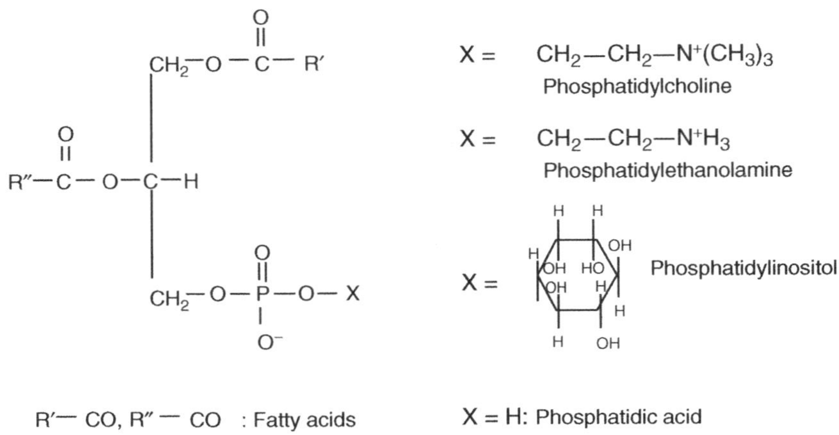

Figure 10.1 *Molecular structure of main phospholipids.*

lecithins are also used as emulsifiers in quite a number of technical, pharmaceutical and cosmetic applications. The sustainable sourcing of lecithin and phospholipids as renewable surfactants for nonfood uses has still a great application potential. Tailor-made modification of phospholipids with enhanced hydrophilic properties may give new value-added commercial outlets. Lecithin is a co-product of the oil seeds milling industry, which has undergone a strong concentration of ownership of oil mills in the last decades. Currently four worldwide operating agricultural commodity companies crush 70% of the oil seeds.

Sourcing of identity preserved (IP) non-GMO (genetically modified organism) soya lecithin for the European food market has changed drastically the lecithin world market supply system since 1996. Traditional non-GMO soya beans availability will become scarcer, which presents a market opportunity for high-quality IP soya, sunflower and rapeseed lecithins.

Companies with oil milling activities and lecithin marketing divisions are Archer Daniels Midland, USA (www.admworld.com), Bunge Inc., USA, through their JV The Solae Company, USA (www.solae.com), Cargill, USA (www.cargill.com), and Imcopa, Brazil (www.imcopa.eu). These companies operate worldwide. Local oil mills may also process lecithins. Medium-sized specialty lecithin marketing companies with a focus on Europe are Stern-Wywiol Group, Germany (www.stern-wywiol-gruppe.de), and Lecico GmbH, Germany (www.lecico.de). Additionally a large number of ingredient and additive companies may have lecithin in their product portfolio, serving various industrial sectors. Producers of purified fractions for the pharmaceutical and cosmetic sectors are Lipoid GmbH, Germany (www.lipoid.com), and Phospholipid GmbH, Germany (www.phospholipid.de).

10.2.1 Production of Soya Lecithin

Hexane extraction of soya beans is carried out in large crushing plants with capacities of 1000–6000 metric tonnes per day (Figure 10.2). Those plants run continuously

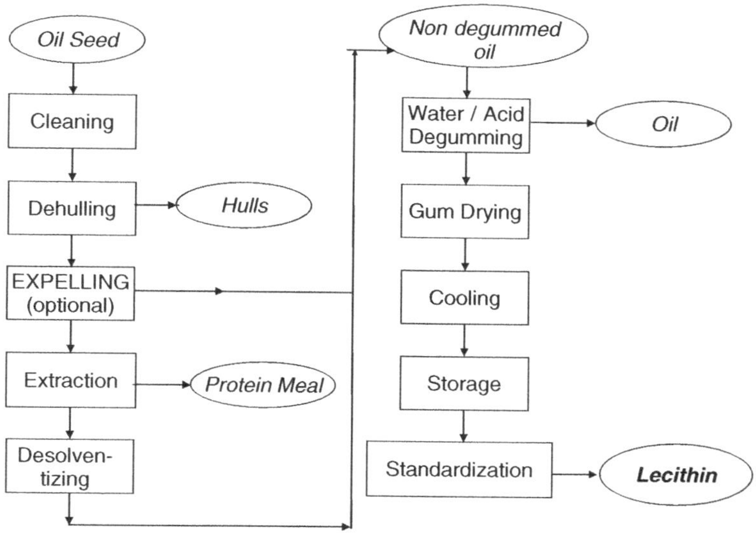

Figure 10.2 *Block diagram of oil seed production and lecithin recovery.*

throughout the year with the scope to extract the 18% soya bean oil from the meal at lowest costs. Extracted soya bean oil contains 800–1200 ppm phosphor, equivalent to 2–3% commercial liquid lecithin. The phospholipids are separated from the oil in the degumming step, since oil refiners need a crude oil with low P content for efficient refining of soya oil with lowest quantities of used chemicals and waste output. The filtration of the extracted oil–hexane miscella or the dried extraction oil is carried out in filtration units with filtration/bleaching earth for optimizing the production of lecithins with low residual dirt (earth, seed protein residue), expressed as hexane insoluble (HI) or toluene insoluble (TI), which must comply with the legal specification maximum of 0.3%. Well-filtered lecithins with a transparent glossy appearance have TI far below 0.1% and a low iron content of Fe < 50 ppm. It is not possible to filter effectively high viscous turbid lecithin at large plant scale operation for achieving a visual clear lecithin.

Lecithin processing is performed in four subsequent unit operations.

10.2.1.1 Hydration of Phospholipids

One to three percent of water is mixed thoroughly with the oil at 50–70 °C. The phospholipids hydrate within 1 hour to form a gum with a higher specific density than the oil. The easily hydratable phospholipids are PC, PI and LPC. PE and PA have low hydrating properties and are therefore marked as nonhydratable phospholipids (NHPs). In practice a mix of these various phospholipids is separated in the lecithin gum. After this water degumming process, sometimes enzymatic degumming is applied for hydrolysing the NHP with phospholipase-A enzymes into hydrophilic lysophosphatidylethanolamine

(LPE) and lysophosphatidic acid (LPA), to be also separated as gum from the oil [1]. However, this gum quality is mostly not used for commercial lecithin production.

10.2.1.2 Separation of the Lecithin Gums

After the hydration period the gums are separated from the oil in continuously operating centrifuges at 50–70 °C. The adjustment of optimal back-pressure conditions in the centrifuge gives efficient degumming yields with a crude soya oil with maximum 0.3% NHP (maximum 100 ppm P) and a lecithin gum with a low oil content (30–39% oil on a dry basis). Gums often contain maximum 50% water, a minimum 33% acetone insoluble (AI) (indication for phospholipids content) and maximum 17% oil. If gums are not dried to lecithins, they are mostly sprayed on the soya meal.

10.2.1.3 Drying

Lecithin is dried to a content of <1% moisture, achieving a long time shelf life and fluidity. For the drying of lecithin gums, batch, semi-batch and continuous drying film evaporators are used. Film evaporators have the advantage of a high performance capacity per unit drying surface of, for example, 70–90 kg gums/m^2 h and a short drying time of 1–2 minutes, which is adequate for achieving a good colour quality [2]. Often a long batch drying time gives severe darkening due to the Maillard reaction of the adherent sugars and the Amadori reaction between sugars and PE. Currently continuous rotating horizontal film evaporators, particularly the Buss-SMS-Canzler Sako KH driers or vertical thin film evaporators (e.g. type Luwa) are used for fast drying.

In Latin America and other continents batch driers are used, claiming that investments are low and that the drying at a maximum of 90 °C is efficient for lecithin with good colour. The De Smet type batch drier is made of a horizontal cylindrical jacketed body with a dome. Hot water circulates in the jacket. A heavy duty, rotating coil mounted on a shaft provides the necessary heat, while at the same time it agitates the lecithin gums.

Natural lecithin has a brown colour. The colour of lecithin can be attributed to the colour components carotenoids, melanoids and porphyrins in addition to the Maillard and Amadori reacted components. Bleaching of preferably the gums prior to drying with a 35% H_2O_2 solution is effective. This bleaching step is limited for food grade lecithins, complying with the EU E322 additive regulation specification peroxide value (POV) maximum of 10 meq/kg. However, lecithin use in technical industries gives freedom for strong bleaching, enabling light yellow lecithins to be processed.

In the USA single bleached grade lecithin (with H_2O_2) with a POV maximum of 100 meq/kg can be legally used in food applications. The US double bleached grade is usually bleached with a high hydrogen peroxide concentration or benzoyl peroxide for technical applications.

10.2.1.4 Cooling

Rapid cooling of the lecithin to below 50 °C in a heat exchanger is effective to prevent post-darkening. Otherwise the results of careful drying will be counteracted by post-darkening during a slow cooling regime from 100 °C to ambient temperature. For bulk storage a temperature of around 40 °C under dry conditions is recommended, using adequate tanks with stirring facilities for keeping stored lecithin homogeneously. At 20–30 °C lecithin can easily be stored for years without significant changes in product quality and functional properties. Standardization of the natural liquid lecithin can be achieved by adjustment of the AI matter with addition of oil or fatty acids to the lecithin gums or the dried lecithin.

10.2.2 Sunflower and Rapeseed Lecithin Production

The lecithin processing from these oil seeds is to a large extent similar to soya lecithin processing. Hence the additional expelling (pressing) operation (see Figure 10.2) is mostly used to squeeze out the seed oil content from 40–50% down to 15–20% in the press cake; the grinded cake is then hexane-extracted for removing the oil until below 0.5%. Both press oil and extraction oil streams are combined for degumming. Since rapeseed contains chlorophyll components, the rapeseed lecithin colour may have a greenish tone. Rapeseed and sunflower lecithin yields are about 0.2–0.3% on a seed basis, while soy lecithin recovery is about 0.5%.

Some European countries, such as Hungary, Ukraine and France, and Argentina grow and process sunflower seeds for oil production. Rapeseed lecithin is produced in some Canadian and European crushing plants. In principle it is possible to process lecithin with good phospholipid composition from expelled seed oils only, but the availability is small.

Using optimal process conditions it is possible to produce standard rapeseed and sunflower lecithins with similar surface-active properties to soya lecithin. Good modification processes of the phospholipids in these lecithins can be applied on a plant scale [3, 4].

10.3 Composition

For decades the four main phospholipids were determined by qualitative thin layer chromatography (TLC). The intensity and size of the coloured spots on the chromatogram could be used for semi-quantitative determination (e.g. American Oil Chemist Society (AOCS) Recommended Practise Ja 7-86). More exact data are obtained by high-performance thin layer chromatography (HPTLC) methods with good repeatability within qualified laboratories. New analytical apparatus allow precise HPTLC determination of all phospholipid and glycolipid classes in vegetable lecithins in a reproducible manner. High-performance liquid chromatography with a light scattering detector (HPLC-LSD) is a common method used to measure six phospholipid classes in standard and fractionated lecithins [5]. This method with good reproducibility is preferred over HPLC with an ultraviolet detector. Based on ring tests organized by the International Lecithin and Phospholipid Society (ILPS) the validated method has been accepted as the official unified German Society for Fat Science (DGF) method (DGF F-I-6a) and AOCS approved method (AOCS Ja 7c-07).

The ultimate most sophisticated method for phospholipidanalysis is the 31P-NMR (nuclear magnetic resonance), analysing quantitatively 15 (lyso-)phospholipid classes with phosphorus as the absolute reference [6, 7].

Table 10.1 *Phospholipid composition in % of liquid vegetable lecithins by 31P-NMR*

Phospholipid	Soya lecithin	Sunflower lecithin	Rapeseed lecithin
Phosphatidylcholine (PC)	15	16	17
Phosphatidylethanolamine (PE)	11	8	9
Phosphatidylinositol (PI)	10	14	10
Phosphatidic acid (PA)	4	3	4
Other phospholipids	7	6	6
All phospholipids	47	47	46

The typical 31P-NMR analyses of commercially produced lecithins from soya beans, rapeseed and sunflower kernels in Table 10.1 show slight differences in phospholipid composition. Differences within one type can be even larger, due to variations in crop harvesting conditions, storage, seed treatment and extraction conditions. Rapeseed and sunflower seed are usually expelled to press out the main part of the oil, after which the remaining cakes are hexane-extracted. This may explain the fact that lecithin, derived from the mixed oils, contains a slightly higher PC and slightly lower PE content [8]. The fatty acid compositions of these three types of lecithin largely follow the typical fatty acid composition of the oils (Table 10.2). Lecithin manufacturers have the expertise to offer different qualities of vegetable lecithins, meeting surfactant properties for specific uses.

10.4 Quality and Analysis of Lecithins

A number of methods are routinely used for determining the specifications during lecithin production, trading and incoming quality control. In Table 10.3 the most used approved methods by the AOCS and the DGF are now given.

- **Acetone-insoluble (AI)**. The amount of AI matter (%AI) is the approximate indication for the amount of phospholipids, glycolipids and carbohydrates, because the oil and fatty acids dissolve in acetone. In crude lecithin AI is synonymous with activity, i.e. functional or nutritional properties. The AI is a commercial and legal specification. However, with NMR and HPLC the more exact amount and composition of true phospholipids can be analysed.
- **Toluene-insoluble (TI)**. The level of TI matter is a measure of the absence of residual fibre and filter processing aid. The level of TI matter in crude lecithin should not exceed 0.3%, and today rarely exceeds 0.1%. In North and South America a similar HI method

Table 10.2 *Fatty acid composition in % of vegetable lecithins*

Fatty acid	Soya lecithin	Sunflower lecithin	Rapeseed lecithin
C16:0	16	11	7
C18:0	4	4	1
C18:1	17	18	56
C18:2	55	63	25
C18:3	7	0	6
Others	1	4	5

Table 10.3 Survey of approved analytical methods

Method	AOCS	DGF
Acetone insoluble	Ja 4-46	F-I 5
Toluene insoluble	–	F-I 4b
Hexane insoluble	Ja 3-87	–
Moisture KF	Ja 2b-87	F-I 4
Acid value	Ja 6-55	F-I 3
Peroxide value	Ja 8-87	F-I 3b
Gardner colour	Ja 9-87	C-IV 4a
Viscosity (rotation)	Ja 10-87	F-I 2a
Viscosity (bubble)	Ja 11-87	–
TLC	Ja 7-86	–
HPLC-UV	Ja 7b-91	–
HPLC-LSD	Ja 7c-07	F-I 6a
HPTLC	–	F-I 6
Iron	Ca 15-75	F-I 4a

is mostly used. Whereas the difference between using the two solvents appears to be negligible, both methods as approved by AOCS and DGF define a different porosity of the used glass filters, which may affect the filtration speed and results.

- **Acid value (AV)**. The acid value (AV) expresses the acidity in mg KOH/g of sample. The AV represents the acidity contributed by phospholipids (in soya lecithin often 18–24 mg KOH/g) and free fatty acids. Usually the free fatty acid content of crude oil is low, so that high AV in liquid lecithins is mostly caused by the addition of distilled free fatty acids for product viscosity reasons. Hydrolysed lecithin will contain free fatty acids from the hydrolysis process. Lecithin exhibits a neutral pH value in aqueous media. To assay for free fatty acids, the correct method is to titrate only the acetone-soluble portion, whereby any contribution from the phospholipids in the AI portion is eliminated.
- **Moisture**. The water content of lecithin products is usually less than 1.0%. As a consequence of lecithin's essentially moisture-free state, lecithins have very low water activity and do not adversely contribute to the microbiological profile of most food systems. Moisture is determined by the Karl Fisher method. A less accurate moisture level can also be determined by azeotropic toluene distillation or drying in an oven at 105 °C.
- **Colour**. Historically, US lecithins have been colour graded as unbleached, single bleached and double bleached. This differentiation is usually not made in Europe. By convention, the amber colour tones of lecithin are measured on the Gardner colour scale or iodine colour scale. The colour range of most clear lecithins is generally in the range of Gardner values 9–17 in the undiluted products, while hydroxylated and strong bleached lecithins may have Gardner colour values of 5–8. If the lecithin is turbid, colour can be measured in a filtered dilution in hexane or toluene, usually in a 1 : 10 ratio, a condition that should be reported in the analysis certificate.
- **Peroxide value (POV)**. The POV of lecithin, produced from fresh or optimal stored beans or seeds, is usually below 2 meq/kg. In contrast to POV values in oil as a result of oxidation, the POV in lecithin is mostly a result of residual hydrogen peroxide from the bleaching process step.

- **Consistency**. Lecithins are available in fluid, paste-like and plastic (solid) forms. Liquid lecithins generally follow Newtonian flow characteristics. The viscosity profile of lecithins is a complex function of AI content, moisture, mineral content, AV and the combined effects of assorted additives such as vegetable oils and surfactants. Generally, a higher AI and/or moisture content yields higher viscosity, whereas an increased AV often decreases viscosity. Certain divalent minerals, such as calcium and magnesium, can also adjust the viscosity level. The sample preparation for the rotation viscosity method eliminates some pseudo-plastic effects. A bubble time method is also used. The measured result is an analytical tool for determining the flow properties of lecithin during pumping and in dosing equipment.
- **Clarity**. In some soya processing plants, high levels of TI or HI matter may partition with the lecithin gums on separation from the oil. This lipid-insoluble material can cause haziness in fluid lecithins. With modern miscella and oil filtration techniques, clear transparent-looking lecithins with very low, or even no, TI contents can be produced. Additionally, moisture of over 1% can also contribute to lack of clarity. Haziness can result in sediments over time; solid particles may appear on the bottom of an otherwise clear liquid lecithin.

10.5 Modification

Standard refined lecithins take the largest share of the lecithin group of emulsifiers. The definition 'refining' is mostly used for the exact adjustment of quality parameters of clear filtered lecithin with guaranteed AI, AV, colour and phospholipid composition. Thus it is a different definition compared to refining of oils.

Modified lecithins have dedicated emulsifying properties due to the increased hydrophilicity. Modification options are: (i) enzymatic and chemical adaptation of the phospholipid molecules, (ii) physical fractionation for separating oil from the phospholipids and (iii) fractionation of phospholipids. A compilation of all typical components in standard liquid soya lecithin, deoiled soya lecithin powder and an alcohol-soluble fraction is given in Table 10.4. Recently Unilever Research scientists have updated this knowledge with new experiments and analytical tools [9].

10.5.1 Modification by Enzymes

Enzymatic hydrolysed lecithins have technological and commercial benefits since they are excellent oil-in-water emulsifiers. Phospholipase A2 enzyme hydrolyses specifically the fatty acid at the ß-position of the phospholipid molecule, which enables the processing of a range of lysophospholipids with different degrees of hydrolysis and emulsifying properties under controlled conditions (Figure 10.3). Since the PE has very weak emulsifying power, the LPE formation with a strong emulsifying property is a focused target of the hydrolysis process. Therefore it is essential to use enzymes with an affinity to strong PE hydrolysis, while PC hydrolysis has a lower priority. Different applications or recipes will require the use of a lecithin with a defined degree of hydrolysis and concentration. The modifications are essential for achieving and adjusting optimal ratios between hydrophilic and lipophilic properties and for ensuring optimal surfactant properties.

Table 10.4 Typical total composition in % of soya lecithin products

Components	Standard liquid	De-oiled lecithin lecithin powder	PC-enriched fraction
Phospholipids			
Phosphatidylcholine (PC)	15	24	38
Phosphatidylethanolamine (PE)	11	17	8
Phosphatidylinositol (PI)	10	16	3
Phosphatidic acid (PA)	4	6	1
Other PL	7	11	1
Glycolipids			
Sterolglycoside	7	11	6
Galactodiglyceride	4	6	3
Complex carbohydrates			
Sucrose	1	1.5	1
Raffinose	1/2	1	1/2
Stachyose	2	3	1
Acetone insolubles (AI)	62	97	62
Neutral oil	37	3	37
Water	<1	<1	<1

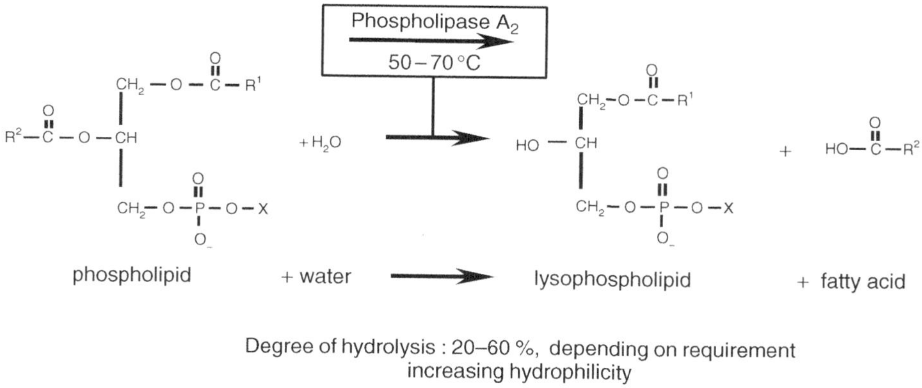

Figure 10.3 Enzymatic hydrolysis of lecithin.

Today, the availability of pure phospholipase A2, A1 and lipase enzymes enables precise degrees of hydrolysis of the various phospholipids. Commercially available lecithins may have a degree of hydrolysis of 30–60% of the phospholipids, which is sufficient for a number of food emulsion and starch interaction applications. However, for technical uses a higher 60–80% degree of hydrolysis may be desired for obtaining the highest possible hydrophilic–lipophilic balance (HLB) value. This requirement can be met by adapting the hydrolysis process conditions accordingly, in particular the reaction balance. In addition to the liquid hydrolysed lecithins, oil-free hydrolysed lecithin powder and special spray-dried products on a carrier are available.

Combinations of phospholipases and lipases represent a tool of enzymatic interesterification of phospholipids with specific fatty acids composition, entering a commercial area for health-related lecithins as well [10, 11]. Hence this interesterification is not primarily focused on the surface-active properties.

10.5.2 Modification by Chemicals

10.5.2.1 Hydrolysis

In principle chemical hydrolysis of the fatty acids from the phospholipid molecules is possible. However, this reaction is not applied on a plant scale, because enzymatic hydrolysis is the preferred state-of-the-art process. Chemical partial hydrolysis can be made with various strong acids, but the modified lecithins usually have a dark brown to black colour, forming an obstacle for successful industrial use in many applications.

10.5.2.2 Acetylation

The amino group in PE reacts with acetic anhydride resulting in acetyl-PE, giving the lecithin enhanced oil-in-water emulsifying properties. The principle of the process is that the 'zwitter-ion' group of the PE is blocked:

$$-NH_2 \rightarrow NH-COCH_3 \text{ (acetylation)}$$

Under plant scale conditions the acetylating reaction with crude lecithin often gives partial acetylated PE. The lecithin is treated with 1.5–5% acetic anhydride at 50–60 °C. The reacted product is neutralized with alkali hydroxide to pH 6.5–8.0. Acetylation can be performed starting from lecithin gums or starting with dried lecithin. Pure fractionated PE is not available as the starting material in large quantities. However, it is useful to acetylate the PE-enriched (PC-depleted) alcohol-insoluble lecithin fraction, in order to upgrade this co-product of the lecithin fractionation [12]. Starting with commercial lecithin with 12% PE, the acetylation degree of 50–80% of PE may result in a lecithin with 6–8% acetylated PE and 4–6% PE in the total phospholipid composition. Residual water has to be removed by drying. For food applications one should take care that no soaps are formed, since they influence the taste negatively. For technical applications a relatively high pH may advantageously give good hydrophilic properties. Therefore the acetylated lecithin has better resistance to browning on heating, because the amine group can no longer react with sugars.

10.5.2.3 Hydroxylation

Hydroxylation of the double bonds in the unsaturated fatty acids of the phospholipids is made in the presence of peroxide and organic acids, resulting in the highest possible lecithin hydrophilicity:

$$-CH = CH- \rightarrow -CH(OOH) - CH(OOH)- \rightarrow -CH(OH) - CH(OH)-$$

Crude lecithin is mixed with 2–14% hydrogen peroxide and 1% lactic acid at 50–60 °C. The lactic acid reduces the pH, which facilitates the reaction. Residual water is removed by drying. During the reaction the coloured components are bleached, which yields a light coloured lecithin. Hydroxylated lecithin is superior for emulsification of oil-in-water emulsions. It easily disperses in cold and water as a whitish emulsion. This is already a good indication that the HLB value increased significantly.

10.5.2.4 Acetylation + Hydroxylation

The amino group of PE has a negative influence on the hydroxylation performance. Therefore it may be appropriate to use acetylated lecithin as the raw material for hydroxylation. These combined modifications yield a lecithin with very hydrophilic properties [12]. The HLB value is reported to reach the number 11 [13].

10.5.2.5 Hydrolysis + Hydroxylation

Interesting products can also be obtained by using enzymatic hydrolysed lecithin as the raw material for producing hydroxylated lecithins [14]. The hydroxylation process conditions are quite similar to those using standard lecithin. Lecithin with slightly more hydrophilic properties is obtained. This was measured in simple dispersion tests. Due to the smaller, more hydrophilic molecules, this product is, for example, a superior fattening agent in leather tanning and fatting.

10.5.2.6 Hydrogenation

Vegetable lecithins contain unsaturated fatty acids in the phospholipid molecules. It is possible to hydrogenate with the use of catalysts such as palladium the unsaturated into saturated fatty acids:

$$-CH{=}CH- \rightarrow -CH_2-CH_2-$$

These products have a higher melting point, are stable against oxidation but less soluble in oils and fats. The process is preferably carried out with quite pure PC fractions. With crude lecithin the reaction is not very well controlled and the catalysts are poisoned by the impurities. Soya lecithins with hydrogenated PC 80–100% contents are commercially available. These fractions are used as polar lipid excipients in pharmaceutical products and cosmetics [15].

10.5.3 Fractionation for Oil Removal

Phospholipids possess polar hydrophilic groups by which they can be separated from the apolar triglycerides. Currently used and potential processes are acetone de-oiling, membrane fractionation and carbon dioxide extraction.

10.5.3.1 Acetone Extraction

Triglycerides dissolve in acetone in contrast to the other more polar components of standard soya lecithin. Therefore this characteristic property is used as a quality control and specification tool in the regulatory definition and marketing of lecithins.

De-oiling with acetone is executed as an efficient continuous process, in which the crude lecithin is mixed and agitated with acetone. The phospholipids, glycolipids and adherent carbohydrates precipitate as sediments, which are centrifuged and/or filtered. A careful drying process is required to eliminate the residual acetone, preventing the formation of undesired off-flavours, particularly mesityloxide. Sophisticated types of belt dryers and fluid bed dryers have proven good results. De-oiled lecithins have only 2–5% remaining oil and are marketed with an AI matter of minimum 95%. Low oil content is beneficial for excellent free flowing properties of the powders or granules. The classification between powders and granules is often made by sieving. The process can also be used for the de-oiling of enzyme and chemically modified lecithins and lecithin fractions.

10.5.3.2 Membrane Technology

Membrane filtration was firstly investigated by Sen Gupta [16] for removing phospholipids from the crude oil miscella feed. The phospholipids (retentate) are bound in large micelles, which do not pass the membrane, while the triglycerides pass with the hexane through the membrane (permeate). The aim of these studies was the replacement of the water degumming step and to produce crude soya oil with low phosphorus content, enabling physical refining. Nowadays hexane-resistant membranes, e.g. ceramic ones, are available in the market and the technology has been improved so far that phospholipid separation from the oil miscella is possible [17]. Now this technology is in use for the production of oil-free lecithin powders [18]. However, the process is cumbersome for deoiling of modified lecithins.

10.5.3.3 Carbon Dioxide Extraction

Supercritical extraction is carried out commercially for the recovery of high-value products such as flavours, oleoresins and decaffeinated coffee beans. In principle CO_2 extraction is an excellent process for de-oiling lecithin without using an organic solvent. On a laboratory and pilot scale the process has resulted in interesting products. Lecithin is sprayed into a process chamber under CO_2 high pressures of 400–700 bar. The triglycerides dissolve in the liquid CO_2, while the phospholipids become available as powder, which falls into the collection vessel. Devices for spraying the viscous lecithin feed with the dense gas are important, because the end product should be a fine oil-free powder [19]. It is difficult to run the unloading of the collection vessel continuously due to the required reduction of the high pressure down to, for example, 50 bar. The advantages of the process are the absence of oxygen and solvent residues. The low oil-dissolving capacity of CO_2, the subsequent high solvent-to-feed ratios and the low yield need further process development before the plant scale de-oiling operation can become economically interesting.

10.5.4 Fractionation of Phospholipids

The phospholipids themselves have different loading and solubility in solvents, so aqueous alcoholic solvents can be used alone or in conjunction with chromatography for separation.

10.5.4.1 Alcohol Fractionation

PC dissolves better and faster than the other phospholipids in alcohols. Therefore the phospholipid mixture in crude soya lecithin can be fractionated into the alcohol-soluble and alcohol-insoluble fractions. Various types of alcohol and concentrations can be used to obtain specific extraction yields and selectivity of the PC/PE ratio. Ethanol-soluble fractions contain a high PC/PE ratio and the insoluble fraction a low PC/PE ratio. Prior to alcohol evaporation, mixtures of refined oil and monodiglycerides or calcium chloride may be added for producing stable liquid PC-enriched fractions [20, 21]. The fractions are used in food emulsions such as a frying agent in margarines, instantizing agent and health supplements. The process engineering aspects and conditions are extensively described in Reference [22].

10.5.4.2 Chromatographic Isolation

More pure PC products can be produced by using column chromatography with adsorbents. On a commercial plant scale oil-free lecithin or its ethanol-soluble PC fraction is treated in a chromatographic column with aluminium oxide or silica gel adsorbents. The PC is concentrated at 70–95% purity [23, 24].

10.6 Emulsifying Properties

10.6.1 Surface Activity

Emulsifiers, including lecithins with hydrophilic and lipophilic segments in their molecular structure, concentrate at the interface between oil and water and subsequently reduce the interfacial tension. The emulsifiers facilitate the formation of the emulsion during energy input with mixers and homogenizers. In food emulsion recipes the low amount (often $<0.5\%$) of emulsifier should support emulsion stability in interaction with the three main food components, fat, protein and carbohydrates, and other texturizing ingredients. The aims of emulsion stability are usually to prevent creaming, coalescence in larger droplets, sedimentation and separation during the shelf life of the product.

The amphiphilic structures of the main phospholipid molecules result in diverse phase behaviour, particularly in aqueous solutions. At neutral pH the head group of the natural phospholipids may be electrically neutral but zwitter ionic or negatively charged. At lower pH the net negative charge facilitates stronger swelling.

PC forms a lamellar layer in the interface between oil and water, different to the reversed hexagonal phase of PE or the hexagonal phase of lysophospholipids (Figure 10.4) [25]. This knowledge is used for the successful application of lecithins with adapted, modified or fractionated phospholipid composition. The different phase structures at the interface influence the emulsion formation and stability.

The enzymatic hydrolysis, acetylation and alcohol fractionation processes all have in common the fact that the ‘zwitter-ion’ group of the PE is modified or removed.

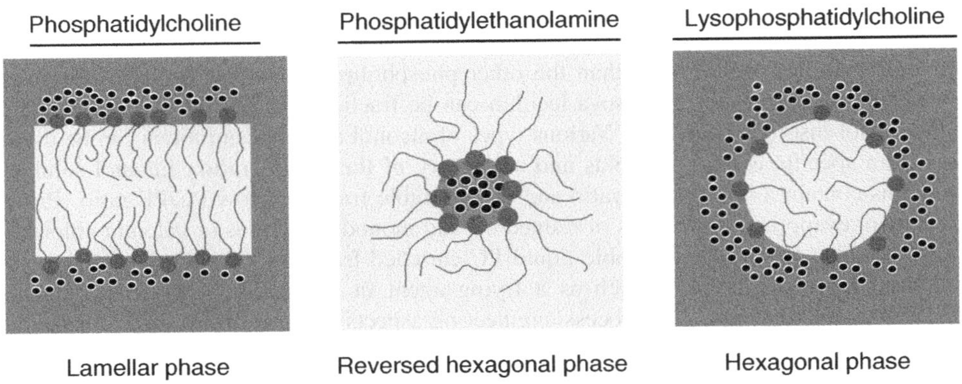

Figure 10.4 *Phospholipid structures at the interface.*

10.6.2 Evaluation of Lecithin Emulsifying Properties

There are manifold emulsifying tests, varying from simple laboratory tests to high sophisticated procedures, requiring specific apparatus and skills. One may start with the assessment of the emulsifying power of a lecithin in simple empirical tests [26]:

- **Dispersion test**. Hydration speed of lecithin in water. Standard liquid lecithins disperse slowly if stirred manually or by a mixer with low-energy input. One may observe different hydration speeds between natural lecithins from various oil mills, due to different production conditions. Modified lecithins usually show fast dispersibility, an indication of stronger surfactant performance. Hydration of lecithin in distilled water is faster than in tap water with a high degree of water hardness.
- **Emulsion capacity**. A lecithin/oil mixture (ratio 1/9 is often used) is emulsified in water and the oil and water separation is measured over a period of time. For example, hydrolysed lecithins give more stable emulsions with low fat creaming compared to the use of a standard liquid lecithin. In Table 10.5 ratings of the dispersing behaviour and oil emulsion capacity of various types of soya lecithins are given.
- **Emulsion stability**. Protein or other ingredients are incorporated in the lecithin/oil/water system and the stability of the emulsion (fat creaming, protein sedimentation) is measured over a period of time. When preparing an oil-in-water emulsion, the modified lecithins with enhanced hydrophilic properties usually give a specific stability performance.

These types of tests can be carried out in a quite simple manner for internal quality control and initial product development. Hence the conditions (mixing time, temperature, stirring equipment, ingredients, time of measurement) should be carefully controlled for obtaining repeatable and reproducible results. Hydrophilic lecithins facilitate the formation of smaller droplets in oil-in-water emulsions.

More detailed scientific information will require sophisticated equipment, experienced test procedures and skilled scientists.

Table 10.5 *Rating of dispersion and emulsifying capacity of soya lecithins*

Type of lecithin	Dispersion rating	Oil emulsifying capacity rating
Standard liquid	1	2
Standard oil-free	3	2
PC-35 fraction	3	4
Acetylated	4	4
Hydrolysed	4	5
Hydroxylated	5	5

Rating 1 = poor, 5 = excellent.

- **Particle size distribution (PSD)**. PSD is a useful technique, giving additional valuable information about the size of droplets in diluted emulsions. The principles of these methods are based on photon correlation spectroscopy and laser diffraction as offered by, for example, Coulter Counter of Malvern Mastersizer instruments. The Malvern Zetasizer gives the additional tool to measure the zeta potential, a repulsing electric force between the ions and the surfactant molecules.
- **Turbidity**. measurements of emulsions are made by, for example, the optical Turbiscan or Beckman QuickSCAN^R vertical scan analyser. The physical evolution of liquid dispersions during resting time in cylindrical glass measurement cells can be followed in the various stages of the destabilization process, discriminating between particle migration (sedimentation, creaming) and particle size variation (flocculation, coalescence) without disturbing the original system and with good sensibility and reproducibility [27, 28].
- **Microscopy**. Particle size, shape and structure of emulsion droplets can be visualized by various microscope techniques, such as phase contrast light microscopy, confocal scanning light microscopy (CSLM), scanning electron microscopy (SEM), transmission electron microscopy (TEM), X-ray microtomography (XRT), atomic force microscopy (AFM) and imaging techniques.

All these supporting data can be used to classify the lecithins in the HLB system by comparing the hydrophilic properties of modified lecithins with known nonionic emulsifiers (Table 10.6). Originally this system was developed as an indicator for selecting appropriate nonionic emulsifiers in a range of HLB values between 0 and 20 [29]. The HLB ranking for lecithins is scientifically problematic, because it provides little information about the emulsion stability properties and synergistic effects. However, it is still a valuable support for selecting emulsifiers for development of emulsions.

Synergy in surface-active properties and emulsion stability can be obtained by combining (modified) lecithins with other surface-active components. Those types of surfactants can be low molecular emulsifiers such as monodiglycerides and their esters, sugar esters or soaps. Polymeric emulsifiers such as certain proteins and hydrocolloids may also have surface-active properties. It is crucial to select the proper conditions in recipe and processing for achieving the desired synergistic properties. Otherwise one surfactant may deplete the other surfactant from the interface. This antagonistic effect will result in emulsion instability [30].

Table 10.6 Hydrophilic–lipophilic balance values of modified lecithins

Lecithin type	HLB value
PE-enriched (PC-depleted) fraction	2–3
Standard liquid lecithin	4
PC-30 fraction	5
Acetylated lecithin	6–7
Enzymatically hydrolysed lecithin	7
Hydroxylated lecithin	9
Acetylated + hydroxylated lecithin	11
Hydrolysed + hydroxylated lecithin	11

Table 10.7 Survey of lecithin application in selected foods

Application	Functionality	Types of lecithin
Baked goods	Volume improvement Fat dispersion, reduction Antistaling Firmness, freshness	Stand., hydrolysed, de-oiled
Chocolate	Rheology, yield value Viscosity modification	Stand.
Chewing gum base	Rheology Tackiness, brittleness	Stand., hydrolysed
Instant drinks dairy/cocoa	Agglomeration wetting Dispersibility	Stand., hydrolysed, PC fraction, de-oiled
Milk protein and replacer	Emulsification, emulsion stabilization Wetting Anti-dusting	Stand., hydrolysed, PC fraction, de-oiled
Margarine	Antispattering in frying Emulsification Mouth feel	Stand., hydrolysed, PC fraction
Flavour	Liposome encapsulation	PC fraction
Pan release agent	Wetting Separation, machineability	Stand., hydrolysed, acetylated

The choice of lecithin is based on European Food additive legislation. Acetylated and hydroxylated lecithins may be applicable in other continents. Stand = standard liquid lecithin, PC fraction = phosphatidylcholine-enriched fraction.

10.7 Applications

The use of lecithins in food, feed and technical products is manifold. Principally, surface-active lecithins may have technological functions in all types of emulsions and suspensions, but the effective use will also depend on the price–performance ratio in comparison to other emulsifiers and surfactants. Food applications have recently been reviewed elsewhere [31, 32]. Table 10.7 gives a short summary of the food uses.

In this chapter only a number of industrial nonfood uses are discussed.

10.7.1 Release and Lubrication Agents

Special refined liquid lecithins in combination with waxes and vegetable oils are used in release agents for food frying. The industrial market for lecithin containing release emulsions is more developed in North America than on other continents. For these applications the hydrophilic acetylated, hydroxylated and fractionated lecithins can be used. A typical pump spray formulation consists of 25% medium-chain triglycerides, 65% vegetable oil and 10% hydrophilic lecithin. Lecithin can be added to oil to prevent food from sticking to oven belts, moulds and cooking surfaces. Less stickiness gives less product loss, less cleaning stops and costs and therefore a higher production throughput. It is beneficial to use heat-resistant lecithins with a low PE content, by which the browning reaction of the PE amino group with sugars can be controlled to remain low. The fatty acid tails of the phospholipids stick as a thin film on the metal surface, while the surface-active heads form a hydrophilic layer.

In a similar way the lubricating properties of lecithins can be explained. Therefore lecithins are used in lubricant formulations in metal and other technical industries.

Whereas hydrophilic lecithins are used as dispersing and wetting agents of lipid-based products, the more hydrophobic standard lecithin and PE/PI phospholipids fraction with low HLB values are used for reducing the hydrophilic properties of powdered plastic intermediates. This leads to more convenient handling and less dusting properties.

10.7.2 Liposome Encapsulation

Liposome encapsulation of high-value functional ingredients is an innovative tool to achieving optimal targeted performance at the lowest concentrations. Liposomes for pharmaceutical/medical purposes enable the encapsulation of combined oil-soluble plus water-soluble drugs and controlled drug transfer and release [33]. Liposomes made with phospholipids are unilamellar and multilamellar vesicles varying from 50 nm up to over 300 nm. The liposome vesicles also have an internal water core, suitable for the inclusion of hydrophilic components. PC can form the desired structured bilayers. The maximum loading of the agent in the bilayer is about 25 wt% based on phospholipid. Due to the configuration of the (poly)unsaturated fatty acids of the PC molecules, the vesicles leak the encapsulated agent. More stable liposomes can be obtained by using PC molecules with hydrogenated saturated fatty acid structures.

The liposomal encapsulation technology is focused on cosmetic and pharmaceutical systems, since the PC fractions are quite costly. Plant scale produced PC phospholipids are used as excipients for targeted drugs transport, encapsulation and cell membrane passage.

For new nanotechnology developments in the technical industries the application of phospholipids as an encapsulation tool may be of challenging interest.

10.7.3 Cosmetics

Lecithin consisting of various phospholipids including polyunsaturated fatty acid has skin softening properties in cosmetic formulations. These preparations include skin creams and lotions, foundations and cleansing creams, sunscreens, soaps, bath oils, shampoos and hair conditioners. Lecithin can also reduce the undesirable oily feeling in cosmetics containing oils and supports moisturizing effect. Phospholipids penetrate the skin and facilitate the penetration of other essential cosmetic compounds. The liposomal encapsulation is a technological challenge to achieve the optimal performance. Cosmetics with advertised slogans as 'liposomes' and 'nanosomes' are marketed. If a surfactant with a high HLB value is needed in the recipe, lecithins should be used in combination with other surfactants.

10.7.4 Leather

The production of leather from animal skins consists of a series of processes, from the slating of fresh skins to tanning and fatting. The tanning is performed with various vegetable, synthetic and mineral tanning agents. After the tanning the skin has to be greased to produce supple leather, suitable for manufacturing goods. Fat liquors consist of sulfonated and sulfated oils with polar groups in the triglycerides. Natural lecithins can be added to the fat blend as a raw material source for the sulfonation and sulfating processes. However, it is also attractive to use hydroxylated lecithins, since these products fulfil the requirements for a good fatting agent: (i) deep penetration into the skin, (ii) fixing in the leather, (iii) good light fastness (resistance) and (iv) low residual fat content in the used liquor. In particular, the very hydrophilic combined hydrolysed and hydroxylated lecithins give deep penetration and good fixation, resulting in leathers with firm grain, good tear and tensile strength [34]. Although leather production has largely moved from European countries to the skin producing, emerging countries with large cattle herds, the world leading leather auxiliary companies still have their roots in Europe. These companies make use of lecithins for their product portfolio.

10.7.5 Paper Coating

Paper and paperboard are often coated to improve properties such as printability and appearance. A coating with superior uniformity, flow properties, stability and ultraviolet (UV) brightness is obtained from a blend of fatty acids and lecithin. The rheological properties of aqueous coatings for rapidly moving webs are enhanced [35]. Hydroxylated or acetylated lecithin is included in the coating recipe, meeting the requirements for high production speed in paper manufacturing. It facilitates excellent properties of tensile strength and UV light absorbency.

10.7.6 Paints

The broad range of functional properties of lecithin makes it highly suitable for many different coating formulations. These include paints, waxes, polishes and wood preservatives. Lecithins function as interfacial agents in paints, lacquers and printing inks, influencing the wetting, dispersing, suspending and stabilizing agents in both oil-base and latex/resin emulsion paints. In paints and lacquers the choice of wetting agents depends on the nature of the pigment, the vehicle and the processing procedure. As a rule, natural grades of lecithin have been recommended up to 1% on a pigment weight basis. Lecithin products facilitate pigment dispersion and redispersion and also reduce thixotropic viscosity properties. In water-based paint systems, water-dispersible lecithins are recommended as such or in combination with other surfactants. Although synthetic surfactants may have taken market share from the old-fashioned standard lecithin, the evaluation of hydroxylated and acetylated lecithins or combined modified lecithins may demonstrate attractive performance.

10.7.7 Plant Crop Protection

Phospholipids facilitate the transport of agrochemicals in the leaves and stems of plants or the organs of insects. It is likely that the surfactant function supports the formation of small particle droplet sizes of the pesticide suspension, making efficient transfer through the cell membranes possible. As a result lower dosages of agrochemicals per spraying period are feasible. The modified vegetable lecithins with enhanced hydrophilicity are preferred. Phospholipid vesicles have been successfully used as model membrane systems to study permeability, ion transport, fluidity and other properties of biological membranes.

10.7.8 Soil Bioremediation

Many soil areas in the world are polluted by chemicals of varying persistency and toxicity. Such wastes may spread out and into the ground and disperse into the ground water. One method for cleaning up contaminants is by enhancing the degradation of them through the use of microorganisms. Phospholipids facilitate the clean-up of contaminated soils by the combined effect of enhancing bacterial growth by supplying essential nutrients while modifying the physical state of oil or wastes [36]. Tests were conducted on soils artificially contaminated with polychlorinated biphenyls (PCBs) using de-oiled lecithin as a natural surfactant. The degradation of PCBs was significantly enhanced at lecithin concentrations in slight excess of the critical micelle concentration.

These few application examples in the nonfood industrial sectors can be easily extended by a long range of specific actual uses. Some other examples are as an antidusting agent on sandy roads, viscosity enhancer in brick production, mosquito control systems and asphalt rheology.

10.8 Legislation and Reach

According to most legislative definitions lecithins are mixtures of phosphatides (= phospholipids) derived from vegetable and animal origins. The EU approved food additive number E322 covers all standard, physical fractionated and enzymatically hydrolysed lecithins for use in the food industry. Hydroxylated and acetylated lecithins do not have an EU approved food additive status.

In the USA lecithins are classified in the Food Chemical Codex (FCC III). The US Food and Drugs Administration (FDA) has regulated under Title 21, part 184 direct food

Table 10.8 Legal purity specifications of food grade lecithin

Purity	FAO/WHO Codex Alimentarius	European Union E322	Food Chemical Codex
Acetone insoluble (%)	>60	>60 Hydrolysed >56	>50
Hexane insoluble (%)	–	–	<0.3
Toluene insoluble (%)	<0.3	<0.3	–
Moisture (%)	–	–	<1.5
Drying loss (%)	<2.0	<2.0	
Acid value (mg KOH/g)	<36	<35 Hydrolysed <45	<36
Peroxide value (meq/kg)	<10	<10	<100
Arsenic (ppm)	<3	<3	–
Lead (ppm)	<10	<5	<1
Mercury (ppm)	–	<1	–
Heavy metals (as Pb) (ppm)	<40	<10	–

substances affirmed as generally recognized as safe (GRAS). Standard lecithin, fractions and acetylated lecithin are listed in § 184.1400, enzyme-modified lecithin in § 184.1063 and hydroxylated lecithin in § 172.814. Table 10.8 lists the various specifications.

For technical uses lecithins are classified in the European Inventory of Existing Commercial Substances (EINECS) (EU) and Chemical Abstract Service (CAS) (US) systems (Table 10.9). The European Union is actively setting up a new system in which all (chemical) substances should have the Registration, Evaluation and Authorization of Chemicals (REACH) status. The new EU law defines the REACH in Regulation 1907/2006/EC. The aims of this regulation are (i) to improve the protection of human health and the environment from the risk that can be posed by chemical substances, (ii) to promote alternative safety testing methods and (iii) to improve the safe handling and use of substances across all sectors of industry. Native lecithin (CAS 8002-43-5), falling under E322, is listed in Annex 4, defining exemptions from the obligation to register. Modified lecithins, even when falling under E322, are not on Annex 4 and must be pre-registered, ensuring that all types of modified lecithins can be applied in technical uses in future. European lecithin manufacturers, organized in the European Lecithin Manufacturers Association (ELMA) may have pre-registered modified lecithins for the REACH inventory list before the deadline of the end of 2008. The effective registration procedure will take a much longer time and can be done later.

Table 10.9 EINECS and CAS classification of lecithins

Lecithin grade and source	EINECS (EU)	CAS (USA)
Lecithin, natural	232-307-2	8002-43-5
Lecithin, acetylated	293-316-5	91053-50-8
Lecithin, hydroxylated	232-440-6	8029-76-3
Lecithin, enzymatically hydrolysed	288-318-8	85711-58-6
Lecithin hydrogenated	295-786-7	92128-87-5
Lecithin, soya bean	310-129-7	8030-76-0
Lecithin, egg yolk	297-639-2	93685-90-6

EINECS = European Inventory of Existing Commercial Substances; CAS = Chemical Abstract Service.

10.9 Conclusion

Lecithin is primarily sourced from the oil-bearing seeds of soya bean, rapeseed and sunflower kernel as a co-product from the seed oil crushing. Phospholipid modification by enzymatic hydrolysis, solvent fractionation, acetylating and hydroxylation processes gives lecithins with specific enhanced hydrophilicity and oil-in-water emulsifying properties. Although lecithin has been known for a long time, phospholipids have still new potentials as tailor-made surfactants in food and nonfood systems. The application knowledge, combined with precise analytical methods to characterize the composition, emulsion particle sizes and emulsion stability, supports an understanding of the functionality of the various phospholipids. Natural and modified lecithins are used as such or in combination with other surfactants in a large range of technical applications. The products are renewable and have good biodegradability. Particularly the potentials for modified hydrophilic lecithins in nonfood uses are worth further exploration.

References

1. Buchold, H., Boensch, R. and Schroeppel, J. (1994) Process for enzymatically degumming vegetable oil, European Patent 0654527.
2. van Nieuwenhuyzen, W. (1976) Lecithin production and properties. *J. Am. Oil Chem. Soc.*, **53**, 425–427.
3. Hollo, J., Peredi, J., Ruzics, A. *et al.* (1993) Sunflower lecithin and possibilities for utilisation. *J. Am. Oil Chem. Soc.*, **70**, 997–1001.
4. Temelli, F. and Dunford, N. T. (1995) Modification of crude canola lecithin for food use. *J. Food Sci.*, **60** (1), 160–163.
5. Balazs, P. E., Schmit, P. L. and Szuhaj, B. F. (1996) High-performance liquid chromatographic separations of soy phospholipids. *J. Am. Oil Chem. Soc.*, **73**, 193–197.
6. Diehl, B. K. W. and Ockels, W. (1995) Quantitative analysis of lecithin: phospholipid analysis with 31-P NMR spectroscopy, in *Phospholipids: Characterization, Metabolism and Novel Biological Application* (eds G. Cevc and F. Paltauf), AOCS Press, Champaign, Illinois, pp. 29–32.
7. Diehl, B. K. W. (2001) High resolution NMR spectroscopy. *Eur. J. Lipid Sci. Technol.*, **103**, 16–20.
8. Wu, Y. and Wang, T. (2003) Phospholipid class and FA compositions of modified soybeans processed with two extraction methods. *J. Am. Oil Chem. Soc.*, **80**, 127–132.
9. Doig, S. M. and Diks, R. M. M. (2003) Toolbox for modification of the lecithin head group. *Eur. J. Lipid Sci. Technol.*, **105**, part 1, 359–367; part 2, 368–376.
10. Xu, X., Vikbjerg, A. F. and Guo, Z. (2008) Enzymatic modification of phospholipids and related polar lipids, in *Phospholipid Technology and Applications* (ed. F. D. Gunstone), The Oily Press, Bridgwater, Massachusetts, pp. 41–82.
11. Vikbjerg, A. F., Mu, H. and Xu, X. (2007) Synthesis of structured phospholipids by immobilized phospholipase A2 catalyzed acidolysis. *J. Biotechnol.*, **128**, 545–554.
12. Szuhaj, B. F. and Vaste, J. R. (1974) Phosphatide preparation process, United States Patent 3 962 292.
13. Bonekamp, A. (2008) Chemical modification, in *Phospholipid Technology and Applications* (ed. F. D. Gunstone), The Oily Press, Bridgwater, Massachusetts, pp. 141–151.
14. Sietz, F. G. (1982) Hydroxyliertes lecithin. *Seife-Öle-Fette-Wächse*, **108**, 637–639.

15. Schneider, M. (2008) Major sources, composition and processing, in *Phospholipid Technology and Applications* (ed. F. D. Gunstone), The Oily Press, Bridgwater, Massachusetts, pp. 21–39.
16. Sen Gupta, A. K. (1982) Separation process, European Patent EP0049914.
17. Raman, L. P., Cheryan, M. and Rajagopalan, N. (1996) Solvent recovery and partial deacification of vegetable oils by membrane technology. *Fett/Lipid*, **98** (1), 10–14.
18. Hutton, H. J. and Guymon, J. S. (2000) Process for producing deoiled phosphatides, World Patent WO 00/34292.
19. Wagner, H. and Eggers, R. (1996) Extraction of spray particles with supercritical fluids in a two-phase flow. *Am. Inst. Chem. Engrs J.*, **42** (7), 1901–1910.
20. Pardun, H. (1971) Verfahren zur Extraktion von Pflanzenrohphosphatiden, German Patent 1692568.
21. Rek, J. H. M., Rijsdijk, A. and Duthilh, C. E. (1985) Phosphate mixtures and process for preparing such mixtures, European Patent 0141442.
22. Liebing, H. and Lau, J. (1976) Über ein Verfahren zur Gewinnung von Lecithin-Fraktionen. *Fette, Seifen, Anstrichm.*, **78**, 123–126.
23. Eikermann, H. (1962) A process for the production of natural phospholipids and substances produced thereby, United States Patent 3,031,478.
24. Günther, B. R. (1984) Process for the separation of oil and/or phosphatidylethanolamine from alcohol soluble phosphatidylcholine products, containing the same, United States Patent 4,425,276.
25. Bergenståhl, B. and Claesson, P. M. (1990) Surface forces in emulsions, in *Food Emulsions* (eds K. Larsson and S. E. Friberg), Marcel Dekker Inc., New York, pp. 41–96.
26. van Nieuwenhuyzen, W. (1997) Funktionalität von Lecithinen. *Fett/Lipid*, **99**, 10–14.
27. Pan, L. C., Tomás, M. C. and Añón, M. C. (2002) Effect of sunflower lecithins on the stability of water in oil (W/O) and oil in water (O/W) emulsions. *J. Surfactants Deterg.*, **5**, 135–143.
28. Pan, L. C., Tomás, M. C. and Añón, M. C. (2004) Oil in water emulsions (O/W) formulated with sunflower lecithins: vesicle formation and stability. *J. Am. Oil Chem. Soc.*, **81**, 241–244.
29. Griffin, W. C. (1949) Classification of surface-active agents by 'HLB'. *J. Soc. Cosmet. Chem.*, **1**, 311–326.
30. Dickinson, E. (1999) Structure, Rheology and Physical Stability of Aggregated Particle Systems Containing Proteins and Lipids, EU Fair Project Contract CT96-1216, Interim Report and Project Dissemination Brochure (May 1999).
31. Hernandez, E. and Quezada, N. (2008) Use of phospholipids as functional ingredients, in *Phospholipid Technology and Applications* (ed. F. D. Gunstone), The Oily Press, Bridgwater, Massachusetts, pp. 83–94.
32. van Nieuwenhuyzen, W. and Tomás, M. C. (2008) Update of vegetable lecithin and phospholipid technologies. *Eur. J. Lipid Sci. Technol.*, **5**, 472–486.
33. Ghyczy, M. (1994) Liposomes for the food industry. *Food Tech. Europe*, **1** (5), 44–46.
34. Pelckmans, H. H. A. and Sietz, F. G. (1981) Lecithin als Fettungsmittel verschiedener Ledersorten. *Leder und Håutemarkt/ Gerbereiwissenschaft und Praxis*, **33**, 256–257.
35. Nikoloff, K. P. (1999) Surfactant-free lubricant for coating moving webs, United States Patent 5,858,933.
36. Fava, F. and Gioia, D. D. (2001) Soya lecithin effects on the aerobic biodegradation of polychlorinated biphenyls in an artificially contaminated soil. *Biotechnol. Bioengng*, **72** (2), 177–184.

11
Sophorolipids and Rhamnolipids

Dirk W. G. Develter and Steve J. J. Fleurackers
Ecover Belgium NV, Malle, Belgium

11.1 Sophorolipids

11.1.1 Introduction

Numerous microorganisms are known to produce and excrete substances with surface-active properties. In general, the function of these compounds is to facilitate the uptake of hydrophobic substrates such as triglycerides or alkanes. The common abbreviation for these biologically produced surface-active agents is biosurfactant. Among these biosurfactants, sophorolipids receive a prominent place in both commercial and academic interests, engendered by a high production yield and finding a variety of applications such as detergency, bioremediation, enhanced oil recovery and therapeutic agents. Furthermore, they are produced by a nonpathogenic organism and exhibit a low toxicity. For a more in-depth review on the subject, see Desai and Banat [1] for microbial surfactants in general and Kitamoto *et al.* [2] for glycolipids in specific.

Sophorolipids were first described in 1961 as being produced by the ascomycetous yeast species *Torulopsis magnoliae* [3], which was discovered actually to be *T. apicola* [4] and got its current name of *Candida apicola* when it was found that the two genera were synonymous. Since then, other species have been found to produce sophorolipids, such as *C. gropengiesseri* [5] and *C. bombicola* [6]. As the latter organism exhibits both the highest yield and productivity when compared to others of the *Candida* genus, this strain has received the most attention when studying sophorolipids. This chapter will therefore focus on those sophorolipids produced by *C. bombicola*.

The biosurfactant molecule consists of sophorose, a diglucose, bound to a hydroxy fatty acid in a glycosidic bond. The sophorose moiety may contain acetyl groups at the 6′ and/or 6″ positions and the molecule itself may form a lactone through linking

Surfactants from Renewable Resources Edited by Mikael Kjellin and Ingegärd Johansson

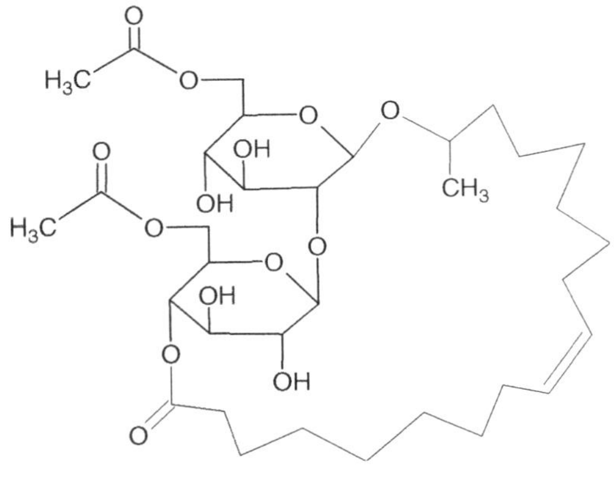

Figure 11.1 *Chemical structure of a sophorolipid.*

the carboxyl group of the hydroxy fatty acid and the 4″-hydroxyl of the sophorose. Figure 11.1 depicts a common sophorolipid, as excreted by *C. bombicola*.

Sophorolipids may be produced on substrates that are, in part, directly incorporated into the molecule as the hydrophobic moiety, as opposed to, for example, rhamnolipids where the biosynthesis occurs with the *de novo* synthesis of the hydrophobic part. Specifically, when providing *C. bombicola* with oleic acid and an otherwise suitable medium, the yeast will incorporate this fatty acid into a sophorolipid. By contrast, when capric acid is provided to *Pseudomonas aeruginosa*, the fatty acid is not directly converted into a rhamnolipid but is metabolized instead. The possibility of direct incorporation is a clear indication as to why sophorolipid production is among the highest of all biosurfactants.

There exists, however, a limitation regarding the kind of hydrophobic substrate that is directly incorporated; only those chains of a specific length are utilized. This limitation generally only allows for chains between palmitic and stearic acid to undergo direct incorporation. Chains below a length of C_{16} are completely metabolized through β-oxidation, whereas chains larger than C_{18} may become shortened by this mechanism and incorporated when they reach a suitable length.

As such, providing *C. bombicola* with a substrate having the correct length will result in a higher production when compared to adding shorter or longer chains because the need for *de novo* synthesis of the hydrophobic moiety is eliminated. This does limit the number of different types, with regard to their lipophilic moiety, of sophorolipids that can be produced on an economically viable scale by the wild-type organism when using commonly available natural fatty acids. Of these fatty acids, six are readily converted into sophorolipids; they are palmitic, palmitoleic, stearic, oleic, linoleic and α-linolenic acid.

11.1.2 Biosynthesis

As of yet, no direct studies have been performed on the biosynthetic pathway of sophorolipids. Several indirect studies have been performed, however, that allow for the construction of a theoretical pathway. These studies were executed on different species of *Candida*, all of which have the capability to produce sophorolipids, and provide an insight as to the different steps that are needed to convert a lipophilic substrate into a sophorolipid. The general pathway is outlined in the next paragraph according to Fleurackers [7] and Van Bogaert [8], followed by a brief explanation of each step.

11.1.2.1 Biosynthetic Pathway

The theoretical pathway of sophorolipid biosynthesis is outlined in Figure 11.2. Steps 1 to 3 take into account the possibility of growing *C. bombicola* and producing sophorolipids on lipophilic substrates other than fatty acids or esters thereof. In this case, an n-alkane is taken up by the yeast and oxidized to an alcohol by cytochrome P450 monooxygenase. It is further oxidized to a fatty acid by alcohol- and aldehyde-dehydrogenase. Alternatively, in step 4, extracellular lipases convert any fatty acid esters into free fatty acids, which are taken up also. In case no lipophilic substrate is supplied, *de novo* synthesis of fatty acids will occur from acetyl-CoA as derived from the glycolytic pathway. Step 5 uses molecular oxygen and NADPH (nicotinamide adenine dinucleotide phosphate, reduced form) to hydroxylate the intracellular fatty acid at the ω or ω-1 position, again through the action of cytochrome P450 monooxygenase. To the resulting hydroxy fatty acid, two glucose units are added by two different glucosyltransferases in steps 6 and 7. The sophorolipid can be excreted as such, or can be modified further. These modifications consist of lactonization by linking the carboxylic function to the 4″-hydroxyl (step 8) and acetylation of the 6′- and/or 6″-hydroxyls (step 9).

11.1.2.2 Metabolite Modification and Uptake

The ability of growing sophorolipid producing strains such as *C. bombicola* and *C. apicola* on alkanes indicates that they possess enzymes that are required for modifying these substrates into fatty acids. In this way, they can be used as a source of carbon and energy when processed by β-oxidation or can be converted to sophorolipids. When using fatty acids as the lipophilic substrate, it may prove advantageous to present them under the form of a triglyceride or (m)ethyl ester as free fatty acids are known to disrupt the electron balance of the cells. As extracellular lipases will gradually release free fatty acids to the medium, this approach is particularly suited when the substrate is added in large quantities as opposed to a continuous feed strategy. Any absorbed lipophilic chains that fall outside the C_{16}–C_{18} fork are either metabolized (short chains) or shorted to a suitable length (long chains) through β-oxidation as they are not accepted in the next step.

11.1.2.3 Lipophilic Metabolite Oxidation

The activation of a fatty acid for sophorolipid synthesis occurs in step 5 through the creation of a hydroxyl group at the ω or ω-1 position, depending on the length of the fatty acid. C_{16} chains will undergo terminal hydroxylation and C_{18} chains are hydroxylated subterminally. This difference in position of the hydroxyl group is a clear indication

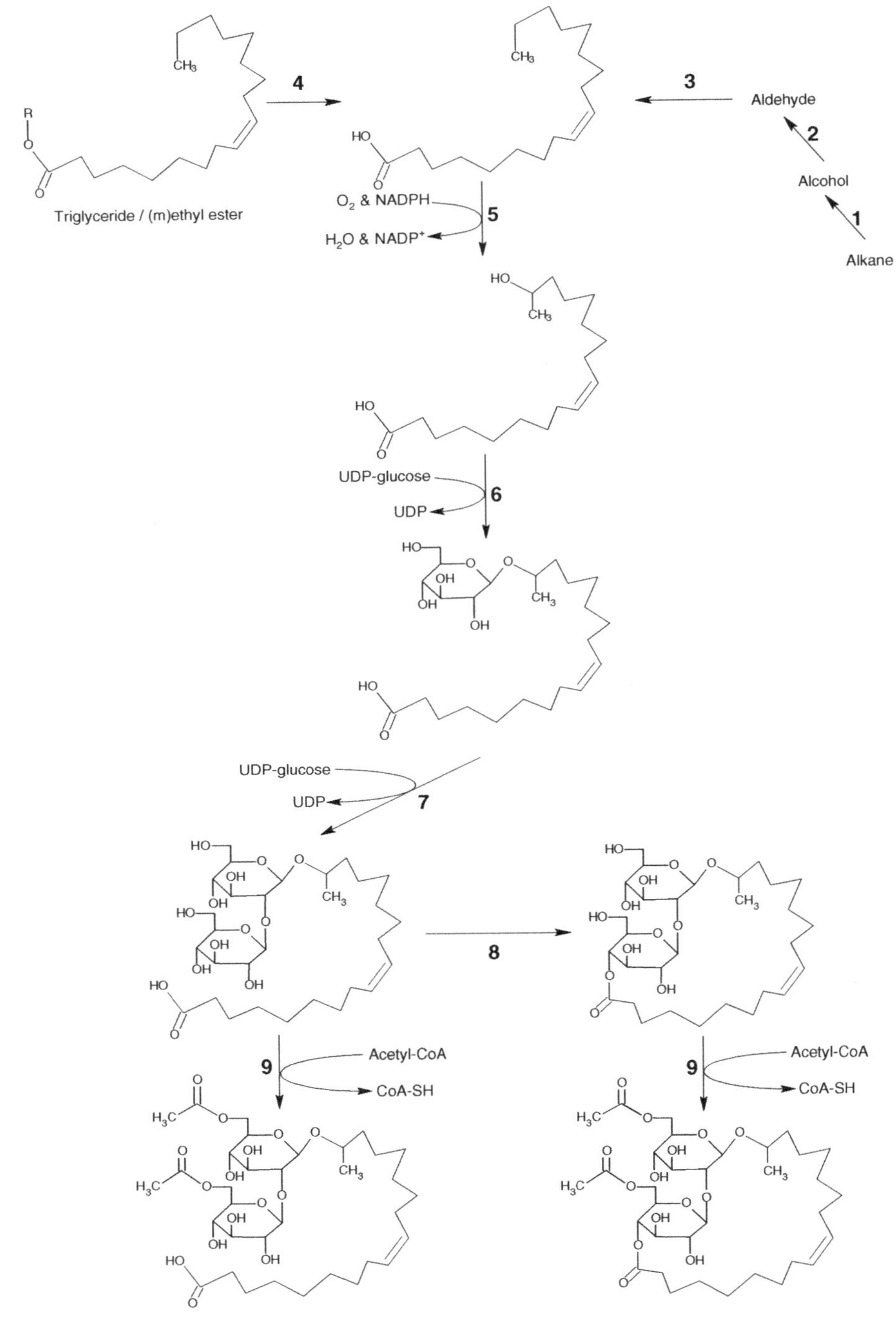

Figure 11.2 *Theoretical pathway of sophorolipid biosynthesis.*

that the ω/ω-1 hydroxylation is carried out by the same enzyme and furthermore is sterically determined by a length of roughly 14 to 15 carbons between the polar carboxylic function and the target carbon. Indeed, when linoleic acid is provided, both ω and ω-1 hydroxylation occur and sophorolipids containing either 17- or 18-hydroxy linoleic acid are produced. Conversely, adding palmitic or stearic acid will result in sophorolipids solely containing 16-hydroxy palmitic acid or 17-hydroxy stearic acid, respectively. Oleic acid predominantly undergoes ω-1 hydroxylation, its length approximating that of stearic acid, with α-linolenic acid mostly being hydroxylated at the terminal carbon, due to the three double bonds that shorten it. Importantly, it is this step that is responsible for the relatively small range of fatty acids that are directly converted into sophorolipids. Providing the yeast with, for example, lauric acid does not lead to the production of C_{12} sophorolipids, save for some trace amounts.

This hydroxylation is performed by cytochrome P450 monooxygenase. Experiments involving the cell-free extract of a *Candida* species and radioactively labelled O_2 have indicated the need for molecular oxygen, with one oxygen atom remaining as part of the hydroxyl group, and NADPH, which is oxidized to $NADP^+$ (nicotinamide adenine dinucleotide phosphate, oxidized form). The involvement of cytochrome P450 monooxygenase in sophorolipid synthesis was first discovered for *C. gropengiesseri* by Jones and Howe [5]. Identification of two P450 genes in *C. apicola* by Lottermoser *et al.* [9] led them to be classified as belonging to the CYP52 family, based on amino acid similarity. In *C. bombicola*, eight different cytochrome P450 monooxygenase genes were recently identified belonging to this CYP52 family, with one having a 91% amino acid identity to the CYP52E2 gene of *C. apicola*, thought to be directly involved in sophorolipid synthesis.

If, alternatively, a substrate were to be provided, which has already been suitably oxidized, say 2-dodecanone, a conversion takes place in reducing the ketone to an alcohol to which sophorose is added, even though it involves an atypical chain length.

11.1.2.4 Adding Glucose and Creating Sophorose

Studies on cell-free enzyme extracts of the sophorolipid producing strain *C. bogoriensis*, later renamed *Rhodotorula bogoriensis*, have identified two different glycosyltransferase actions. The first creates a glycosidic bond between the 1′ carbon of glucose and the hydroxylated carbon of the hydroxy fatty acid. The second transferase adds another molecule of glucose through the formation of a second glycosidic bond between positions 2′ and 1′. Both reactions depend on the nucleotide-activated glucose, uridine diphosphate glucose (UDPG) [10]. The question remains whether these reactions are catalysed by either one or two activities of the same (multi)enzyme or if two different enzymes are involved. Both enzyme activities could not be separated and purification yielded only one electrophoretic band. However, the different nature of the substrate in both reactions – a hydroxylated fatty acid as opposed to a glycosyl hydroxy fatty acid – and the fact that the monoglucose intermediate was detected would support the notion of two distinct reaction sites, either located on one enzyme or on two very similar ones. It is further assumed that the sophorolipid synthesis in *C. bombicola* occurs in a similar fashion.

The hydroxylation of the lipophilic substrate may be regarded as activating the substrate before glucose can be added. When providing substrates that are already suitably oxidized, the stereochemical configuration of the activated carbon must be taken into account. If the yeast itself performs the activation through ω-1 hydroxylation, a chiral carbon arises having an *S* configuration and it is this configuration that also must be present if an otherwise activated and chiral substrate is presented. Indeed, experiments have shown that providing the yeast with ricinoleic acid, already having a hydroxyl group in the *R* configuration, does not lead to the addition of glucose to this hydroxy fatty acid. Furthermore, when using either nonterminal dodecanols or dodecanones, it has been observed that the yields for the ketones are actually higher than those for the alcohol. Having an additional step of reducing the ketone would *prima facie* imply a lower yield, but seeing that the alcohols were presented as a racemic mixture, only half were able to accept the addition of glucose, whereas the ketones could all be converted into the correct stereoisomer.

11.1.2.5 Acetylation and Lactonization

The predominant form of ultimately secreted sophorolipid contains two acetyl esters at the 6′ and 6″ positions and a lactonic link between the carboxyl group and the 4″-hydroxyl [11]. In *R. bogoriensis*, it was found by Esders and Light [12] and Bucholtz and Light [13] that an acetyl-coenzyme A dependent acetyl transferase is involved in sophorolipid synthesis and has been partially purified. It is assumed that the acetylation occurs in a similar fashion for *C. bombicola*. Formation of the lactonic bond is believed to be catalysed by a specific lactone esterase, but as yet no such enzyme has been identified for the production of sophorolipids. The addition of these three esteric bonds is not required for the molecule to be excreted as all eight possible combinations of the three esters are detected in the final sophorolipids. They do, however, play a role in the efficiency of the molecule as the diacetylated lactonic sophorolipid exhibits, for instance, the best hard surface cleaning when compared to sophorolipids that have less of these esteric bonds but are otherwise identical with regard to the lipophilic moiety.

11.1.2.6 Production Conditions

A typical medium for the production of sophorolipids will consist of a nitrogen source, small amounts of various minerals and over 10% of glucose as the main source of energy and carbon. For the nitrogen source, yeast extract is mainly used on a laboratory scale and processed waste products such as corn steep liquor are used on an industrial scale. Small amounts of mineral nitrogen may be used as a supplement. An example of such a medium is shown in Table 11.1, according to Lang *et al.* [14].

To the compounds shown above, an amount of lipophilic substrate is added either wholly, stepwise or continuously. Additional glucose is usually added when its concentration drops below 1%. The pH stabilizes to around 3–3.5 after the growth phase, so very little, if any, pH correction is needed. Some researchers do prefer to keep the pH at 3.5. The hydroxylation of the lipophilic substrate requires molecular oxygen; therefore the medium needs be aerated sufficiently in order to achieve proper production levels. In order for sophorolipid production to commence, the carbon-to-nitrogen ratio must be sufficiently high. Production therefore occurs almost exclusively in the stationary

Table 11.1 Example of a medium composition for sophorolipid production

Ingredient	Concentration (g/l)
Glucose	120
Yeast extract	5
Sodium citrate.3H_2O	5
Ammonium chloride	1.5
Potassium dihydrogen phosphate	1
Dipotassium hydrogen phosphate	1.16
Magnesium sulphate.7H_2O	0.7
Sodium chloride	0.5
Calcium chloride.2H_2O	0.27

phase when almost all available nitrogen has been depleted [15]. Consequently, feeding lipophilic substrate during the growth phase of the organism will either cause it to be metabolized as a secondary energy source or accumulate and potentially interfere with cell growth, as is the case for free fatty acids. With regard to temperature, 21 °C was found by Gobbert, Lang and Wagner [16] to be the optimal temperature for sophorolipid formation and 28.8 °C for the optimal growth of *C. bombicola*. Most productions are performed, however, at either 25 or 30 °C [17]. The former yields a lower biomass but has a higher rate of consumption of glucose when compared to the latter, resulting in a negligible difference of sophorolipid production between those two temperatures. Furthermore, performing the process at slightly elevated temperatures facilitates handling, sampling and availability to the organism of the lipophilic substrate. The duration of a production cycle is typically 10 days. Experiments involving the recovery of the yeast after production for use in a subsequent cycle without the growth phase have indicated that significantly lower yields are obtained. However, it is generally preferred to grow a new culture for each new production.

11.1.3 Life Cycle Analysis

An unpublished cradle-to-gate life cycle analysis (LCA) study subcontracted to Vito (Mol, Belgium) tried to identify environmental bottlenecks of rhamnolipid and sophorolipid production and compare them with those of market reference surfactants according to the ISO 1404x series. Experimental data for the pilot plant scale were complemented with LCA data from the Vito database as well as literature data for the upscaled production scenario. In order to compare the environmental profile of these biosurfactants with market reference surfactants data were used from Zah and Hischier [18]. The Eco-Indicator '99 methodology [19] was used to determine environmental impacts.

11.1.3.1 Sophorolipids

The environmental profile of sophorolipid synthesis in Figure 11.3 indicates that the biosynthetic phase and in particular the required electricity consumption accounts for 45–90% of all cradle-to-gate impact categories, whereas the sterilization of the reactor

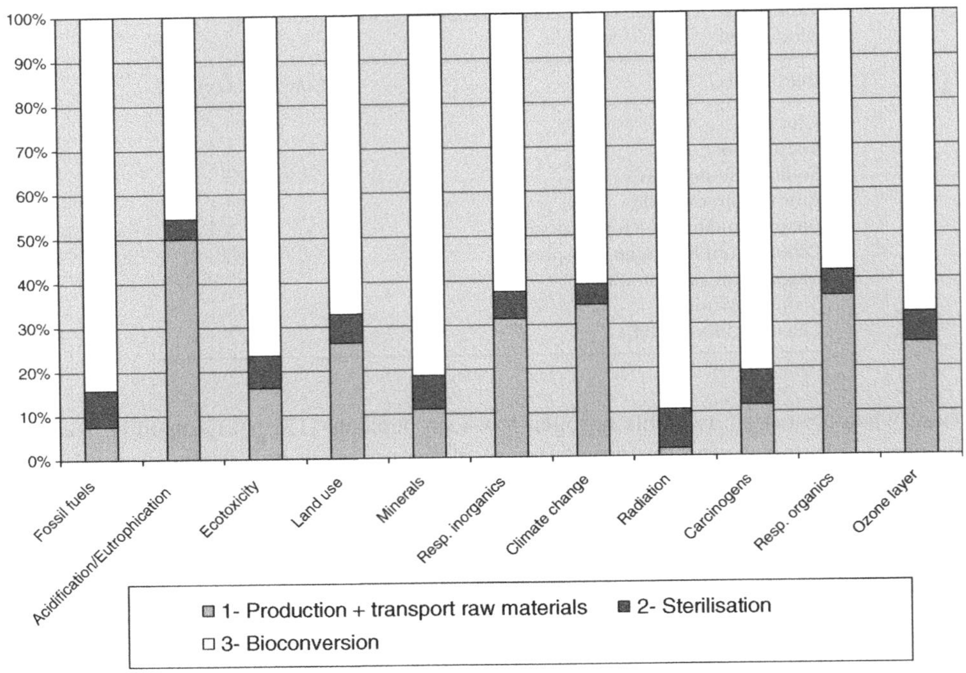

Figure 11.3 *Environmental profile of sophorolipid biosynthesis.*

accounts for less than 10%. Production and transport of raw materials account for the rest of the environmental impact since the downstream processing of the surfactants is negligible. For rhamnolipid synthesis, the distribution across the categories is similar.

Sophorolipids cause less or equal environmental damage prior to their use and disposal as compared to other surfactants, in particular when compared to other oleochemicals. The synthesis of rhamnolipids at present scores significantly worse than that of most reference surfactants due to its low yield. Impact categories where sophorolipids score worse than the reference surfactants are almost entirely due to the electricity generation needed for the fermentation phase by either nuclear ('radiation') or fossil fuel plants ('ozone' and 'acidification'). Producing synthetic fertilizer for growing the canola feedstock requires natural gas, which significantly contributes to the radiation impact of sophorolipids. Acidification and eutrophication are mainly related to the atmospheric emission of sulfur and nitrogen oxides during electricity production as well as fertilizer production and overseas transport. Renkin [20] demonstrated both rhamnolipids and sophorolipids to be an order of magnitude less toxic to aquatic life than several reference surfactants upon their discharge in the environment.

The climate change caused by sophorolipid synthesis is mainly due to CO_2 emissions, 74% of which is fossil CO_2 and again is almost entirely related to electricity generation. The biogenic nonfossil CO_2 emissions account for 24% of greenhouse total CO_2 emissions and are mainly (90%) caused by the fermentation process itself due to the metabolic conversion of either glucose or lipophilic substrate to CO_2. An important conclusion of this LCA study is that any measure to save electricity would be a major opportunity for improvement.

11.1.3.2 Comparison to Other Surfactants

Using renewable feedstock as the starting material for surfactant synthesis implies an obvious climate saving advantage of carbon capture and temporary storage in feedstock as opposed to setting free fossil carbon stocks. The extent to which the carbon loop is closed, however, largely depends on the energetic requirements and carbon efficiency of the conversion process. The Zah and Hischier [18] study clearly demonstrates the impact of sourcing with, for example, coconut scoring worse than palm kernel. An important observation is the importance of overall process efficiency which is good for the fully petrochemical and nonrenewable linear alkylbenzene sulfonate (LAS). It must be noted that this is obviously an oversimplification of the full carbon cycle as this cycle would also contain the carbon inefficient conversion of plankton and associated biodegraded compounds to the precursor of crude oil, kerogen. Stalmans *et al.* [21] and Hirsinger [22] compare an oleochemical surfactant with a petrochemical of the same structure and note an environmental benefit of approximately 25–75% in terms of process energy. The surfactants on which these studies were based include fatty alcohol ethoxylates (FAEO), lauryl sulfate (LS) and lauryl ether sulfate (LES). Figure 11.4 shows the comparison between sophorolipids and several previously mentioned surfactants.

Whereas biosurfactants are also produced at ambient temperature and pressure from renewable feedstocks with little to no manipulation, the LCA study performed by Vito demonstrated that the electricity needed for providing the optimal growth conditions is quite substantial. Moreover, when using a bioreactor for the conversion of feedstock to surfactants, part of this feedstock is unavoidably lost through assimilation by the production organism for cell maintenance and growth. *C. bombicola*, for example, typically achieves an overall substrate conversion of 65–68%, which is good when compared to many other production organisms but poor in purely chemical terms. The remainder of the substrates (mostly glucose) is converted into biomass and CO_2, which partly explains why sophorolipids do not perform better on the 'climate change' bar in Figure 11.4. In producing rhamnolipids, *Pseudomonas aeruginosa* reaches even higher conversion rates of up to 88%, but exhibits a productivity of 10–20% relative to that of *Candida*, which heavily affects the life cycle data. Carbon inefficiency inherent to fermentation processes can be minimized by optimization of both the production strain and conditions, increasing the productivity, for example, by decreasing the duration of production for the same yield, since this is correlated to energy consumption and/or the substrate conversion efficiency.

11.1.3.3 The 'Bi-cycle' Approach

Next to this, entirely different mechanisms for minimizing CO_2 loss by closed cycles and a cradle-to-cradle approach are currently being intensively explored. A prominent example entails coupling heterotrophic with autotrophic (usually photosynthetic) production processes. The Powerfarms! Project, for example, is an initiative of the Dutch

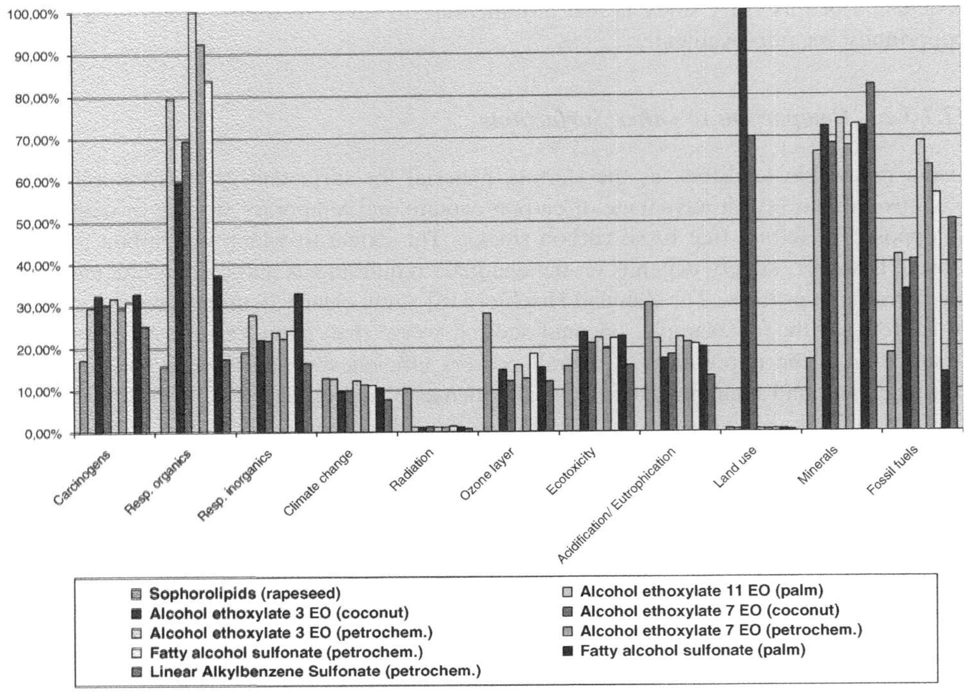

Figure 11.4 *Environmental profiles of sophorolipids and chemically produced surfactants.*

InnovatieNetwerk, Ingrepro en Sustec, in which the dung of dairy cattle is converted into biomass, biogas, single cell oil and clean water by anaerobic digestion and algae cultivation in open raceway ponds or photobioreactors (see also www.oilgae.com). Similarly, the Belgian–German research project ‘Bi-Cycle’ studies the admittedly challenging subject of coupling two fermentation processes producing bioethanol and single cell oil for biodiesel, using the otherwise wasted CO_2 to boost algal growth. ‘Holding on to carbon’ in such a way can drastically improve overall carbon efficiency. This concept can likewise be applied to biosurfactant synthesis as visualized in Figure 11.5, where carbon otherwise wasted to CO_2 is transformed into single-cell oil for use as a substrate in subsequent fermentations to sophorolipids.

After use, the surfactant will be disposed of in the environment. Here again the extent to which the carbon cycle can be closed will depend on the biodegradability of the obtained surfactants. In fact, Renkin [20] demonstrated both sophorolipids and rhamnolipids to be fully degraded without leaving traces of stable metabolites. This is in contrast to LAS and other certain branched surfactants, of which 5–10% of their molecular structure remains as nondegraded material and lingers in the environment as molecular garbage. The degradable part of LAS adds up carbon that was captured hundreds of millions of years ago to the cycling biogenic carbon, and it is exactly this lapse in time that further contributes to climate change.

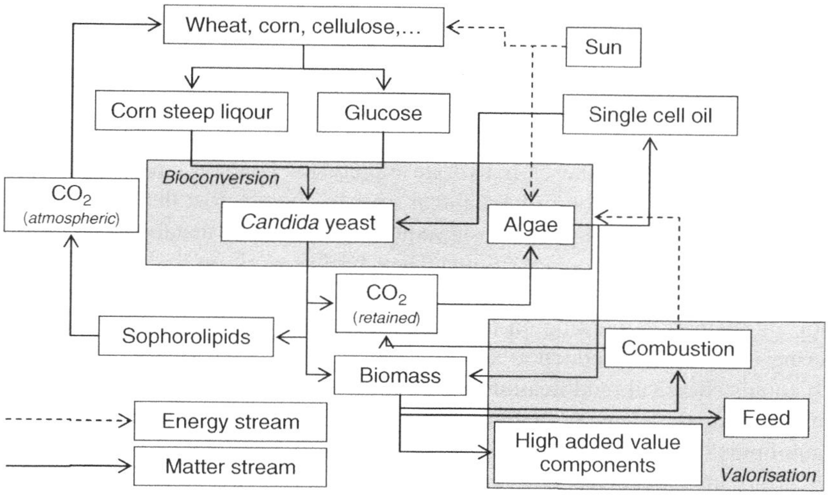

Figure 11.5 Schematic overview of the 'bi-cycle' concept.

11.1.4 Properties of Native Sophorolipids

Native sophorolipids are understood to be produced by the wild-type organism and contain a C_{16}–C_{18} hydroxy fatty acid. Two types of variations occur: the nature of the lipophilic moiety and the amount of esterification. For the lipophilic moiety, predominantly hydroxy stearic acid is observed in the case of *de novo* synthesis [23] and when feeding a fatty acid of a suitable length, the concurrent hydroxy fatty acid is found. The extent of the esterification, either acetylation or lactonization, determines the solubility of the final product and is influenced by the pH. A value of 3–3.5 predominantly yields the poorly soluble diacetyl lactone and biosynthesis using a pH of 2 results in a more soluble and thus less esterified product. It is this last parameter that most influences the emulsifying properties of the sophorolipids. Adapting these properties to specific applications is further explored in Section 11.2.

11.1.5 Brief Market Description

Sophorolipids are currently marketed by the French company Soliance for use in cosmetics. This relates to their attributed antimicrobial properties, among others against *Propionibacterium acnes* and *Corynebacterium xerosis*. The company Ecover Belgium NV uses rhamnolipids and sophorolipids in its hard surface cleaners. The required sophorolipids are produced using proprietary technology focused on solvent-free product extraction and purification. Another market player is the Korean MG Intobio Co. Ltd offering 'Sopholine' cosmetics. The Japanese company Saraya issued a patent application on the use of sophorolipids in their 'Sophoron' automatic dishwashing powder.

Three prominent academic groups are currently developing sophorolipid applications and technology of special interest are those of Richard Gross, of Daniel Solaiman and of Wim Soetaert. The first group as well as the group of Solaiman worked on the use of various surplus and low-cost agricultural co-products such as animal fats, plant oils and soya molasses as fermentative substrates. The group of Gross currently focus on chemical derivatives of the carbohydrate moiety for medical, specifically virucidal, applications, whereas the group of Solaiman tries to characterize the genes involved in the biosynthesis of sophorolipids for subsequent genetic and metabolic engineering to improve product yields and compositions. They further work on the chemical modification of sophorolipids to improve their solubility and surface-active properties. Finally, the group of Soetaert is working on the genetics of sophorolipid synthesis and genetical engineering of *Candida* production strains yielding tailor-made glycolipids. Additional research entails chemical modifications of the fatty moiety and exploring new application areas in the cosmetic, cleaning, agricultural, petroleum and chemical industries.

Rhamnolipids have been commercialized by Jeneil Biosurfactant Company, which put a lot of effort into achieving administrative approval of rhamnolipid use as pesticide adjuvants and for poultry hygiene. Aurora Advanced Beauty Labs Inc. has a substantial intellectual property (IP) portfolio in the domain of rhamnolipid use for pharmaceutical applications and announced that it was introducing various cosmetic products such as anti-wrinkle creams, moisturisers, beauty treatment products and skin protection ointments.

11.2 Derivatives of Native Sophorolipids

With regard to medium composition and feeding strategy, the production of native sophorolipids has an established optimum at which a relatively fixed mixture of product is obtained. The nature of this mixture does not exhibit good water solubility and therefore sophorolipids that are produced on an industrially viable scale are unsuited for a number of detergency applications, where clear or homogeneous formulations are required. The reason behind this poor solubility lies in the large fraction of diacetylated lactonic sophorolipids that are formed and in which the three esteric bonds increase the hydrophobic nature of the molecule. Removing these bonds will increase the solubility of the surfactant and shortening the hydroxy fatty acid moiety increases it further. Alternatively, the hydroxy fatty acid can be released from the sophorolipid entirely and can be used as a starting material for the perfume industry or as a precursor for polyfunctional fatty amines and other polymeric applications. Finally, transesterification of the lactone produces antibacterial and antiviral agents.

11.2.1 Hydrolysis

Diacetylated lactonic sophorolipids have both esteric and glycosidic bonds which exhibit different sensitivities to either acid or alkaline hydrolysis. The latter will only hydrolyse the esters whereas acid hydrolysis at first cleaves the esters and, under more harsh reaction conditions, removes the sophorose moiety to yield a hydroxy fatty acid. In performing alkaline hydrolysis, the gradual removal of the esters transforms the solid

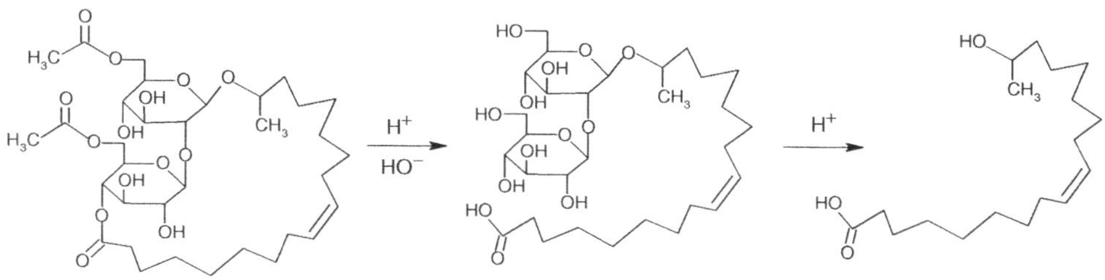

Figure 11.6 *Alkaline (first step) and acid (both steps) hydrolysis of a diacetylated lactonic sophorolipid.*

and insoluble diacetyl lactone into a heavier-than-water and immiscible liquid phase, which in itself contains 40% of water and a mixture of sophorolipids with a varying degree of acetylation and lactonization. In this process, both the sensitivity of the esteric bonds to an alkaline pH and the pH itself play a role. The esters will hydrolyse above a pH of 7 and subsequently release carboxylic groups, lowering the pH, which decreases the solubility of the sophorolipids. Further hydrolysis decreases the hydrophobicity of the mixture and makes it water soluble in increasingly more acidic environments. Ultimately, only the deacetylated free acid sophorolipid is found, which dissolves at a moderately acidic pH or higher, indicating that at least a fraction of the free carboxylic function must be ionized in order for it to dissolve. Acid hydrolysis occurs more slowly due to the decreased solubility and first results in a similar liquid heavier-than-water phase from ester cleavage and is ultimately followed by a top phase of liquid hydroxy fatty acid after hydrolysing the glycosidic bonds. Figure 11.6 shows the ultimate result of both alkaline and acidic hydrolysis [7].

11.2.2 Ozonolysis

The availability and liquid state at room temperature of oleic acid makes it one of the most used lipophilic substrates in sophorolipid synthesis. The resulting surfactant, even after complete alkaline hydrolysis, is still too hydrophobic to dissolve in water when the free carboxylic functions are completely protonated. In having hydroxy oleic acid as the lipophilic moiety, the double bond may be cleaved using ozone without affecting the sugar and thus retaining the amphiphilic nature. The two most common ways to perform ozonolysis include reductive and oxidative workup, resulting in a terminal aldehyde or carboxyl group, respectively. The resulting products are shown in Figure 11.7.

Product A has been shown to exhibit excellent wetting properties and, due the increased solubility and free carboxyl group, has the potential to act as a calcium sequestering agent while the aldehyde function of product B is a likely starting point for further chemical derivatization [24].

11.2.3 Transesterification

Treatment of the diacetylated lactonic sophorolipid with an alcohol will result in removing both acetyl groups and opening the lactone ring to the concurrent alcohol ester. Figure 11.8 depicts the transesterification of a sophorolipid with methanol.

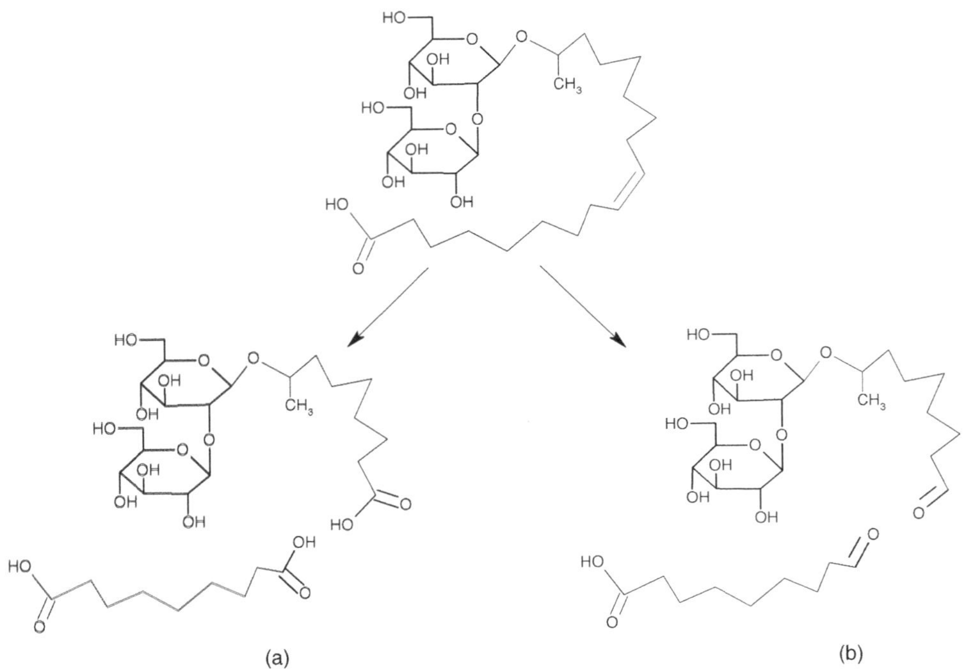

Figure 11.7 *Resulting products of sophorolipid ozonolysis with oxidative (A) and reductive (B) workup.*

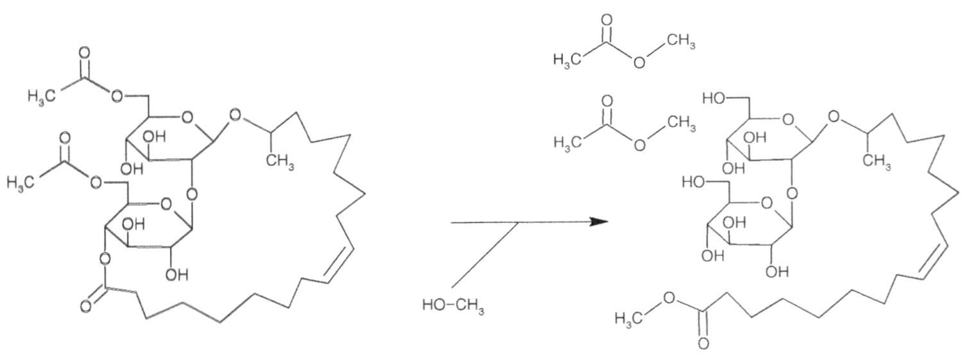

Figure 11.8 *Transesterification of sophorolipids with methanol.*

The resulting product is subjected to enzymatic acetylation using vinyl acetate to reattach acetyl groups at positions 6′ and 6″. Different esters, as well as the naturally produced sophorolipids, were shown to have inhibitory effects on bacteria such as *E. coli* and a species of *Salmonella*, in particular from the ethyl ester, which for those two bacteria has a minimal inhibitory (50%) concentration of below 10 mg/l. Experiments on the human immunodeficiency virus showed a log reduction of four units in the virus titre when exposed to 0.3% of the hexyl ester [25].

11.3 Biosynthesis of Novel Sophorolipids

11.3.1 General Outline

The limited range of lipophilic substrate associated with wild-type sophorolipids, either from direct incorporation or through *de novo* synthesis, also limits the various chemical derivatives that can be obtained without eclipsing their renewable character or dramatically increasing their cost. In an attempt to increase the range of the lipophilic moiety, various approaches have been developed to convert renewable lipophilic substrates with shorter chains into surface-active molecules by the use of *C. bombicola*. Initial research revealed that subterminally oxidized shorter chains such as 2-dodecanone were readily converted into the corresponding alkyl sophoroside under standard fermentation conditions, but that the corresponding primary isomers were mostly metabolized. Figure 11.9 shows the resulting surfactant from the bioconversion of 2-dodecanone.

Bioconversion of primary alcohols and aldehydes presents the difficulty of the substrate having been oxidized at a position that makes it easily accessible for degradation via the β-oxidation pathway. Indeed, initial research indicated that they were metabolized and used in the *de novo* production of native sophorolipids instead [26]. Several solutions to this problem were developed afterwards and three approaches are presented that allow for the direct use of primary fatty alcohols in converting them into detergents using *C. bombicola*. As the organism will add sophorose to a lipophilic alcohol having a wider range of chain lengths when compared to fatty acids, it can be concluded that the activation through hydroxylation is also the limiting step that prevents direct incorporation of fatty acids outside the C_{16}–C_{18} range.

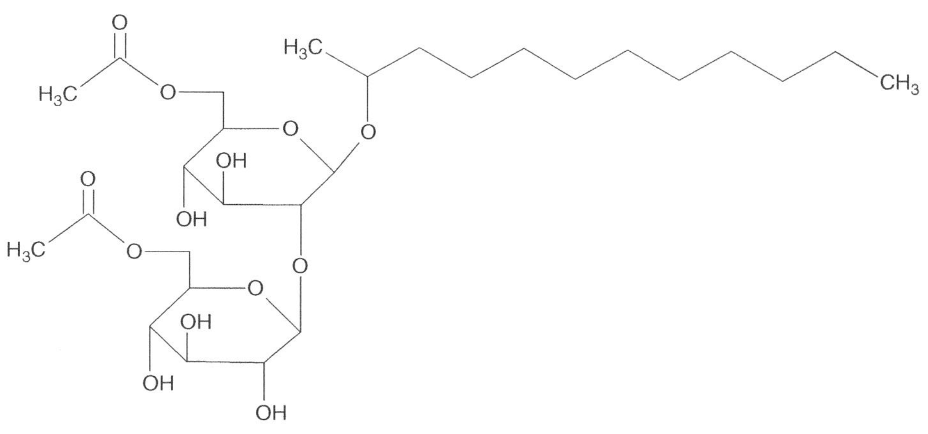

Figure 11.9 Sophorolipid from 2-dodecanone.

11.3.2 Changing the Medium Conditions

The conversion of primary fatty alcohols was first achieved though modification of the production medium [14]. Both glucose and available nitrogen were increased without a corresponding increase in aeration. As a result, biomass after the growth phase was 50% more than under the previous conditions and thus relatively less oxygen is available. It is supposed that this relative decrease precludes the oxidative breakdown of the substrate and allows for sufficient time to add the two molecules of glucose. When using this approach, around 70% of the 1-dodecanol used is converted into the concurrent sophorolipid, 10% of the alcohol was recovered as lauric acid and the remaining 20% were metabolized. No sophorolipids containing hydroxy lauric acid are formed. It is worth noting that the product of this bioconversion, after alkaline hydrolysis to remove any acetyl residues, bears a striking structural resemblance to the synthetically obtained alkyl-polyglycoside. Figure 11.10 shows the structures (a) of the sophorolipid based on 1-dodecanol and (b) of a lauryl polyglycoside with a degree of polymerization of two.

A comparison of the surface-active properties of those two compounds revealed only a minor difference: the lauryl sophoroside lowered the surface tension of water from 72 to 33 mN/m whereas the commercially available alkyl polyglycoside (APG) 1200 Plantaren®, which on average has 1.3 glucose units per molecule, performed somewhat better and lowered the surface tension to 27 mN/m.

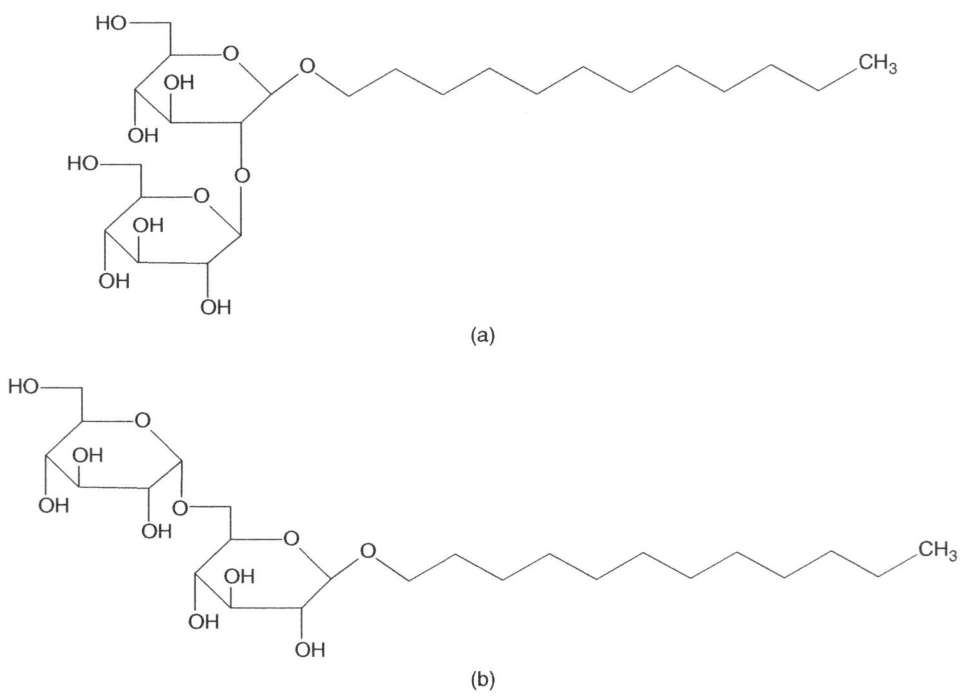

Figure 11.10 Structural comparison of sophorolipids from (a) 1-dodecanol and (b) APG.

11.3.3 Elongating the Substrate

As described by Develter and Fleurackers [24], a different approach is taken when 1-dodecanol is elongated by way of esterification with glutaric acid. The resulting molecule has a chain length which approximates that of oleic acid and is assimilated by the yeast as such. Figure 11.11 offers a comparison between the structures of oleic acid and the elongated 1-dodecanol, for which dodecyl glutarate is shown as an example.

The free carboxylic group of the glutarate residue is recognized as the polar end of a fatty acid and on the opposite side the usual ω-1 hydroxylation occurs. Two molecules of glucose are added and normal acetylation and lactonization occur. Subjecting the resulting molecule to alkaline hydrolysis will not only remove the acetyl residues and open the lactone but will also remove the glutaric acid elongator. As a result, a novel sophorolipid is obtained which has (11*S*)-1,11-dodecadiol as its lipophilic moiety. Figure 11.12 depicts this sophorolipid before and after alkaline hydrolysis.

This technique allows for the conversion into sophorolipids of a host of substrates, as long as they can be elongated to approximate a final chain length in the C_{16}–C_{18} range. As the dodecyl ester is wholly incorporated, it was concluded that extracellular lipases did not affect the bond. This is the opposite to when (m)ethyl or glyceryl esters are fed and cleavage of the ester occurs. It was furthermore found that a polar terminus is not required for converting the elongated substrate as long as the chain length is

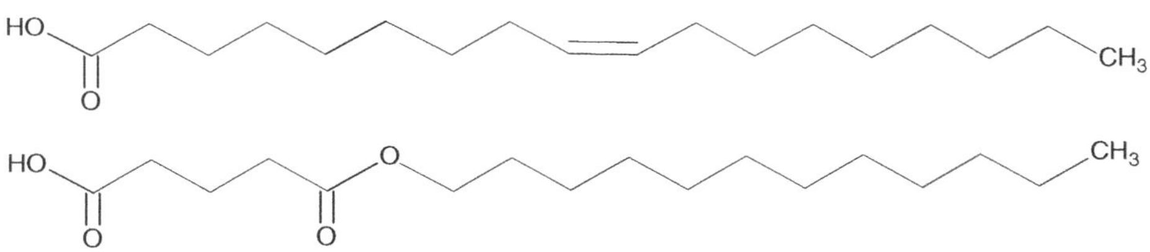

Figure 11.11 Structural comparison between oleic acid (top) and dodecyl glutarate (bottom).

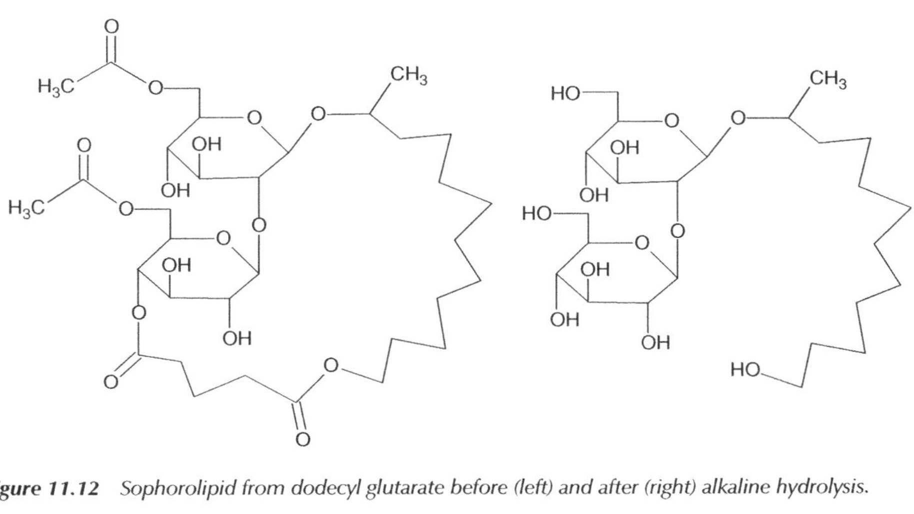

Figure 11.12 Sophorolipid from dodecyl glutarate before (left) and after (right) alkaline hydrolysis.

suitable. Absence of such a polar terminus does, however, result in the hydroxylation and subsequent addition of sophorose at either end of the molecule. For example, when dodecyl pentanoate is used as substrate and the resulting sophorolipids are subjected to alkaline hydrolysis, two sophorolipids with different lipophilic moieties will emerge; one having (11*S*)-1,11-dodecadiol and the other (*S*)-4-hydroxy pentanoic acid.

It is interesting to note that this method not only increases the range of obtainable sophorolipids but also provides a biological way to perform the stereospecific hydroxylation of a number of substrates. All that is required of such a substrate is that it contains a function that allows an elongator molecule to be attached and removed and that the target site for hydroxylation is sterically accessible.

11.3.4 Disabling the β-Oxidation

In order to increase the yield obtained from primary fatty alcohols, the β-oxidation of the organism can be disabled to minimize the loss of substrate due to energy harvesting. Favourable results were obtained when the multifunctional enzyme type 2 (MFE-2) was disrupted. As this enzyme catalyses the second and third step of the β-oxidation and no evidence exists of isozymes, it was determined to be the optimal target for creating a knock-out mutant. When using 1-dodecanol as the lipophilic substrate, various $MFE\text{-}2^-$ mutant strains where this enzyme has been disabled cause a two- to threefold increase in yield when compared to the wild-type organism's production of sophorolipids under the same conditions. Surprisingly, the $MFE\text{-}2^-$ mutant produces significantly less of the normal sophorolipid when rapeseed is used and a suitable amount of glucose is present to act as a source for carbon and energy. Disabling the β-oxidation would therefore promote the conversion of fatty alcohols into sophorolipids but not stimulate the production of wild-type sophorolipids through keeping the fatty acids from being metabolized. A suggested explanation for this discrepancy entails the accumulation of dehydrogenated acyl-CoA derivatives resulting from the (uninhibited) first step of the β-oxidation [8].

11.4 Rhamnolipids

11.4.1 Structure of Rhamnolipids

Biosurfactants containing L-rhamnose and β-hydroxydecanoic acid are called rhamnolipids. *Pseudomonas* sp., especially *Pseudomonas aeruginosa*, have been known since 1949 to produce rhamnolipids under growth-limiting conditions on hydrocarbons and other hydrophobic C-sources, such as seed oils or n-hexane. Different kinds of rhamnolipids are produced [27], as illustrated in Figure 11.13 [28].

Rhamnolipid 1 and rhamnolipid 3 are the major rhamnolipids produced by using resting cells of *Pseudomonas aeruginosa* DSM 2874. Two further rhamnolipids that are similar in structure but contain only one hydroxydecanoic acid unit, rhamnolipids 2 and 4, have also been detected. The rhamnolipid production with resting cells is a two-step process. In a first step *Pseudomonas aeruginosa* cells are produced and harvested. In a second step this biomass is used for the rhamnolipid production under growth-limiting conditions [29].

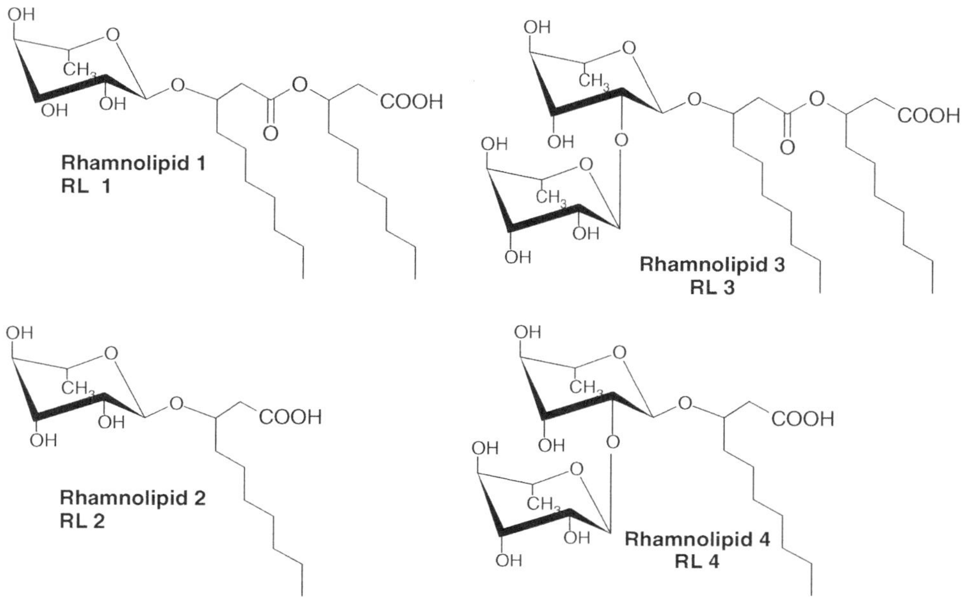

Figure 11.13 *Structures of the most common rhamnolipids.*

11.4.2 Challenges in Biosynthesis

Although rhamnolipids are commercially available from the company Jeneil Biosurfactant Inc., the main problems in rhamnolipid production continue to be the pathogenic nature of the producing organism, a relatively low productivity, an unreliable fermentation process prone to substrate and/or product inhibition and strong foam formation, difficult biomass separation from the product, product impurity with various subtypes and alginate side product and the use of organic solvents for downstream processing. Several research groups have tried to tackle some of these problems, among others, but not limited to, Gruber and co-workers [30], Wullbrandt and co-workers from Hoechst [31], as well as Syldatk and co-workers from Braunschweig, Stuttgart and Karlsruhe University (initially resulting in EP0153634 from Wintershall Aktiengesellschaft). The latter research group is still actively researching process optimization, product purification and screening for nonpathogenic production organisms.

Attempts have been described to reduce foam formation by removing CO_2 by membrane filtration. Gruber *et al.* [30] in particular describe such a process using glycerine as substrate. A European funded project coordinated by Ecover concluded, however, that these membranes are prone to swelling when plant oil and particularly oleic acid is used as the substrate, which coincidently gives the highest productivity. Additionally, the project tried to control foam formation using alternative mechanical foam control strategies focusing on the mixing and aeration technology among others, with bubble-free aeration by gas exchange membranes and/or microsparging with oxygen-enriched air), as well as on foam measurement and control, among others, with ultrasonic foam destruction and product separation, rather than on chemical foam control.

Another approach exemplified by a patent application from the University of Karlsruhe is to operate at high concentrations of sunflower oil and at low oxygen concentrations, i.e. at lower working volumes and high foam levels that are controlled by mechanical foam destruction. This results in a product yield of about 50 g/l, which is still only half the yield once reported by Giani and co-workers [31]. On the other hand, a high volumetric productivity is achieved while undesired side products like alginate only occur to a minor extent. After harvesting, the cells can be reused in the next fermentation without significant loss of product yield. This procedure reduces the amount of organic waste and an unnecessary complete sterilization can be avoided.

Another aspect prone to improvement is the product separation and purification step. This is particularly challenging due to high protein and alginate concentrations that tend to clog membrane filters, resulting in a drastically reduced transmembrane flux. In addition to this, rhamnolipids aggregate in micelles of about the same size as *Pseudomonas* cells, which obviously complicates solvent-free product separation.

11.4.3 Physicochemical Properties

Of the cell-free supernatants of strains grown on frying oils, those of *P. aeruginosa* 47T2 and of eight other *Pseudomonas* strains showed surface tension values from 32 to 36 mN/m [32]. An emulsion with kerosene remained stable for three months. Abalos and co-workers [33] cultured *P. aeruginosa* AT10 in a mineral salts medium with 5% of waste-free fatty acids in a food-oil refinery. The final production of rhamnolipids was 9.5 g/l after 96 hours. The surface tension of the supernatant reached 28 mN/m at this time.

The dynamic surface activity of the commercial rhamnolipid mixture JBR425 from Jeneil was determined as a function of concentration and time with the maximum bubble pressure method using an online bubble tensiometer (Sita T60). Figure 11.14 highlights the good surfactancy properties of rhamnolipids, with low minimum surface tension and moderate dynamics, meaning a relatively fast decrease of surface tension at new surfaces and low bubble lifetimes.

After logarithmation of concentration (ppm) and bubble lifetime (ms), and taking the logarithm of a min/max transformed tensioactivity $\log((72.5 - 25.5)/(72.5 - \sigma) - 1)$ a multivariate regression was performed that allowed a good predictive model to be selected (adjusted $R^2 = 0.92$; $F(7.122) = 219.07$; $p < 0.000$), the theoretical values correlating well with the experimental data.

Likewise for sophorolipids (Figure 11.15), a good model was compiled with surface tension as low as 33 mN/m, but dropping slower than rhamnolipids, both in terms of surfactant concentration and surface age. The minimum surface tension values are in good agreement with the findings of Abalos *et al.* [33] and Lang *et al.* [14], although the lower critical micelle concentration (CMC) of 40 ppm reported for sophorolipids would suggest their surface activity to start at lower concentrations than with rhamnolipids, which these data cannot confirm.

The metal complexing and mineral oil emulsifying properties of rhamnolipids make them particularly suitable candidates for in situ bioremediation, as has been described in numerous publications. Monorhamnolipid (RL1) seems to be the most effective subtype

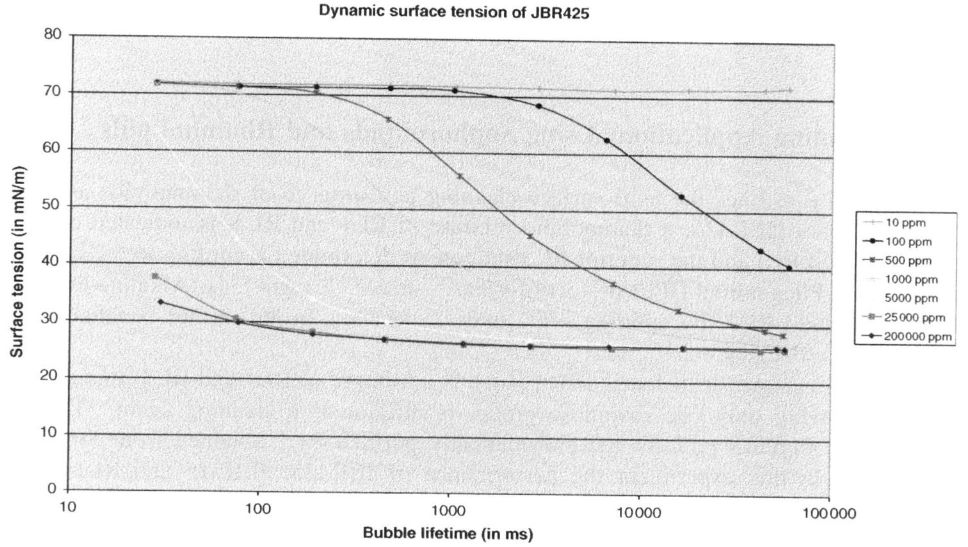

Figure 11.14 *Dynamic surface tension (mN/m) of the rhamnolipid mixture JBR425 as a function of concentration and bubble lifetime.*

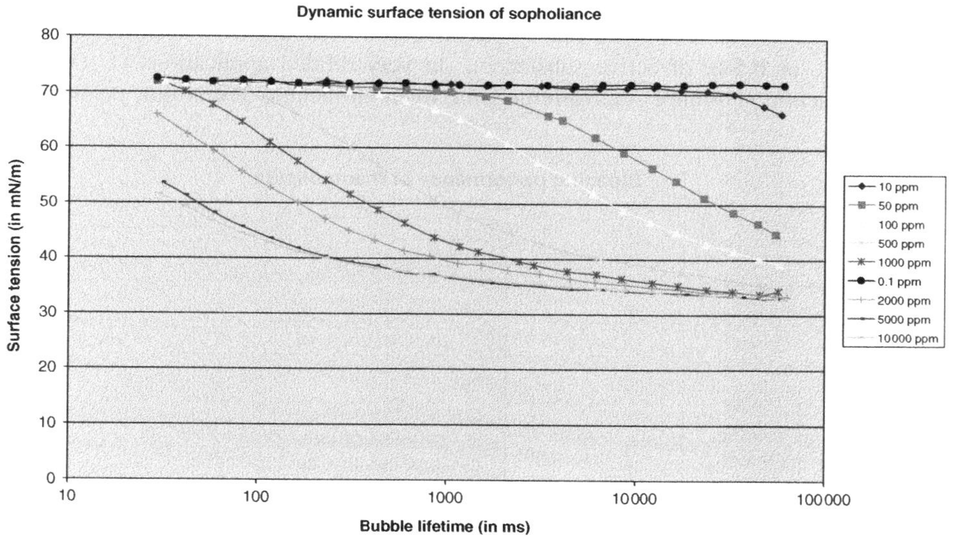

Figure 11.15 *Dynamic surface tension (mN/m) of sophorolipids as a function of concentration and bubble lifetime.*

against lead, cadmium and mercury. The rhamnolipids M7 described by Abalos *et al.* [33] also showed excellent antifungal properties against phytopathogenic fungi *Botrytis cinerea* and *Rhizotecnia solani* (minimum inhibitory concentration (MIC): 18 ppm).

11.5 Cleaning Applications Using Sophorolipids and Rhamnolipids

Figure 11.16 illustrates the hard surface cleaning performance of rhamnolipids as produced by Jeneil (JBR425, a rhamnolipid mixture of RL1 and RL3, is indicated on the graph as Jeneil RL) and the superior RL3 subtype as determined according to the Institut für Putz und Pflegemittel (IPP) protocol for hard surface cleaners [34]. Alkaline hydrolysis to RL2 and RL4 (by splitting off a hydroxydecanoic group) is not beneficial for rhamnolipid efficiency.

Since RL3 is superior to Jeneil JBR (which is a mixture of RL1 and RL3) this means that RL1, having only one rhamnose group, is inferior as a cleaning agent. This is confirmed in Figure 11.17 for the enzymatically purified RL1 obtained from Stuttgart University. In this experiment the performance of JBR425, JBR515 and RL1 were tested in concentrated form (1/10, or 0.5%, of active substance of surfactant) as well as in diluted form (1/100, or 0.05%, of active substance). RL1 in fact proved to be a very sticky material, which produced little foam and with hardly any surface activity (data not shown). Another interesting feature of Figure 11.17 is the effect of rhamnolipid purification (JBR515 being a purified version of JBR425), which is clearly beneficial for the performance.

Figure 11.18 illustrates the excellent hard surface cleaning properties of biosurfactants as compared to two commonly used surfactants LES and APG650. Especially sophorolipids ('Sopholiance') are highly effective and certainly in concentrated applications (1/10, or 0.5%, of active substance). In very diluted applications (1/1000, or 0.005%, of active substance) the difference between surfactants levels out, contrary to

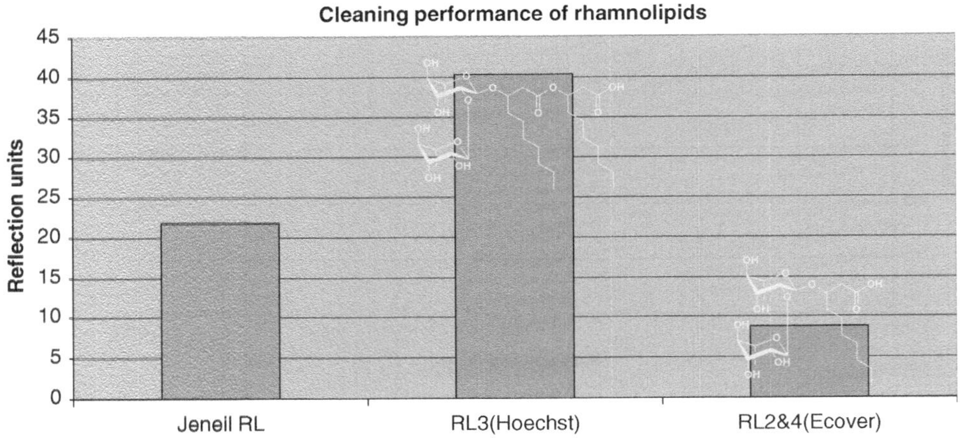

Figure 11.16 *Cleaning performance of rhamnolipids.*

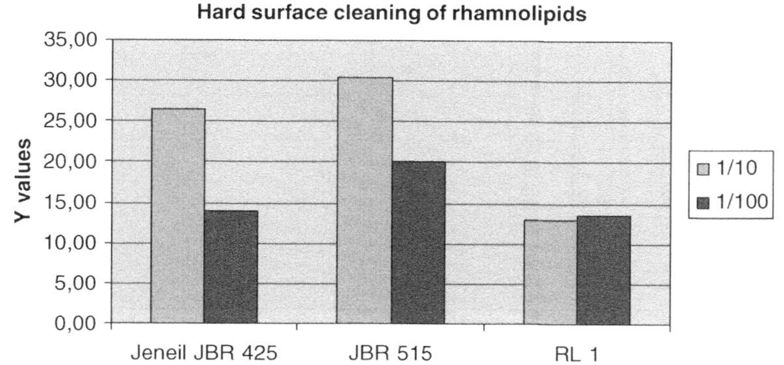

Figure 11.17 *Hard surface cleaning performance of JBR425, JBR515 and RL1 (Stuttgart) in concentrated and diluted applications.*

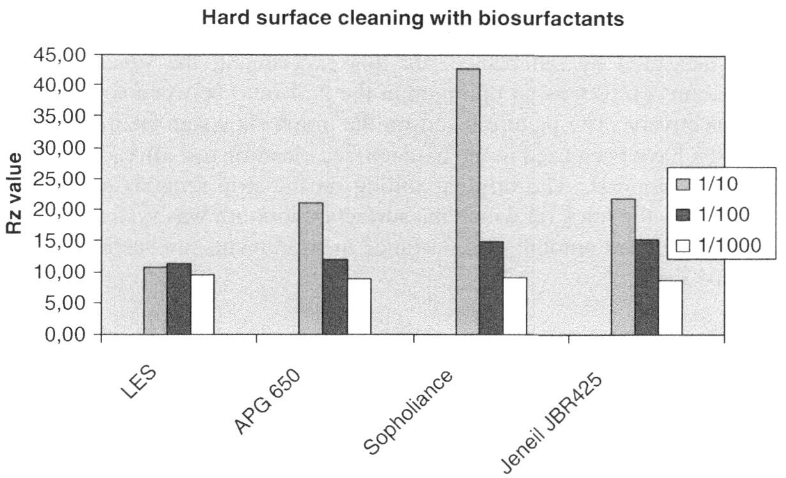

Figure 11.18 *Hard surface cleaning performance of LES (FAES, Ifrapon LOS 70N from Ifrachem), APG650® (Glucopon from Cognis), Sopholiance (from Soliance) and JBR425 (from Jeneil) in concentrated and diluted applications.*

some claims that biosurfactants remain effective at extremely low concentrations. Efficiency in hard surface cleaning is measured through reflectance, as indicated by its concurrent Rz value. This value effectively measures how 'white' a polyvinyl chloride (PVC) test strip was after a standard (black) soiling mixture was applied; according to the set conditions for the test, this mixture was removed using surfactants, which are the subject of the experiment. A high Rz value indicates good reflectance and therefore good soil removal.

As elaborated by Develter, Renkin and Jacobs [28], sophorolipids are complementary to alkylpolyglucosides (using Glucopon600®), with mixtures of the two performing well in both concentrated and diluted applications in both aqueous and nonaqueous products.

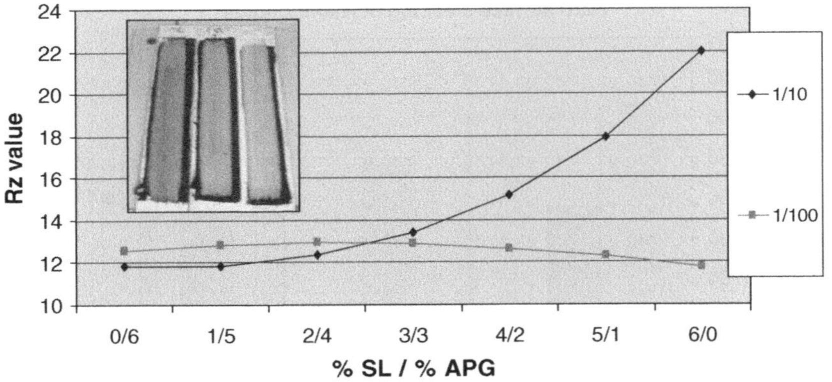

Figure 11.19 Cleaning efficiency as measured by the Rz value for sophorolipid and APG mixtures.

Figure 11.19 shows the synergy between these two surfactants as tested for hard surface cleaning and measured by reflectance. In this experiment, the synergy is shown for diluted applications (1/100) as an optimum at the 2 : 4 ratio between sophorolipids (SLs) and APG, respectively. The picture insert on the graph shows an example of three PVC test strips, which have been used in the hard surface cleaning test after a certain surfactant mixture had been applied. The original soiling on the strip remains as a black border, which encompasses the track on which the surfactant mixture was systematically applied. A statistically significant amount of reflectance measurements are carried out on random positions in this track.

References

1. Desai, J. D. and Banat, I. M. (1997) Microbial production of surfactants and their commercial potential. *Microbiol. Molec. Biol. Rev.*, **61**, 47–59.
2. Kitamoto, D., Isoda, H. and Nakahara, T. (2002) Functions and potential application of glycolipids biosurfactants – from energy-saving materials to gene delivery carriers. *J. Biosci. Bioengng*, **94**, 187–201.
3. Gorin, P., Spencer, J. and Tulloch, A. (1961) Hydroxy fatty acid glycosides of sophorose from *Torulopsis magnolia*. *Can. J. Chem.*, **39**, 846–855.
4. Tulloch, A. and Spencer, J. (1968) Fermentation of long-chain compounds by *Torulopsis apicola* IV. Products from esters and hydrocarbons with 14 and 15 carbon atoms and from methyl palmitoleate. *Can. J. Chem.*, **46**, 1523–1528.
5. Jones, D. and Howe, R. (1968) Microbiological oxidation of long-chain aliphatic compounds. Part I. Alkanes and alk-1-enes. *J. Chem. Soc. C*, **257** (16), 2801–2808.
6. Spencer, J., Spencer, D. and Tulloch, A. (1979) Extracellular glycolipids of yeasts, *J. Biol. Chem.*, **3**, 523–540.
7. Fleurackers, S. (2006) On the use of waste frying oil in the synthesis of sophorolipids. *Eur. J. Lipid Sci. Technol.*, **108** (1), 5–12.
8. Van Bogaert, I. (2008) Microbial production of sophorolipids by the yeast *Candida bombicola*, PhD thesis, Faculty of Bioscience Engineering, Ghent University, Ghent, Belgium.

9. Lottermoser, K., Schunck, W. H. and Asperger, O. (1996) Cytochromes P450 of the sophorose lipid-producing yeast *Candida apicola*: heterogeneity and polymerase chain reaction-mediated cloning of two genes. *Yeast*, **12**, 565–575.
10. Breithaupt, T. B. and Light, R. J. (1982) Affinity-chromatography and further characterization of the glucosyltransferases involved in hydroxy-docosanoic acid sophoroside production in *Candida bogoriensis*. *J. Biol. Chem.*, **257**, 9622–9628.
11. Asmer, H. J., Lang, S., Wagner, F. and Wray, V. (1988) Microbial-production, structure elucidation and bioconversion of sophorose lipids. *J. Am. Oil Chem. Soc.*, **65**, 1460–1466.
12. Esders, T. W. and Light, R. J. (1972) Glucosyl- and acetyltransferases involved in the biosynthesis of glycolipids from *Candida bogoriensis*. *J. Biol. Chem.*, **247**, 1375–1386.
13. Bucholtz, M. L. and Light, R. J. (1976) Acetylation of 13-sophorosyloxydocosanoic acid by an acetyltransferase purified from *Candida bogoriensis*. *J. Biol. Chem.*, **251**, 424–430.
14. Lang, S., Brakemeier, A., Heckmann, R. *et al.* (2000) Production of native and modified sophorose lipids. *Chim. Oggi*, **18**, 76–79.
15. Davila, A. M., Marchal, R. and Vandecasteele, J. P. (1992) Kinetics and balance of a fermentation free from product inhibition-sophorose lipid production by *Candida bombicola*. *Appl. Microbiol. Biotechnol.*, **38**, 6–11.
16. Gobbert, U., Lang, S. and Wagner, F. (1984) Sophorose lipid formation by resting cells of *Torulopsis bombicola*. *Biotechnol. Lett.*, **6**, 225–230.
17. Casas, J. A. and Garcia-Ochoa, F. (1999) Sophorolipid production by *Candida bombicola*: medium composition and culture methods. *J. Biosci. Bioengng*, **88**, 488–494.
18. Zah, R. and Hischier, R. (2003) *Life Cycle Inventories of Detergents – Final Report Ecoinvent 2000*, vol. 12, Swiss Centre for LCI, EMPA-SG, Dübendorf.
19. Goedkoop, M., Effting, S. and Collingo, M. (2000) *The Eco-Indicator'99 – A Damage Oriented Method for Life Cycle Impact Assessment – Manual for Designers*, 2nd edn, Pre Consultants B.V.
20. Renkin, M. (2003) Environmental profile of sophorolipid and rhamnolipid biosurfactants. Convention Giornate CID No10, Milan, Italy, vol. 80, no. 4, pp. 249–252.
21. Stalmans, M., Berenbold, H., Berna, J. L. and Cavalli, L. (1995) European life-cycle inventory for detergent surfactants production. *Tenside Surfactant Deterg.*, **32**, 84–109.
22. Hirsinger, F. (1998) Comparing oleochemical and petrochemical surfactants with the life cycle inventory method (LCI). *Chim. Oggi*, **16** (10), 62–66.
23. Stüwer, O., Hommel, R., Haferburg, D. and Kleber, H.-P. (1987) Production of crystalline surface-active glycolipids by a strain of *Torulopsis apicola*. *J. Biotechnol.*, **6**, 259–269.
24. Develter, D. and Fleurackers, S. (2007) A method for the production of short chained glycolipids, Patent EP07101581.2.
25. Shah, V., Doncel, G. F., Seyoum, T. *et al.* (2005) Sophorolipids, microbial glycolipids with anti-human immunodeficiency virus and sperm-immobilizing activities. *Antimicrob. Agents Chemother.*, **49**, 4093–4100.
26. Brakemeier, A., Lang, S., Wullbrandt, D. *et al.* (1995) Novel sophorose lipids from microbial conversion of 2-alkanols. *Biotechnol. Lett.*, **17** (11), 1183–1188.
27. Lang, S. and Wullbrandt, D. (1999) Rhamnose lipids biosynthesis, microbial production and application potential. *Appl. Microbiol. Biotechnol.*, **51**, 22–32.
28. Develter, D., Renkin, M. and Jacobs, I. (2003) Detergent compositions, Patent EP1445302A2.
29. Trummler, K., Effenberger, F. and Syldatk, C. (2003) An integrated microbial/enzymatic process for production of rhamnolipids and L-(+)-rhamnose from rapeseed oil with *Pseudomonas* sp. DSM 2874. *Eur. J. Lipid Sci. Technol.*, **105** (10), 563–571.
30. Gruber, T., Chmiel, H., Käppeli, O. *et al.* (1993) Integrated process forcontinuous rhamnolipid biosynthesis, in *Biosurfactants, Production, Properties, Applications*, Surfactant Science Series, vol. 48 (ed. N. Kosaric), Elsevier, Amsterdam, The Netherlands.

31. Giani, C., Wullbrandt, D., Rothert, R. and Meiwes, J. (1997) *Pseudomonas aeruginosa* and its use in a process for the biotechnological preparation of L-rhamnose, Patent US5658793.
32. Haba, E., Espuny, M. J., Busquets, M. and Manresa, A. (2000) Screening and production of rhamnolipids by *Pseudomonas aeruginosa* 47T2 NCIB 40044 from waste frying oils. *J. Appl. Microbiol.*, **88**, 379–387.
33. Abalos, A., Pinazo, A., Infante, M. R. *et al.* (2001) Physicochemical and antimicrobial properties of new rhamnolipids produced by *Pseudomonas aeruginosa* AT10 from soybean oil refinery wastes. *Langmuir*, **17**, 1367–1371.
34. Anonymous (1986) Qualitätsnormen für Fussbodenpflege- und Reinigungsmittel. *Seife, Öle, Fette, Wachse*, **112** (10), 371–372.

12
Saponin-Based Surfactants

Wieslaw Oleszek[1] **and Arafa Hamed**[2]

[1]*Institute of Soil Science and Plant Cultivation, State Research Institute, Pulawy, Poland*
[2]*Department of Botany, Aswan Faculty of Science, South Valley University, Aswan, Egypt*

12.1 Introduction

Saponins belong to a diverse group of naturally occurring surface-active compounds. They occur in different tissues of a large number of plant species belonging to nearly 100 families. Although predominant in angiosperms, saponins also occur in some ferns (species of *Polypodium* and *Cyclamen*) and possibly algae, and in certain lower marine organisms. They can be found in different plant parts including roots, shoots, flowers and seeds. Their common feature is the formation of a soapy lather when shaken in water solution. This property has for a long time been used in the identification of saponin-containing plant species as well as for their quantification. The height of the froth when shaken in a glass tube and the time of its disappearance was a semiquantitative test. Since some other plant components may also form froth when shaken, a number of species were misclassified as saponin-containing plants, but in general most literature concerning saponin distribution in nature was based on this test.

In the Orient plants rich in saponins were used as a soap substitute in folk medicine. Thus, in many cases the common names of saponin-rich plant species were derived from this feature, e.g. soaproot (*Chlorogalum pomeidianum*), soapbark (*Quillaya saponaria*), soapwort (*Saponaria officinalis*), soapberry (*Sapindus saponaria*), soapnut (*Sapindus mukurossi*), soapjacob or ra-ra Ayoub – 'bring back youth and health' (*Glinus lotoides*), which is based on the folklore belief that the Prophet Jacob (Arabic = Ayoub), having dermal disease, cured himself by rubbing his skin with this plant. For these reasons saponins found some industrial interest mainly as surface-active or foaming agents. When occurring in food, saponins were for a long time considered as antinutritional compounds

Surfactants from Renewable Resources Edited by Mikael Kjellin and Ingegärd Johansson

due to their throat irritating properties and some bitterness; therefore food processing was targeted to remove these undesirable chemicals from the diet. Also the goal of breeding programmes of cultivated species was directed into the removal of these compounds, e.g. low-saponin alfalfa varieties. Nowadays saponins are considered in some cases as health beneficial food components due to their cholesterol lowering and anticancer properties, e.g. soya bean, garlic and onion.

12.2 Molecular Properties

Saponins occurring in plants are predominantly glycosides possessing one, two or three sugar chains attached to the aglycone, and the terms monodesmosides, bidesmosides or tridesmosides have been given to them, respectively (Greek *desmos* = chain). The aglycones, also called sapogenins, which are the nonpolar parts of the molecule, may have a steroidal or triterpene backbone (Figure 12.1). The monodesmosidic saponins have a sugar chain attached usually at C-3 of the aglycone. The bidesmosides have two sugar chains, most often with one attached through the ether linkage at C-3 and one attached through the ester linkage at C-28 in triterpene saponins or an ether linkage at C-26 in steroidal saponins. The tridesmosides may have a third sugar chain linked through an ether or ester link at one of the OH or COOH functional groups occurring on the aglycone. Some authors also include in the saponin family steroidal glycoalkaloids occurring predominantly in the Solanaceae family and cucurbitacines, the bitter principles of some Cucurbitaceae species. The most common monosaccharides appearing in the sugar chains include: D-glucose (glc), D-galactose (gal), D-glucuronic acid (glcA), L-rhamnose (rha), L-arabinose (ara), D-xylose (xyl), D-apiose (api) and D-fucose (fuc). The steroidal skeletons have in most cases furostanol or spirostanol form; furostanol glycosides usually have a bidesmosidic and spirostanol monodesmosidic nature. Both steroidal and triterpene sapogenins may have a number of functional groups (−OH, −COOH, $-CH_3$) causing great natural diversity only because of the aglycone nature. This diversity has been multiplied by the number and composition of sugar chains. Thus, in most cases when the term 'saponins' is used, this should be understood as a complex mixture of glycosides with the same sapogenin or with different sapogenins. Their composition and the concentration can be different in plant parts (roots, shoots, leaves, flowers, fruits) and can be strongly influenced by environmental factors and the state of development of the plant.

The complex structure of saponins may undergo chemical transformations during storage or processing, which in turn may modify their properties and biological activity. The glycosidic linkage (between the sugar chain and the aglycone) and the interglycosidic linkages between the sugar residues can undergo hydrolysis in the presence of acids/alkali, hydrothermolysis (heating in the presence of water) or enzymatic/microbial transformations, resulting in the formation of aglycones, prosapogenins (partially hydrolysed saponins), sugar residues or monosaccharides depending on the hydrolysis method and conditions [1]. Complete acid hydrolysis yields the constituent aglycone and monosaccharides, whereas under basic hydrolysis conditions, cleavage of ester-linked sugar chains results in the formation of prosapogenins [2]. The solubility behaviour of the parent aglycone can be markedly different from the saponin due to its lipophilic nature.

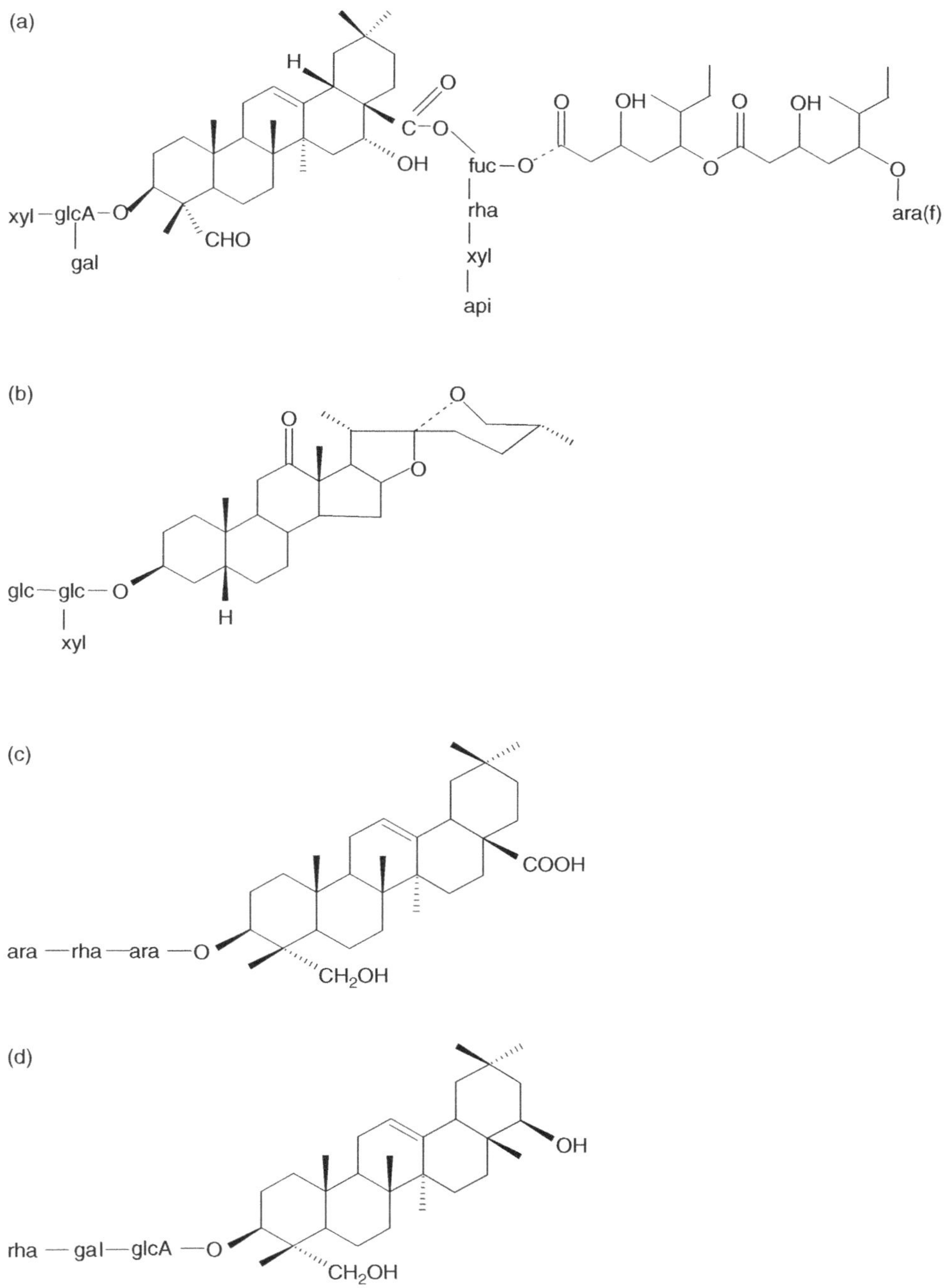

Figure 12.1 *Basic structures of saponins: (a) triterpene saponin QS-21 from Quillaya saponaria; (b) steroidal saponin from Yucca schidigera; (c) hederagenin glycoside from Sapindus mukurossi; (d) soyasaponin I from soybean (Glycine max).*

12.3 Sources of Saponins

The triterpene saponins are predominantly present in dicotyledons (Leguminosae, Araliacae and Caryophyllaceae) while steroidal saponins occur predominantly in monocotyledons (Liliaceae, Dioscoreaceae and Agavaceae). The main dietary sources of saponins are leguminous plants, which include soya bean, chickpea, mungbean, peanuts, lentils and beans, but they are also present in some other edible plant species like oats, leek, garlic, asparagus, tea, spinach, sugar beet, sesame and yam. The concentrations of saponins in these products are not very high, ranging from 0.1 up to 2% in dry matter [3], but they are recognized as antinutritional factors, which should be eliminated by breeding or processing. Recent trends in nutrition showing the beneficial effect of some natural products on human health stimulated some research also on saponins. They were shown to express different kinds of activities including anticarcinogenic, antioxidant, hypocholesterolemic, hepatoprotective, antiviral, antifungal and antibacterial functions *in vitro* or *in vivo* animal tests. Once consumed they may increase the intestinal permeability and interact with bile acids. The largest number of nutritional experiments has been performed on soya bean saponins, but saponins from oat, quinoa, sunflower, garlic and tea have also been studied.

The main nonfood sources that have been commercially used in food and cosmetic industries include: soap bark tree (*Quillaya saponaria*), Mohave yucca (*Yucca schidigera*), fenugreek (*Trigoonella foenum-graceum*), horse chestnut (*Aesculus hippocastanum*), licorice (*Glycyrrhiza glabra*), soapwort (*Saponaria officinalis*), gypsophylla (*Gypsophylla paniculata*), sarsaparilla (*Smilax regelli*) and some others (Table 12.1).

The concentration of saponins in some of these species is high. The best sources of these compounds seem to be *Yucca schidigera* and *Quillaya saponaria*, in which concentration of saponins reaches 10% of dry matter. Moreover, these two plants have been given by the Food and Drug Administration (FDA) a generally recognized as safe (GRAS) label, and are accepted as safe food, feedstuff and cosmetic ingredients in the United States. Since 1962 these species have been accepted as emulsifiers and foaming agents at a maximum concentration of 20 ppm in the UK and Japan. In Japan the saponins from two other sources, e.g. soya bean and Enju, have been accepted for food applications.

12.4 Saponins as Emulsifiers and Surfactants

The ability of a saponin to foam is caused by the combination of the nonpolar sapogenin and water-soluble side chain (see Figure 12.1(b) to (d)), which is similar to the structure of most synthetic surfactants having lipophilic and hydrophilic molecular parts. In synthetic surfactants lipophiles are usually similar from one surfactant to another (i.e. straight or branched alkyl chains) but hydrophiles show a range of chemical types. This has been the basis for surfactant classification as anionic, cationic, nonionic and amphoteric. In the case of saponins, hydrophiles are built of a sugar chain, which can differ in the length, branching, substitution and composition (glucose, galactose, rhamnose, arabinose, xylose, apiose and uronic acid), while the lipophiles may have a steroidal or triterpene structure. Hence, saponins of this composition are nonionic surfactants. Saponins with one sugar

Table 12.1 *Plant sources for industrially utilized saponins*

Plant source	Common name	Use	Plant parts	Concentration (%)
Aesculus hippocastanum	Horse chestnut	Aescin source	Seeds	10
Agave sisalana	Sisal, henequen, hemp plant	Hecoginin source	Leaves	>12
Balanites aegyptiaca	Heglig, lalob, desert date	Diosgenin and yamogenin source	Fruits, seeds, bark	22–27
Chlorogalum pomeridianum	Soaproot, California soap plant	Amolonin source	Bulbs	19–22
Costus speciosus	Crape ginger, Malay ginger	Diosgenin source	Rhizomes	3.86
Digitalis lanata	Woolly foxglove, Grecian foxglove	Digoxigenin source	Leaves	–
Digitalis purpurea	Purple foxglove, lady's foxglove	Digoxigenin source	Leaves, seeds	1.25
Dioscorea composite	Yams, barbasco	Diosgenin source	Rhizomes, roots	4–6
Glinus lotoides	Soap Jacob, lotus sweetjuice	Hopane and oleanane source	Roots, leaves, seeds	16.5
Glycine max	Soya bean	Soyasaponin source	Sprouts, seeds	0.5
Gypsophilla paniculata	Baby's breath	Gypsogenin source	Roots	>10
Quillaya saponaria	Quaillaya, Murillo's-bark	Soaps, foaming agents	Bark	≥25
Smillax spp.	Sarsaparilla	Smilagegnin source	Rhizomes, roots	2
Solanum spp.	Bitter nightshade, black nightshade	Solasodine source	Fruits, stems	0.3–0.8
Sapindus mukurossi	Soapnut	Hedragenin source	Fruits and roots	20
Sapindus saponaria	Soapberry	Hedragenin source	Fruits	11
Tribulus terrestris	Puncturevine, yellow vine and goathead	Protodioscin source	Fruits	>20
Trigonella faenum graecum	Fenugreek	Diosgenin source	Seeds and leaves	8–10
Yucca schidigera	Mohave yucca, Joshua tree	Soaps, foaming agents	Stalk and roots	10

Adapted from W. Oleszek, Saponins, *Natural Food Antimicrobial Systems*, 2000, CRC Press.

chain have the best foaming characteristics. For saponins with two or three sugar chains the foaming ability decreases and in some saponins no foaming in water solution has been observed, but due to their chemical structure they are still considered as saponins. The emulsifying properties of saponins are due to the fact that they have a salt-free nature, making them less likely to be affected by alkaline or acid conditions.

In water solution saponins form micelle-like aggregates [4] and show a critical micelle concentration (CMC). Below this concentration, molecules remain unassociated. An abrupt change in physical properties appears when the concentration surpasses the CMC and the solute starts to form micelles. Soya bean saponins, the saponins from *Saponaria officinalis*, and *Quillaya saponaria*, form micelles in aqueous solutions, the size and structure of which are dependent on the type of saponin. Commercial 'saponin white' from *Saponaria officinalis* and soya bean saponins form small micelles consisting of only two molecules, while the aggregates of saponin of *Quillaya saponaria* consist of 50 molecules, and appear to be significantly less hydrated [5]. These differences are quite unexpected as aglycones of quillaya saponin differ from the aglycone of 'saponin white' only by one hydroxyl group. Presumably they aggregate by hydrophobic interaction of their aglycones (as for other surfactants), leaving the hydrophilic sugar groups exposed to the water. The micelle-forming properties and the aggregation number (number of monomers in a micelle) of *Quillaya* saponins were affected by temperature, salt concentration and pH of the aqueous phase. At 25 °C, the values of the CMC of *Quillaya* saponins were in the range of 0.5 to 0.8 g/l. The CMC increased with temperature and pH but decreased with increasing salt concentration [5]. The shape of the micelles depends also on the saponin structure. The micelles formed by 'saponin white' and *Quillaya* saponins appear as elongated and filamentous, while those formed by soya bean saponins appear spherical [5]. Probably the reason for these differences is in the aglycone structures. The soya bean aglycones, unlike the aglycones of *Saponaria officinalis* and *Quillaya saponaria* saponins, do not possess carboxylic functions and therefore they are more uniformly hydrophobic.

The presence of carboxylic acid in the saponin molecule may strongly influence the surface activity, emulsion stability or zeta potential of the emulsion droplets. Not only the presence but also the location in the molecule has been extremely important. This can be shown by the comparison of surface activity of soyasaponin I, the dominant saponin of soya bean and monodesmosodic saponins of *Sapindus mukurossi* [6]. Soyasaponin I contains the carboxylic group in the hydrophilic sugar chain part of the molecule. The carboxylic groups dissociate in the aqueous phase and form free carboxyl anion, which increases the solubility of a molecule in water. In contrast, the saponins of *Sapindus* contain the carboxylic group attached to the aglycone part of the molecule, which is hydrophobic, and the dissociation of this carboxylic group is very low (Figure 12.1(c) and (d)). Due to these differences the surface activity, emulsion stability and foamability were higher and the surface and interfacial tensions were lower for *Sapidsus* saponins as compared to soyasaponin I. However, the creaming stability of soyasaponin I was higher than that of *Sapindus* saponins.

Saponins can also form mixed 'sandwich-like' or 'pile of coins like' micelles with bile acids. These are much larger than the micelles of saponins alone and again they differ depending on the structure of the aglycone. Saponin white and *Quillaya* saponin form filamentous structures with bile acids, while the soya bean saponins have a loose, open structure with considerable penetration of water [5]. The ability of saponin to form these stable micelles with bile acids has very important nutritional consequences. The food and feedstuff containing saponins increase fecal excretion of bile acids, which in consequence leads to their reduced reabsorption and to lowering of the plasma cholesterol concentration.

Saponins also affect the permeability of intestinal cells by forming addition complexes with sterols (e.g. cholesterol) in mucosal cell membranes [7]. This leads to destabilization of the membranes and an increase in the permeability of intestinal mucosal cells, which inhibit active nutrient transport. Thus this facilitates the uptake of substances to which the gut would normally be impermeable, e.g. milk alergen α-lactoglobulin.

12.5 Application of Saponins as Surfactants and Emulsifiers

In the early days of mankind development of different plants and plant extracts were explored for personal hygiene purposes. However, this was not an efficient way to meet demands of a growing human population. Thus, the natural surfactants were replaced with different synthetic products. The synthetic chemicals that enter our environment have usually no 'ecological history' and have been very slowly degraded by microorganisms. Those synthetic chemicals may have detrimental effects on human health, being the main factor of so-called 'civilization diseases'. This includes also the common synthetic surfactants that can have a negative effect such a skin irritation. Thus, there has been an increasing interest in once again using natural products that could substitute for synthetic ones. Current pressure to move away from nonrenewable petroleum feedstocks and towards plants as a source of raw materials has led to much effort in developing surfactants from oleochemical feedstocks. Many recently developed surfactants are the result of an attempt to satisfy modern consumers' desire for products to be 'more natural'. Thus, the history of surfactant development has turned full circle, as demonstrated by soap production (Figure 12.2).

Two major natural surfactants being used include lecithin, an emulsifier used in chocolate and ice cream manufacture, and plant saponins. However, the use of saponins has some limitations. First of all, these compounds are not as potent surfactants as the synthetic ones, and, second, the supply from plant sources has been quite limited, which makes these products less available and rather expensive.

One of the most important sources of saponins has been *Yucca schidigera*, which is native to the southwestern United States and Mexico. Native Americans used this plant to make soap. Currently, most commercial production of *Y. schidigera* products takes place in Mexico (DesertKing Int., ChulaVista, California, USA). The trunk of the plant is the part used. The logs are mechanically macerated and the macerated material is subjected to mechanical squeezing in a press, producing yucca juice. The juice is concentrated by evaporation, with the concentrated product referred to as an extract. The term 'yucca extract' is slightly misleading, in that the plant juice is removed by mechanical means, rather than by solvent extraction. Their antifungal and antibacterial properties are also important in cosmetic applications, in addition to their emollient effects.

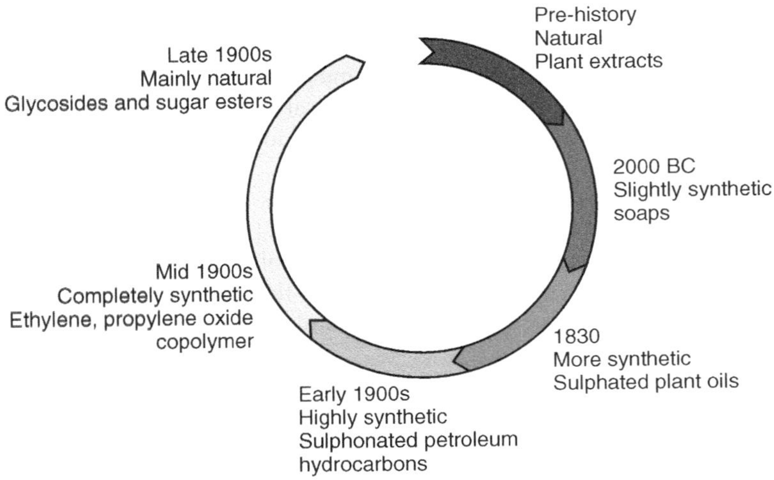

Figure 12.2 *Soap development going full circle.*
From T. Hargreaves, *Chem. Br.*, 2003, **39** (7), 38–41 (http://www.rsc.org/chemistryworld/Issues/2003/July/amphiphiles.asp). Reproduced by permission of The Royal Society of Chemistry.

The second saponin source of commercial value is *Quillaya saponaria*, found in the arid areas of Chile. The bark of the tree is the part used and has been used as shampoo in Chile for hundreds of years. Yucca and quillaya extracts are commonly used as foaming agents for beverages such as root beer (6–7 fluid ounces of yucca and 3.5 ounces of quillaya per 100 gallons of syrup). These have also been used in other soft dinks (1–2 fluid ounces per 100 gallons of syrup) and cocktail mixes. These phytochemicals have also commercial applications such as ore separation in industrial and mining operations, and are useful as components in products such as photographic emulsions, cosmetics and shampoos.

As a natural emulsifier *Quillaya* and *Yucca* saponins have been used in the following ways for products patented in Japan (Natural foaming agents quillaya and yucca, DK Int. commercial leaflet):

- to emulsify oil-based flavours for candy,
- to prevent precipitation in a protein containing liquid composition,
- to help prevent oil separation in mayonnaise,
- for use as a leavening agent in the bakery industry,
- to produce an oil in a water-type emulsion composition,
- to increase stability of cream when added to coffee,
- as a natural dispersing agent for waxes used in food coatings.

A unique organic toothpaste characterized by the use of *Quillaya* and *Yucca* saponins as the cleansing and foaming agent has been developed. The content of saponins may form up to 10% by weight of the toothpaste [9].

The crude saponin fraction from the pericarp of *Sapindus mucurossi* Gaertn., which grows in China and Japan, has been used as a natural detergent and as foam-stabilizing agents in chemical fire extinguishers in Japan. The saponin extract from this plant has been listed in the Japanese Cosmetic Ingredient Codex, and is authorized as an ingredient in cosmetics by the Ministry of Health and Welfare of Japan. This saponin cannot be used as a food ingredient, as hederagenin glycosides and sesquiterpene oligoglycosides, which are components of 'crude saponins' from this plant, can be toxic. Due to their antidermatophytic activity they have been very promising as raw materials for ingredients to be used in cosmetics. Dermal toxicity tests did not show primary dermal irritation, sensitization, phytotoxicity or photosensitization effects [10].

Other sources of saponins that have been studied for possible commercial application include soapwort *Saponaria officianalis*, whose foliage yields a glycoside capable of wetting, foaming and grease dispersion – the very qualities that we recognize in a modern detergent. These natural glycosides have been used for specialized processes, such as washing delicate fabrics. A new quaternary ammonium compound, hydrolysed ginseng saponin quaternary (HGSQ), from Korean ginseng (*Panax ginseng*) saponin and 2,3-epoxypropyltrimethyl ammonium chloride, has been developed as a conditioning agent for hair care products. This structure has a hydrophobic group from the aglycone of ginseng saponin, which is biologically active and considered as the most important component of Korean ginseng [11].

The best hair loss shampoo formulations can also incorporate saponins, from soap bark, soapwort, sarsaparilla and ivy. These saponins make a very good lather but have low cleansing properties. To get appropriate detergency for washing hair, a high concentration of saponins is required, but a high concentration can be harsh to the hair fibre. Therefore, while formulating shampoo, these natural surfactants are generally combined with synthetic ones to ensure good cleansing and satisfactory cosmetic qualities.

The triterpene saponin has been patented in the USA as an effective component of fire-fighting foams [12]. Saponins were used in very early fire-fighting foams as foaming agents, but have long since been discontinued due to their high cost, which makes them unsuitable for use as foaming agents, given the existence of cheaper alternatives. It was invented under the patent that the use of low levels of a saponin (maximum 2%, by weight) gives a synergistic relationship between the saponins and surfactants present in the formulations, which provides a surprising and significant reduction in the amount of surfactants needed for effective fire-fighting performance. A considerable improvement in the heat resistance of the foam formulations is also observed.

Due to the foaming abilities saponins may also be used as emulsifiers helping in degradation of xenobiotics like polycyclic aromatic hydrocarbons (PAHs). The increased solubility of PAH in the presence of saponins make them easily available for degrading bacteria [13]. An aqueous preparation containing vitamin E, prepared by emulsifying or solubilizing vitamin E in an aqueous phase in the presence of a saponin, shows excellent transparency and thermal stability, and can be widely used in the fields of medicines, cosmetics, foodstuffs and animal nutrition [14]. The tea saponin paraffin emulsifier (TS-80 emulsifier) has been widely used in the building board industry. Because of its small oil droplet size, level of degree and good stability, the emulsion's character is much better than general emulsifier such as: alkyl sodium sulfonate, sodium oleate and ammonium oleate.

The addition of a small amount of a saponin in an aqueous enzyme-based composition containing a bacterial inhibiting stabilizer provides a product that is an effective water clarifier and solid surface cleanser. These compositions may be used to clean metals and metal-plated surfaces, such as stainless steel and chrome plating, plastics, plastic composites, ceramics, painted surfaces, wood, glass, textiles, carpeting, animal hair and skin, and the like, to remove a wide variety of food, animal and cosmetic induced stains, dirt and grime, oil, grease and the like. The composition provides superior deodorizing capabilities of such soiled surfaces [15].

The surfactant activity of saponins also finds some application in animal production. Feed grains such as barley, wheat and oats contain nonstarch polysaccharides (NSPs) such as α-glucans, which are viscous gums that are poorly water soluble. They cause a 'plugging-up' of the intestinal mucosa in poultry because of their high viscosity. Saponins via their surfactant activity might be effective in improving the water solubility of NSP and in consequence the feeding value of barley, wheat and oats for poultry. However, this concept needs further study. Saponin-based surfactants may also influence starch characteristics and ruminal dry matter and starch degradability of steam-flaked grain. The mechanism by which the surfactant enhanced the degradability was not known. Ageing of the hot flakes results in a quadratic decrease in dry matter and starch ruminal degradability. The ageing process affects starch gelatinization enthalpy values of flaked grain in a manner opposite to that observed for ruminal dry matter and starch degradation. This phenomenon is most likely explained by increased starch intramolecular associations or crystallinity associated with starch annealing, or both. Because the rate of degradation was not affected by the surfactant, the increase in degradability was attributed mainly to increases in dry matter and starch solubility [16].

Acknowledgements

This work was financed in part by the European Community (FP6-2005-FOOD-4-C) project Feed-Seg, Contract No. 043077.

References

1. Oleszek, W. (2000) Saponins, in *Natural Food Antimicrobial Systems* (ed. A. S. Naidu), CRC Press, LLC, pp. 295–324.
2. Oleszek, W. (2002) Chromatographic determination of plant saponins – review. *J. Chromatogr. A*, **967** (1), 147–162.
3. Rao, A. V. and Gurfinkel, D. M. (2000) Dietary saponins and human health, in *Saponins in Food Feedstuffs and Medicinal Plants* (eds W. Oleszek and A. Marston), Kluwer Academic Publishers, Dordrecht, pp. 255–270.
4. Oakenful, D. and Sidhu, G. S. (1989) Saponins, in *Toxicants of Plant Origin* (ed. R. R. Cheeke), CRC Press, Boca Raton, Florida, pp. 97–142.
5. Oakenful, D. (1986) Aggregation of saponins and bile acids in aqueous solution. *Aust. J. Chem.*, **39**, 1671–1683.

6. Gohtani, S., Murakami, A. and Yamano, Y. (1996) Physico-chemical properties of mon-odesmoside saponins of Sapindaceae (*Sapindus mukurossi* Gaertn.) at air/water and oil/water interfaces. *Food Sci. Technol. Int.*, **2**, 34–37.
7. Johnson, I. T., Gee, J. M., Price, K. *et al.* (1986) Influence of saponins on gut permeability and active nutrient transport *in vivo*. *J. Nutr.*, **116**, 2270–2277.
8. Hargreaves, T. (2003) Surfactants: the ubiquitous amphiphiles. *Chem. Br.*, **39** (7), 38–41.
9. Olmstead, M. J. (1997) Organic toothpaste containing saponin, US Patent 6485711.
10. Tanaka, O., Tamura, Y., Masuda, H. and Mizutani, K. (1996) Application of saponins in foods and cosmetics: saponins from Mohave yucca and *Sapindus mukurossi*, in *Saponins Used in Food and Agriculture* (eds G. R. Waller and K. Yamasaki), Plenum Press, New York, pp. 1–11.
11. Kim, Y.-D., Kim, C.-K., Lee, C.-N. and Ha, B.-J. (1988) Hydrolysed ginseng-saponin quaternary: a novel conditioning agent for hair care products, in XVth Congress of the International Federation of the Societies of Cosmetic Chemists, 26–29 September, 1988, London.
12. Meyer, D. J. (2000) Fire fighting foams utilizing saponins, US Patent 6051154.
13. Soeder, C. J., Papaderos, A., Kleespies, M. *et al.* (1996) Influence of phytocenic surfactants (*Quillaja* saponin and soya lecithin) on bioelimination of fluoranthrene and fluoranthrene by three bacteria. *Appl. Microbiol. Biotechnol.*, **44**, 654–659.
14. Shirakawa, Y., Itoh, M., Koyama, K. and Minowa, Y. (2001) Aqueous preparation containing vitamin E and saponins, US Patent 4568667.
15. Kulperger, R. J. (1996) Enzymatic solutions containing saponins and stabilizers, US Patent 5503766.
16. Hristov, A. N., Zaman, S., VanderPol, M. *et al.* (2007) Effect of a saponin-based surfactant and aging time on ruminal degradability of flakedcorn grain dry matter and starch. *J. Anim. Sci.*, **85**, 1459–1466.

Part 5

Polymeric Surfactants/Surface-Active Polymers

13
Surface-Active Polymers from Cellulose

Leif Karlson
AkzoNobel Functional Chemicals, Stenungsund, Sweden

13.1 Introduction

Water-soluble cellulose ethers (cellulose derivatives, or CDs) have found many applications. The major Application areas for CDs are [1–3]:

- **Water-borne paint**. Mainly nonionic cellulose ethers are used as rheology modifiers for water-borne paint. The rheology control of the paint influences such properties as paint consistency, brush load, levelling, sagging and hiding power. Besides the thickening, the polymer takes an active part in the particle stabilization in the paint. This is by far the most important application for a hydrophobically modified cellulose derivative (HM-CD).
- **Construction**. Many cement- and gypsum-based adhesives, mortars, fillers joint compound, and so on, contain cellulose ethers. The main contribution from the polymer is increased water retention. This means that the polymer helps to prevent water from escaping from the formulation and into the underlying material, which is often porous. The water content in the formulation during the hardening process of the cement or gypsum is important for the strength. The polymer also contributes with improved consistency and sag resistance, increased open time and adjustability.
- **Pharmacutical formulations**. Carboxy methyl cellulose (CMC), hydroxypropyl cellulose (HPC) and methyl hydroxypropyl cellulose (MHPC) are commonly used ingredients in tablet formulations of pharmaceuticals. They can either be used as one

Surfactants from Renewable Resources Edited by Mikael Kjellin and Ingegärd Johansson

component of the excipient or as a coating layer on the outside of the tablet. During the dissolution process of the tablet the polymer forms a gel, resulting in an extended release of the active drug.
- **Oilfield**. In oilfield applications CMC and hydroxyethyl cellulose (HEC) are used as rheology modifiers for drilling fluids. The main contribution is a reduction of fluid loss but they can also make an important contribution to the shale inhibitive properties. High viscosity of the drilling fluid helps the transport of drilled solids and cuttings to the surface. HEC is also present in products for oil well cementing.
- **Food**. The ability of cellulose ethers to retain moisture evaporation is used for the extended shelf life of bread and other baked goods. Methyl cellulose (MC) forms strong gel structures that trap volume-enhancing gases and retain moisture. CDs are used in ice cream to give the right rheology but also to prevent ice crystal growth. Diary products as well as sauces and soups can contain CD for rheology control and stabilization of fat and protein.
- **Cleaning**. Cellulose ethers are used as rheology modifiers in many household products, e.g. in liquid detergents and cleaners. They also contribute to stabilization and binding.
- **Personal care**. CDs are added to many personal care products like shampoos, hair conditioners and skin care products as thickeners. The ability to stabilize suspensions and emulsions is also utilized for this application. The film-forming properties of CDs are important for hair conditioning products.

In 2007 the total worldwide capacity for nonionic CDs was 363 000 tonnes per year. The annual capacity for technical CMC was 192 000 and for purified CMC 271 000 tonnes [4].

13.2 Structure and Synthesis of Cellulose Ether

Cellulose ether is a generic term for polymers synthesized via an etherification reaction on cellulose. CMC, HEC, HPC and ethyl hydroxyethyl cellulose (EHEC) are some examples of cellulose ethers. Cellulose is a polysaccharide built up from 1,4-anhydroglucose (AHG) units. The cellulose molecules in native cellulose form large crystalline regions, and therefore cellulose is insoluble in water. To make cellulose soluble it has to be modified to split the crystalline packing. The process for making cellulose derivatives starts with an alkalization step (Figure 13.1). The alkalization has two purposes. Firstly, by introducing charges into the molecules the cellulose swells. This makes individual cellulose chains available for the chemical reaction. Secondly, it also acts as catalysation for the modification reactions.

All the common reagents for the etherification reaction of cellulose are either epoxides or halides. Ethylene oxide (EO), propylene oxide (PO) and alkyl glycidyl ethers are examples of epoxides and monochloroacetic acid (MCA), methyl chloride (MC), ethyl chloride (EC) and long-chain alkyl bromides are examples of halides that are commonly used. The reactions are performed at elevated temperature. Reactions of volatile compounds such as EO, PO, MC and EC require a pressurized reaction vessel.

Each AHG has three hydroxyl groups available for reaction. The reaction of one EO or PO molecule to one of the hydroxyl groups on an AHG results in a new hydroxyl group

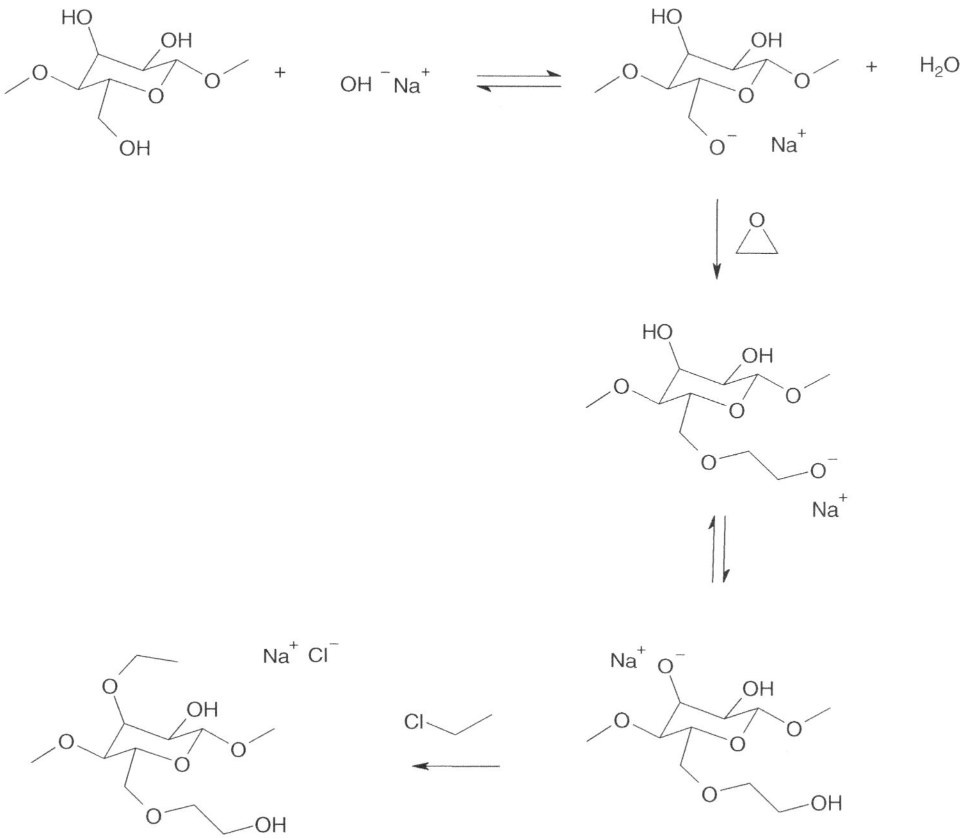

Figure 13.1 *The reaction scheme for production of ethyl hydroxyethyl cellulose (EHEC), which includes the reaction steps alkalization, ethoxylation and alkylation with ethyl chloride.*

that is also reactive (Figure 13.1). The reactivity of the newly formed hydroxyl group is comparable to that of the hydroxyl groups on the AHG, which means that besides the reaction of the hydroxyl groups on the AHG there is also a chain growth reaction going on. The outcome is that short oligo (EO or PO) chains are formed [5]. The molar substitution of EO (MS_{EO}) or PO (MS_{PO}) is the average total number of EO or PO groups, respectively, per AHG (Figure 13.2). For practical reasons the upper limit for MS_{EO} is about 2.5–3 since the efficiency of the reaction decreases dramatically above that level due to side reactions. Glycols, ethers and salt are the main by-products.

In contrast to the reaction with EO or PO where new hydroxyl groups form, the reaction with MCA, MC or EC consumes sodium hydroxide, and the hydroxyl group that has reacted with any of these reagents is terminated for any further reaction (Figure 13.1). The number of hydroxyl groups per AHG that has reacted is expressed as the degree of substitution (DS) and the figure ranges from 0 to 3. Practically the upper limit for DS_{ethyl} is about 1 since the water solubility of the final EHEC polymer decreases dramatically with increasing DS_{ethyl} [6]. The effect on water solubility of methyl groups is much

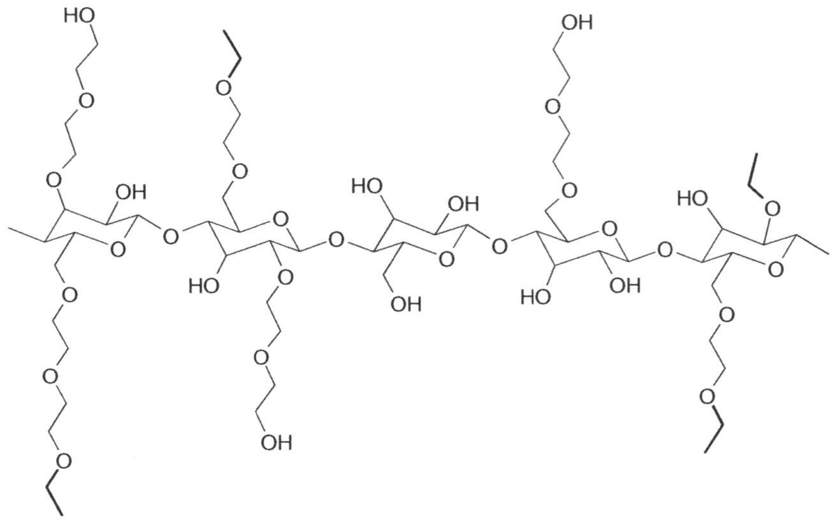

Figure 13.2 Possible structure element of an EHEC molecule, where ⁄⁀O represents a hydroxyethyl group and ethyl groups are represented by bold lines. In this example $MS_{EO} = (4 + 3 + 0 + 2 + 1)/5 = 2$ and $DS_{ethyl} = (1 + 1 + 0 + 0 + 2)/5 = 0.8$.

less pronounced and MC with $DS_{methyl} = 1.8–2$ is still soluble in water. Of course the reaction does not give a perfectly homogeneous substituent distribution over all AHGs. It is likely that the synthesis process for cellulose ethers gives an uneven distribution of the substituents. Therefore the numbers of DS and MS are average values.

Segments of AHG units that have a high degree of methyl or ethyl substituents (or other short hydrophobic groups $< C_{12}$) are slightly hydrophobic. In water solution the ethyl groups can give rise to hydrophobic interactions provided that they are situated in long sequences. This is the origin of the backbone associations and the reason why, for example, methyl hydroxyethyl cellulose (MHEC), EHEC and MHPC are surface active and show an associative behaviour [7–13]. An AHG unit bearing hydrophobic groups can be seen as a hydrophobic monomer unit of a copolymer. The cellulose backbone is relatively stiff and the associations from the short hydrophobic groups are too weak to force the polymer backbone to bend into a loop where the hydrophobic groups could intraaggregate. Instead, the result is interassociations between hydrophobic segments on different polymer chains, which can be detected as increased solution viscosity [14–16].

By reacting aliphatic groups, with a chain length equal to or longer than C_{12}, to the cellulose ether an HM-CD is obtained (Figure 13.3). The HM-CDs obtained in this way are examples of comb-like hydrophobically modifiedpolymers (HM-Ps). They have hydrophobic groups grafted along the water-soluble polymer backbone. Only a small amount of hydrophobic groups is required to totally change the properties of the polymer [14, 15]. Less than 1% of the glucose units of the cellulose ether backbone need to carry hydrophobic groups in order to change substantially the solution properties as compared to those of the corresponding unmodified CD (Figure 13.3). Hydrophobically modified

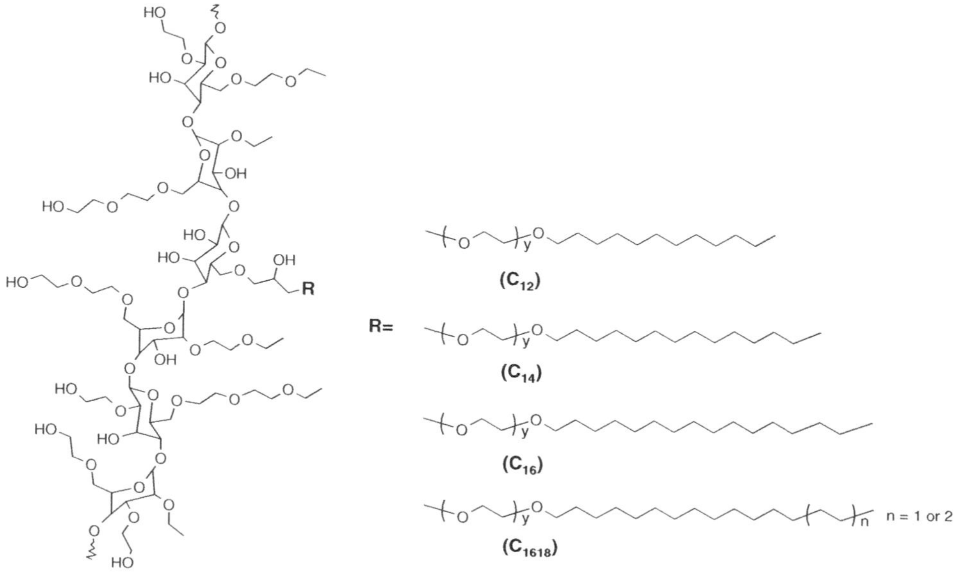

Figure 13.3 *Possible structure segment of the HM-EHECs studied in Reference [12]. R = (C_{12}) for HM-(C_{12})-EHEC, R = (C_{14}) for HM-(C_{14})-EHEC, R = (C_{16}) for HM-(C_{16})-EHEC and R = a blend of (C_{16}) and (C_{18}) for HM-($C_{16,18}$)-EHEC.*

hydroxyethyl cellulose (HM-HEC) and hydrophobically modified ethyl hydroxyethyl cellulose (HM-EHEC) are examples of commercially available HM-CDs.

Cellulose ethers can be found with both technical or purified qualities. Technical CMC contains about 65% CD. The remaining 35% consists mainly of salt resulting from the etherification reaction. The purification of CDs is done by extraction of the technical products. The products with a cloud point are often extracted by hot water whereas alcohol or alcohol/water mixtures are used for the CDs that are readily soluble in hot water. For most applications CD of about 90% purity is used. The main components of the remaining 10% are glycols that are formed as by-products during the reaction and salt. The salt originates either from the etherification reaction, neutralization of unconsumed alkali with acid or from salt that is added to the water during the washing procedure in order to reduce the solubility of the CD. Special grades of CMC, hydroxypropylmethylcellulose (HPMC), MC and HEC of >99% purity are available for food and pharmaceutical applications.

13.3 Cellulose Ethers in Aqueous Solution

The behaviour of polymer molecules in solution depends to a large extent on the polymer concentration, c. Most applications where CDs are used are in the semidilute region. In drug delivery systems for pharmaceutical applications the polymer is likely to be in the highly concentrated regime where the polymers form a gel.

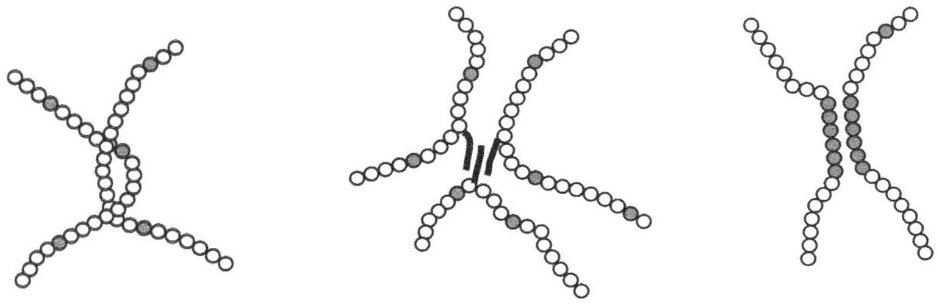

Figure 13.4 *Schematic illustrations of the different types of interpolymeric cross-links in HM-EHEC solution: (a) chain entanglements, (b) associations between hydrophobic side chains and (c) associations between hydrophobic segments of the polymer backbone. Unfilled rings represent hydrophilic glucose units, filled rings represent hydrophobic glucose units and bold lines represent hydrophobic side chains.*

In the semidilute regime the polymer concentration is above the overlap concentration (c^*) and the chain of one polymer molecule will become entangled with other polymer molecules (Figure 13.4), resulting in a transient polymer network, which can be detected by a dramatic increase in the viscosity of the polymer solution. The overlap concentration can roughly be estimated as the reciprocal of the intrinsic viscosity, $c^* \approx 1/(\eta)$, and is for most polymers in the region of 0.1–10% w/w. The chemical structure of the polymer is very important for the coil size and thereby for the behaviour of the polymer in solution. An increased molecular weight (MW) results in larger coils and more chain entanglements, which can be seen as increased viscosity [17].

When describing the behaviour of hydrophobically modified polymers it is important to notice that contrary to the unmodified analogue the HM-P molecule also has the possibility of associating with other HM-P molecules. The association of the hydrophobic groups is very similar to the self-association of surfactants. To minimize the contact between water and hydrophobic groups the hydrophobic groups associate with each other and form a water-poor domain, which is the interior of a micelle. The surface of the micelle is covered by the hydrophilic polymer backbone. In aqueous solution the hydrophobic groups of a hydrophobically modified polymer associate with each other, resulting in physical bonds that hold different parts of the polymer chains together (Figure 13.4). In a snapshot picture it can be described as a cross-linked gel, but in contrast to covalent bonds the physical bonds are reversible. They break and reform continuously. A hydrophobic group on one polymer molecule can either take part of an intramolecular association, that is it interacts with another hydrophobic group on the same polymer chain, or interacts with a hydrophobic group on another polymer molecule (intermolecular association) (Figure 13.5). At low concentrations the probability for interaction between different HM-P molecules is small. Upon increasing polymer concentration intermolecular associations become more important and the three-dimensional network is formed. This gives rise to a dramatic increase in the solution viscosity. The onset concentration of intermolecular association is often well below the overlap concentration, c^*, of the corresponding unmodified polymer with the same MW [18, 19].

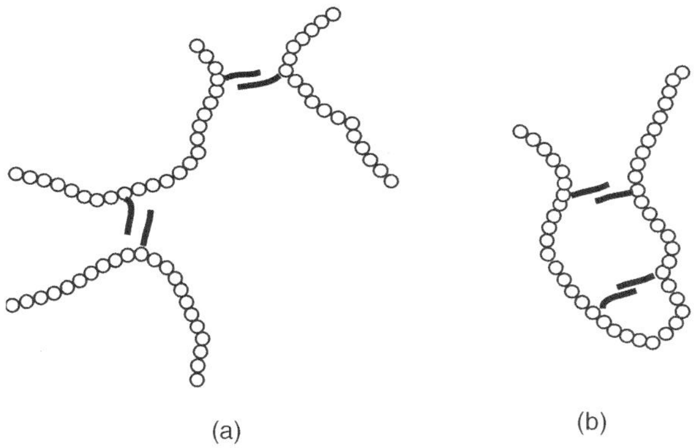

Figure 13.5 *Schematic illustrations of (a) intermolecular associations and (b) intramolecular associations.*

The strength of the hydrophobic interactions between polymer chains is influenced by the length of the hydrophobic groups and the molar substitution of hydrophobic groups ($MS_{hydrophobe}$).

It has been shown in several studies that the length of hydrophobic groups has a dramatic effect on the low shear viscosity of aqueous solutions of HM-CDs [12, 20]. The hydrophobic groups should have at least 12 carbon atoms to have any noticeable effect on the viscosity. By increasing the length of the hydrophobic chains from C_{12} to C_{16} the viscosity increased by two orders of magnitude (Figure 13.6). This effect is ascribed to the residence time of the hydrophobic chains in the 'polymer micelles', which increases for longer hydrophobic groups and results in slower motions of the polymer molecules and thereby a higher viscosity [21, 22].

The influence of $MS_{hydrophobe}$ on the solution viscosity can be divided into three different regions. At low $MS_{hydrophobe}$ values there is a positive correlation between $MS_{hydrophobe}$ and viscosity. This can be explained by an increased number of interconnection points holding the polymer network together. Depending on the structure of the

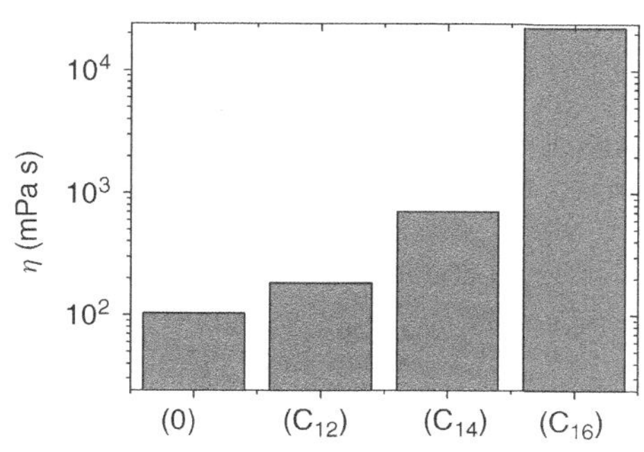

Figure 13.6 *Zero shear viscosity (η) of 1% w/w solutions of HM-EHEC with varying lengths of the hydrophobic groups, where (0) represents unmodified EHEC.*

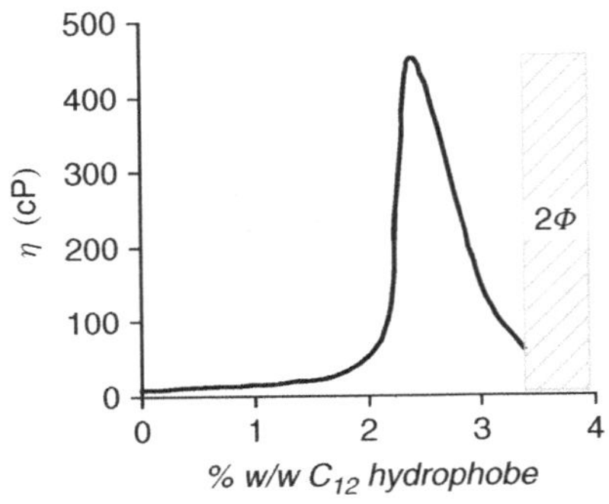

Figure 13.7 Brookfield viscosity of 2% w/w solution of HM-HEC substituted with 1,2-epoxydodecane as a function of degree of hydrophobic modification.
Reproduced from *J. Pol. Sci.*, 1982, **20**, 443–455.

polymer backbone and the length of the hydrophobic groups there is a viscosity maximum somewhere typically in the range of 1–5 hydrophobic groups per 100 repeating units of the polymer backbone if a comb-like polymer is investigated (Figure 13.7) [14]. The reason for the decrease is a conversion of intermolecular associations to intramolecular association and a gradual degradation of the polymer network [23, 24]. At even higher $MS_{hydrophobe}$ the HM-P becomes insoluble in water.

For grafted HM-P with low $MS_{hydrophobe}$ values, like HM-HEC and HM-EHEC, the average number of hydrophobic groups per micelle (N_R) is low. The low aggregation number is likely to result from the polymer chain being a very large head group. The relatively stiff backbone from cellulose ether prevents formation of loops and the consequence is that only a small number of hydrophobic groups can take part in the formation of each micelle. N_R for HM(NP)-EHEC (HM-EHEC modified with nonylphenol (NP) groups) and HM-HEC micelles have been determined to be about 5–10 [7, 22] compared to 60–80 for surfactants forming spherical micelles [17]. The consequence is that rather poor micellar structures are formed with a high degree of contact between water and hydrophobic groups.

Shorter hydrophobic groups can give rise to hydrophobic associations if they are situated in an uneven way along the polymer backbone. Segments of the backbone with high density of short-chain hydrophobic groups can associate with each other. Small hydrophobic groups such as methyl, ethyl and hydroxypropyl groups can give rise to these effects.

From what has been discussed above, it follows that there are at least three types of interpolymer cross-links that contribute to the formation of the three-dimensional network of an HM-CD solution (Figure 13.4). Apart from chain entanglements and associations between hydrophobic side groups, associations of hydrophobic segments of the polymer backbone also play an important role even if the contribution from hydrophobic backbone interactions is relatively small compared with the other two.

The rheology profiles of cellulose ethers differ depending on MW and type of substituents. When comparing two cellulose ethers with the same substitution pattern but with different MWs the shear rate dependence of the viscosity is much more pronounced

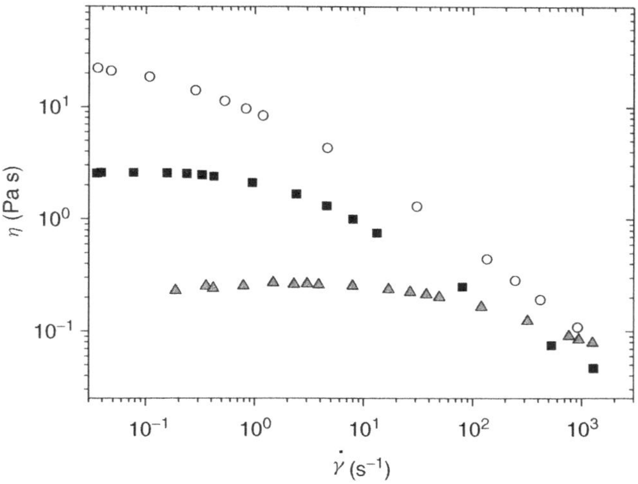

Figure 13.8 *Viscosity (η) as a function of shear rate (γ̇) for solutions of high MW EHEC 1% in water (○), low MW EHEC 2% in water (▲) and HM-EHEC 1% in water (■).*

for the high MW CD (Figure 13.8). As described above, the thickening effect of an HM-CD depends both on chain entanglements and hydrophobic interactions. The result is that, even though the MW of the HM-CD is much lower than for the corresponding unmodified CD, the zero shear viscosity of the solution of HM-CD can be as high as or even higher than the viscosity of the solution of the unmodified CD.

Water-soluble polymers based on cellulose can be more or less surface-active depending on the substituents. In aqueous solutions cellulose ethers that only contain hydrophilic substituents like CMC and HEC show hardly any surface activity at all, as can be seen from the surface tension figures of Table 13.1. The surface tension of solutions of cellulose ethers containing hydrophobic substituents like MC, HPC and combinations of hydrophilic and hydrophobic substituents like EHEC and MHEC show considerably lower figures for surface tension. As exemplified by HM-HEC and HM-EHEC, addition of a small amount of long-chain alkyl groups to a cellulose ether only give rise to a small change in surface tension compared to their unmodified analogues.

Table 13.1 *Surface tension of 1% solutions of different cellulose ethers in water*

Polymer	Surface tension (mN m)	Reference
CMC	71	[25]
HEC	67	[25]
HPC	44	[25]
MHEC	51	[26]
MC	52	[26]
HM-HEC	62	[27]
MHPC	50	[27]
EHEC	52	
HM-EHEC	51	

13.4 Interaction with Surfactants

Several studies have investigated the interaction between different types of surfactants and cellulose ethers (CDs). The relatively hydrophilic HEC interacts with the anionic surfactants sodium dodecyl sulfate (SDS) and sodium octyl benzene sulfate (SOBS) but not with cationic surfactants [29]. In contrast both anionic and cationic surfactants bind to the relatively more hydrophobic polymer EHEC [30–32]. The binding of ionic surfactants leads to an electrostatic repulsion between the polymer molecules. For the EHEC system there is a viscosity increase at low surfactant additions due to the increased strength of the hydrophobic interactions. With increasing surfactant concentration the electrostatic repulsion becomes increasingly more important, resulting in a breakdown of the polymer network of EHEC held together by weak interactions from hydrophobic segments of the polymer backbone [33]. This can be detected as a decrease in the viscosity to a level that is even below the viscosity of the EHEC solution without any surfactant added. For ionic surfactants the binding to hydrophilic nonionic CDs, e.g. HEC, has been attributed to screening of the repulsive forces between the charged head groups of the surfactant molecules in the micelle. The hydrophilic polymers interact mainly with the hydrophilic corona of the micelles. With increasing hydrophobicity of the polymer the interactions with the hydrophobic core becomes increasingly more favourable [22, 29, 34].

In contrast to their unmodified analogues, HM-CDs interact with all major types of surfactants. The extent of binding is larger for the surfactants that bind both to the polymer backbone and the hydrophobic side chains. It is well known that, depending on the concentration, addition of surfactant can either increase or decrease viscosity of a solution of an HM polymer [35–49]. At high surfactant concentrations associations between hydrophobic parts of the polymer chains are disrupted.

In aqueous solution HM-CDs interact strongly with surfactants, leading to the formation of mixed micelles. It has been found that the viscosity of the polymer solution passes via a pronounced maximum when the surfactant concentration is gradually increased (Figure 13.9). The degree of interaction is determined both by the structure of the surfactant and the nature of the polymer. As described above, the micellar structure of the HM-CD normally has low aggregation numbers compared to surfactant micelles and the consequence is a quite large degree of contact between water and the hydrophobic groups. At low surfactant concentrations, already far below the critical micellization concentration of the surfactant, the surfactant molecules are incorporated into the existing micelles from the HM-CD. Incorporation of surfactant molecules into the micelles reduces the water hydrocarbon contact. This increases the activation energy for detachment of a hydrophobic group from the micelle, thereby increasing the residence time of the hydrophobic groups within the micelles and thus leading to stronger associations [7, 42]. For HM-CDs it is suggested that the effect of increased viscosity upon addition of a surfactant is caused mainly by increased relaxation times [7, 22, 47].

At surfactant concentrations above the viscosity maximum the number of micelles in the solution increases. This results in an increased ratio between micelles and hydrophobic groups of the polymer. In this process the decreased viscosity is a consequence of the physical network losing some of its connectivity. At high surfactant concentrations

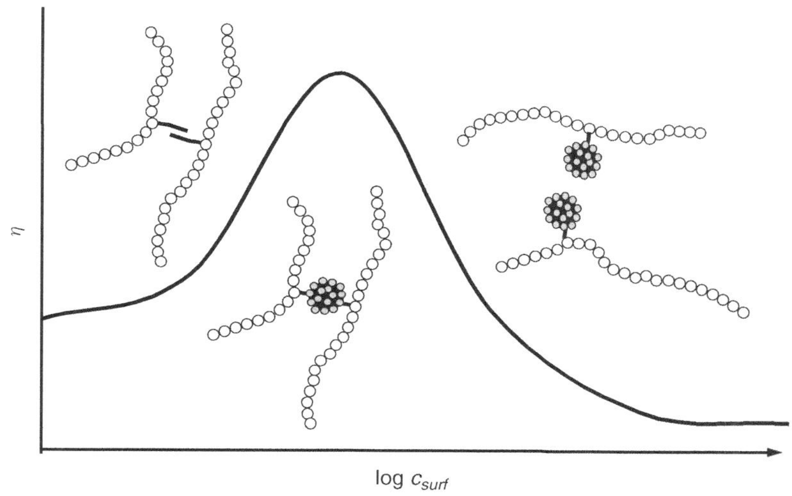

Figure 13.9 *Schematic illustration of the influence of surfactant concentration on the viscosity of solutions of HM polymers.*

where the number of micelles exceeds the number of polymer hydrophobic groups in the system there is only one polymer hydrophobic group in each micelle. At this stage the viscosity is independent of the surfactant concentration and has a value that is even lower than for the HM-CD solution before addition of the surfactant. How strong the effect is depends on the structure of the surfactant. Normally nonionic polymers interact more strongly with anionic surfactants than with nonionic or cationic surfactants. In line with this, it has been found that anionic surfactants give the most pronounced viscosity increase and also the largest reduction of the viscosity at excess surfactant [50].

13.5 Clouding

For most substances the solubility increases with increasing temperature. This is not the case for some cellulose ethers. They belong to a family of substances that have a reversed relationship between solubility and temperature [9–12, 51–59]. The solubility of these substances decreases with increasing temperature. HPC, HPMC, MC and EHEC all have this type of behaviour. At temperatures above a critical value a water solution containing any of these polymers phase separates into one polymer-rich phase and one phase depleted in polymer. The phase separation can be detected by the scattering of light, resulting in a cloudy appearance of the solution. The temperature where the solution first becomes hazy is referred to as the cloud point temperature, T_{Cp}. The process is reversible and decreasing the temperature below T_{Cp} results in a one-phase situation and a transparent solution.

T_{Cp} is correlated to figures for MS_{EO}, MS_{PO}, $\text{DS}_{\text{methyl}}$ and DS_{ethyl}. T_{Cp} increases with increasing MS_{EO} and decreases with increasing $\text{DS}_{\text{methyl}}$ and DS_{ethyl} [6]. The cloud point is dramatically influenced by the introduction of hydrophobic groups on the polymer. As

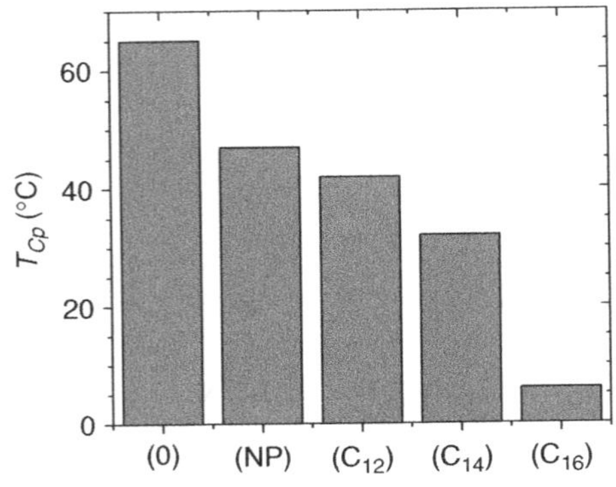

Figure 13.10 *T_{Cp} of 1% w/w solutions of HM-EHEC with varying lengths of the hydrophobic groups, where (0) represents unmodified EHEC.*

can be seen in Figure 13.10, T_{Cp} is shifted dramatically when an EHEC is modified with hydrophobic groups. The longer the hydrophobic group, the more pronounced is the shift in T_{Cp}. The large difference in T_{Cp} between the polymers indicates that the strength of the hydrophobic association is much larger for longer hydrophobic groups. As discussed above, the effect on the solution viscosity of the polymers also reveals large variations in the strength of the associations [55].

Surfactants strongly influence the phase separation temperature of cellulose derivative solutions. Depending on the surfactant concentration, c_{surf}, and type of surfactant, addition of surfactants can either increase or decrease T_{Cp}. Upon progressively increasing the surfactant concentration, c_{surf}, of the ionic surfactant $C_{12}SO_4Na$ (SDS), T_{Cp} is found to decrease initially (Figure 13.11). At slightly higher c_{surf} the T_{Cp} passes through a minimum and at even higher c_{surf}, T_{Cp} increases. The trend is similar for addition of other micelle-forming surfactants provided that the surfactant molecules associate with the polymer. If there is no association between the polymer and surfactant the result can be a segregative phase separation with the polymer enriched in one phase and the surfactant in the other phase.

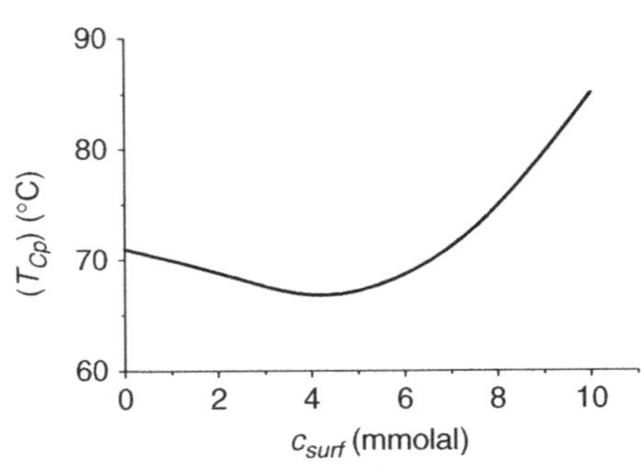

Figure 13.11 *T_{Cp} as a function of sodium dodecyl sulfate concentration for 0.9% w/w solution of EHEC.*
Reproduced from *J. Phys. Chem.*, 1990, **94** (12), 5005–5015.

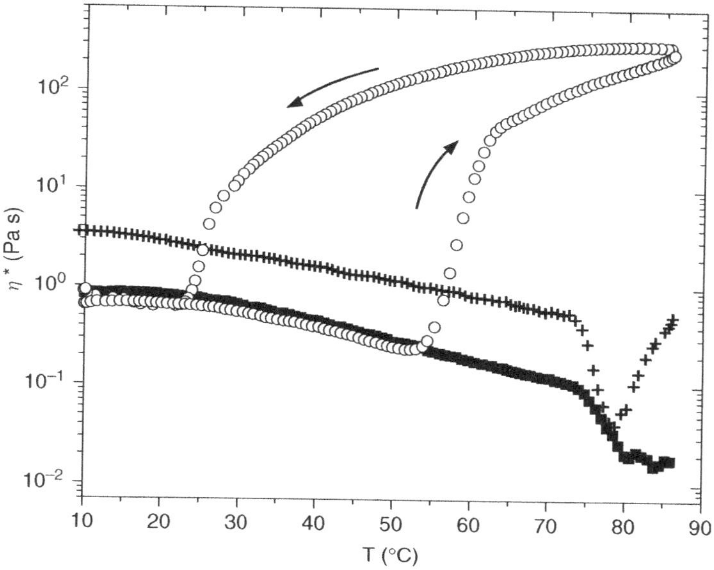

Figure 13.12 *Complex viscosity (η*) as a function of temperature for 1% w/w polymer solutions of (+) Bermocoll E481FQ (EHEC), (○) Methocel A16M (MC) and (■) Methocel K15M (HPMC). The heating and cooling rates are 1.5 °/min.*

The viscosity above the cloud point of aqueous solutions of different CDs depends a lot on the type of substituents on the polymer. When the phase separation occurs for a solution of EHEC or HPMC this results in dramatic reduction of the solution viscosity (Figure 13.12). In contrast a solution of MC shows a huge increase in viscosity above T_{Cp}. As can be seen in Figure 13.12, the viscosity increases by three orders of magnitude and the result is a stiff gel. The viscosity upon heating and cooling of the MC solution shows a well-pronounced hysteresis loop. The size of the loop depends on the heating and cooling rates, indicating that the system is not in equilibrium and that the relaxation processes are very slow. Several mechanisms for the gel formation have been proposed and are still being debated [9–11, 51].

References

1. Ashland Aqualon Functional Ingredients, http://www.herc.com/aqualon/.
2. Dow Products and Services, http://www.dow.com/products_services/.
3. AkzoNobel Cellulosic Specialities, www.akzonobel.com/cs.
4. Will, R., Löchner, U. and Yokese, K. (2007) *CEH Marketing Research Report: Cellulose Ethers*, SRI Consulting.
5. Brandt, L. (2002) Cellulose ethers – cellulose ethers, *Ullmann's Encyclopedia of Industrial Chemistry*, Wiley-VCH Verlag GmbH, Weinheim.
6. Thuresson, K., Karlström, G. and Lindman, B. (1995) Phase diagrams of mixtures of a non-ionic polymer, hexanol, and water. An experimental and theoretical study of the effect of hydrophobic modification. *J. Phys. Chem.*, **99** (11), 3823–3831.

7. Thuresson, K., Soderman, O., Hansson, P. and Wang, G. (1996) Binding of SDS to ethyl(hydroxyethyl)cellulose. Effect of hydrophobic modification of the polymer. *J. Phys. Chem.*, **100** (12), 4909–4918.
8. Cabane, B., Lindell, K., Engström, S. and Lindman, B. (1996) Microphase separation in polymer + surfactant systems. *Macromolecules*, **29** (9), 3188–3197.
9. Kobayashi, K., Huang, C. and Lodge, T. P. (1999) Thermoreversible gelation of aqueous methylcellulose solutions. *Macromolecules*, **32** (21), 7070–7077.
10. Li, L., Wang, Q. and Xu, Y. (2003) Thermoreversible association and gelation of methylcellulose in aqueous solutions. *Nihon Reorji Gakkaishi*, **31** (5), 287–296.
11. Desbrières, J., Hirrien, M. and Ross-Murphy, S. B. (2000) Thermogelation of methylcellulose: rheological considerations. *Polymer*, **41**, 2451–2461.
12. Karlson, L., Joabsson, F. and Thuresson, K. (2000) Phase behavior and rheology in water and in model paint formulations thickened with HM-EHEC: influence of chemical structure and the distribution of hydrophobic tails. *Carbohydr. Polym.*, **41**, 25–35.
13. Karlson, L., Thuresson, K. and Lindman, B. (2002) A rheological investigation of the complex formation between hydrophobically modified ethyl (hydroxy ethyl) cellulose and cyclodextrin. *Carbohydr. Polym.*, **50** (3), 219–226.
14. Landoll, L. M. (1982) Nonionic polymer surfactants. *J. Polym. Sci.*, **20**, 443–455.
15. Wang, K. T., Iliopoulos, I. and Audebert, R. (1988) Viscometric behaviour of hydrophobically modified poly(sodium acrylate). *Polym. Bull.*, **20**, 577–582.
16. Hill, A., Candau, F. and Selb, J. (1991) Aqueous solution properties of hydrophobically associating copolymers. *Prog. Colloid Polym. Sci.*, **84**, 61–65.
17. Evans, D. F. and Wennerström, H. (1999) *The Colloidal Domain. Where Physics, Chemistry, Biology, and Technology Meet*, 2nd edn, Wiley-VCH Verlag GmbH.
18. Schaller, E. (1985) Rheology modifiers for water-borne paints. *Surf. Coat. Aust.*, **22** (10), 6–13.
19. Semenov, A. N., Joanny, J.-F. and Khokhlov, A. R. (1995) Associating polymers: equilibrium and linear viscoelasticity. *Macromolecules*, **28** (4), 1066–1075.
20. Sau, A. C. and Landoll, L. M. (1989) Synthesis and solution properties of hydrophobically modified (hydroxyethyl) cellulose, in *Polymers in Aqueous Media*, vol. 223 (ed. J. E. Glass), American Chemical Society, Washington, DC, pp. 343–364.
21. Annable, T., Buscall, R., Ettelaie, R. and Whittlestone, D. (1993) The rheology of solutions of associating polymers: comparison of experimental behavior with transient network theory. *J. Rheol.*, **37** (4), 695–726.
22. Piculell, L., Nilsson, S., Sjöström, J. and Thuresson, K. (2000) How much surfactant binds to an associating polymer? The HMHEC/SDS case revisited, in *Associative Polymers in Aqueous Media*, vol. 765 (ed. J. E. Glass), American Chemical Society, Washington, DC, pp. 317–335.
23. Volpert, E., Selb, J. and Candau, F. (1996) Influence of the hydrophobe structure on the composition, microstructure and rheology in associating polyacrylamides prepared by micellar copolymerization. *Macromolecules*, **29** (5), 1452–1463.
24. Tam, K. C., Farmer, M. L., Jenkins, R. D. and Bassett, D. R. (1998) Rheological properties of hydrophobically modified alkali-soluble polymers – effects of ethylene-oxide chain length. *J. Polym. Sci. B. Polym. Phys.*, **36**, 2275–2290.
25. Ashland Aqualon Functional Ingredients, http://www.herc.com/aqualon/product_data/brochures/250_49.pdf#lsf.
26. Ashland Aqualon Functional Ingredients, http://www.herc.com/aqualon/construction/index.html.
27. Product Data: NATROSOL® *PLUS* CS, Grade 330, HCMC, http://www.in-cosmetics.com/ExhibitorLibrary/108/4032-3.PDF.

28. METHOCEL cellulose ethers: High-performance Thickeners for Paint and Coatings, http://www.dow.com/PublishedLiterature/dh_003d/0901b8038003df8a.pdf?filepath=meth ocel/pdfs/noreg/192-01098.pdf&fromPage=GetDoc
29. Sjöström, J. and Piculell, L. (2001) Simple gel swelling experiments distinguish between associating and nonassociating polymer-surfactant pairs. *Langmuir*, **17** (13), 3836–3843.
30. Carlsson, A., Karlström, G. and Lindman, B. (1986) Synergistic surfactant-electrolyte effect in polymer solutions. *Langmuir*, **2** (4), 536–537.
31. Rosén, O., Sjöström, J. and Piculell, L. (1998) Responsive polymer gels based on hydrophobically modified cellulose ethers and their interactions with ionic surfactants. *Langmuir*, **14** (20), 5795–5801.
32. Wang, G. and Olofsson, G. (1995) Ethyl (hydroxyethyl) cellulose and ionic surfactants in dilute solutions. Calorimetric and viscosity study of the interaction with SDS and some cationic surfactants. *J. Phys. Chem.*, **99** (15), 5588–5596.
33. Carlsson, A., Karlström, G., Lindman, B. and Stenberg, O. (1988) Interaction between ethyl (hydroxyethyl) cellulose and sodium dodecyl sulphate in aqueous solution. *Colloid Polym. Sci.*, **266** (11), 1031–1036.
34. Cabane, B. (1977) Structure of some polymer–detergent aggregates in water. *J. Phys. Chem.*, **81**, 1639–1645.
35. Aubry, T. and Moan, M. (1996) Influence of a nonionic surfactant on the rheology of a hydrophobically associating water soluble polymer. *J. Rheol.*, **40** (3), 441–448.
36. Chen, M. and Glass, J. E. (1995) The synthesis and solution rheology of branched water-soluble associative polymers. *Polym. Mater. Sci. Engng*, **73**, 449–450.
37. Chronakis, I. S. and Alexandridis, P. (2001) Rheological properties of oppositely charged polyelectrolyte–surfactant mixtures: effect of polymer molecular weight and surfactant architecture. *Macromolecules*, **34** (14), 5005–5018.
38. Iliopoulos, I., Wang, T. K. and Audebert, R. (1991) Viscometric evidence of interactions between hydrophobically modified poly(sodium acrylate) and dodecyl sulfate. *Langmuir*, **7** (4), 617–619.
39. Jiménez-Rigaldo, E., Selb, J. and Candau, F. (2000) Effects of surfactant on viscoelastic behavior of semidilute solutions of multisticker associating polyacrylamides. *Langmuir*, **16** (23), 8611–8621.
40. Loyen, K., Iliopoulos, I., Olsson, U. and Audebert, R. (1995) Association between hydrophobic polyelectrolytes and nonionic surfactants. Phase behavior and rheology. *Prog. Colloid Polym. Sci.*, **98**, 42–46.
41. Magny, B., Iliopoulos, I., Audebert, R. *et al.* (1992) Interactions between hydrophobically modified polymers and surfactants. *Prog. Colloid Polym. Sci.*, **89**, 118–121.
42. Nilsson, S., Thuresson, K., Hansson, P. and Lindman, B. (1998) Mixed solutions of surfactant and hydrophobically modified polymer. Controlling viscosity with micellar size. *J. Phys. Chem.*, **102** (37), 7099–7105.
43. Olesen, K. R., Bassett, D. R. and Wilkerson, C. L. (1998) Surfactant co-thickening in model associative polymers. *Prog. Org. Coat.*, **35**, 161–170.
44. Panmai, S., Prud'homme, R. K., Peiffer, D. G. *et al.* (1998) Rheology of solutions of hydrophobically modified polymers with spherical and rod-like surfactant micelles. *Polym. Mater. Sci. Engng*, **79**, 419–420.
45. Persson, K., Wang, G. and Olofsson, G. (1994) Self-diffusion, thermal effects and viscosity of monodisperse associative polymer: self-association and interaction with surfactants. *J. Chem. Soc. Faraday Trans.*, **90** (23), 3555–3562.
46. Piculell, L., Guillemet, F., Thuresson, K. *et al.* (1996) Binding of surfactants to hydrophobically modified polymers. *Adv. Colloid Interface Sci.*, **63**, 1–21.

47. Piculell, L., Thuresson, K. and Ericsson, O. (1995) Surfactant binding and micellisation in polymer solutions and gels: binding isotherms and their consequences. *Faraday Discuss.*, **101**, 307–318.
48. Steffenhagen, M. J., Xing, L.-L., Elliott, P. T. *et al.* (2001) The influence of HEUR architectures and surfactant types on viscosity build behavior. *Polym. Mater. Sci. Engng*, **85**, 217–218.
49. Williams, P. A., Meadows, J., Phillips, G. O. and Senan, C. (1990) Rheological characteristics of cellulosic and non-cellulosic associative thickeners. *Cellul. Sources Explor.*, **37**, 295–302.
50. Thuresson, K. and Lindman, B. (1997) Effect of hydrophobic modification of a nonionic cellulose derivative on the interaction with surfactants. Phase behaviour and association. *J. Phys. Chem. B*, **101**, 6460–6468.
51. Philip, K. and Ibbett, R. (1993) Thermogelation i cellulose ethers in *Cellulosics: Chemical, Biochemical and Material Aspects*, (eds Kennedy, J. F., Phillips, G. O. and Williams, P. A.), Ellis Horwood, New York, pp. 313–318.
52. Lee, S. C., Cho, Y. W. and Park, K. (2005) Control of thermogelation properties of hydrophobically-modified methyl cellulose. *J. Bioact. Compat. Polym.*, **20**, 5–13.
53. Hourdet, D., L'Alloret, F. and Audebert, R. (1997) Synthesis of thermoassociative copolymers. *Polymer*, **38**, 2535–2547.
54. Ford, J. L. (1999) Thermal analysis of hydroxypropy methyl cellulose and methyl cellulose: powders, gels and matrix tablets. *Int. J. Pharm.*, **179**, 209–228.
55. Thuresson, K. and Joabsson, F. (1999) Phase separation induced by dilution in a system of a hydrophobically modified polymer. *Colloids and Surf. A: Physicochem. Engng Aspects*, **151**, 513–523.
56. Karlström, G., Carlsson, A. and Lindman, B. (1990) Phase diagrams of nonionic polymer–water systems. Experimental and theoretical studies of the effects of surfactants and other cosolutes. *J. Phys. Chem.*, **94** (12), 5005–5015.
57. Thuresson, K., Nilsson, S. and Lindman, B. (1995) Effects on phase behaviour and viscosity of hydrophobically modification of a nonionic cellulose ether. Influence of cosolutes, in *Cellulose and Cellulose Derivatives: Physico-chemical Aspects and Industrial Applications* (eds J. F. Kennedy, G. O. Phillips, P. A. Williams and L. Piculell), Woodhead Publishing Ltd, Cambridge, pp. 323–329.
58. Thuresson, K. and Lindman, B. (1999) Association in nonionic cellulose ether solutions due to microcrystallites. *Colloids Surf. A: Physicochem. Engng Aspects*, **159** (1), 219–226.
59. Carlsson, A., Lindman, B., Nilsson, P. G. and Karlsson, G. (1986) Hydration of ethyl hydroxyethyl cellulose. *Polymer*, **27**, 431–436.

14

New Developments in the Commercial Utilization of Lignosulfonates

Rolf Andreas Lauten[1], Bernt O. Myrvold[1] and Stig Are Gundersen[2]

[1]*Borregaard Corporate R&D, LignoTech Group, Sarpsborg, Norway*
[2]*Borregaard LignoTech, Sarpsborg, Norway*

14.1 Introduction

Lignosulfonates or sulfonated lignins specify the complex anionic polymers that are recovered from liquors from the pulping industry. Over the years this product has evolved from being conceived as a waste material to become a refined chemical that constitutes a viable alternative to synthetic chemicals in many applications. A general overview of typical applications for lignosulfonates is given by Gargulak and Lebo [1] and their summary is still valid. Recently, a review [2] summarizing some advances on chemical characterization of lignosulfonates was published. In this chapter we will present recent advances in typical speciality applications within the agrochemical industry. Development of a lignosulfonate-based superplasticizer for the construction industry is also mentioned. We also discuss the production process of lignosulfonates and review literature on the fate of lignosulfonates released into the environment.

14.2 Lignosulfonates

Lignosulfonates are used in a wide range of different applications; common to most of them is that lignosulfonates are used as a dispersant. In suspensions, lignosulfonate

Surfactants from Renewable Resources Edited by Mikael Kjellin and Ingegärd Johansson

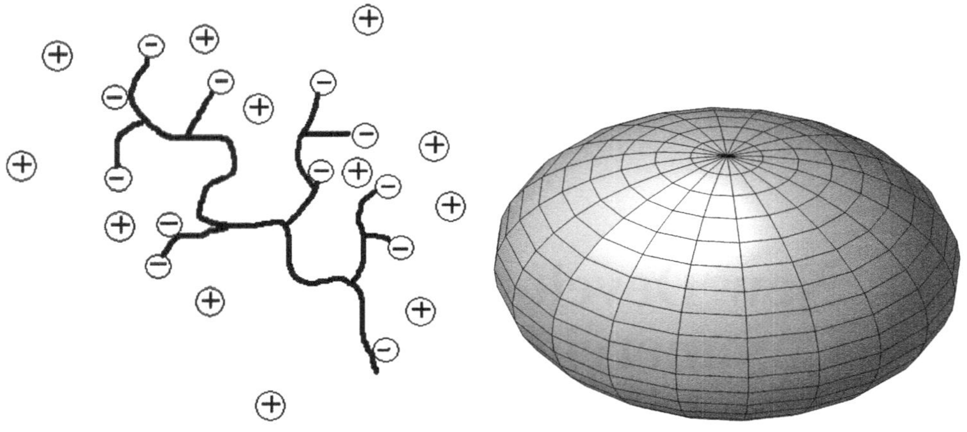

(a) (b)

Figure 14.1 A schematic figure illustrating a possible lignosulfonate structure. The longest chain fragment constitutes a lignosulfonate backbone and shorter chains are grafted to this backbone (a). According to Vainio et al. [5], the shape of a lignosulfonate molecule can be approximated by an ellipsoid (b), as shown to the right. The two representations are not to scale.

adsorbs onto an interface. In lignosulfonate-based emulsions lignosulfonate is enriched on the oil–water interface. The poor solubility in oil-based solvents requires more energy-intensive preparation methods than typical emulsions based upon synthetic emulsifiers. For both classes of dispersed systems it is of interest with a physicochemical understanding of lignosulfonate properties. The established idea of lignosulfonate structure is the microgel concept introduced by Goring [3]. Myrvold [4] has recently modified this model, suggesting that lignosulfonates are better described as randomly branched polyelectrolytes. This model is based upon a scaling analysis of how the intrinsic viscosity varies with molecular weight. One of the arguments for branched structure instead of a microgel is related to the flexibility of lignosulfonate molecules. A schematic drawing of the suggested branched structure is given in Figure 14.1. That lignosulfonates indeed are compact and rigid molecules was further elaborated by Vainio *et al.* [5], who suggested that lignosulfonates are reminiscent of hyperbranched polymers that behave in a similar way to particles in solution. Similar conjectures have been published earlier [6, 7].

A new characterization method that is of particular interest within formulations for the agrochemical industry is hydrophobic interaction chromatography applied to lignosulfonates [8]. As described below, during pulping the lignin molecule is cut randomly at some locations, especially at benzylic positions where sulfonic groups are introduced. The molecular weight distribution is typically wide [9] and the degree of sulfonation varies throughout the distribution [10]. If the degree of sulfonation varies with molecular weight the hydrophilic character of a lignosulfonate molecule will also vary. Depending on the pulping conditions and post-processing applied to sulfite liquor, the hydrophilic character of lignosulfonate may be varied and to a certain extent tailored to specific applications.

14.3 Lignosulfonate Production

The sulfite pulping process was invented by B. C. Tilghman and the first commercially successful mill was constructed and operated by C. D Ekman in Sweden in 1874 [11]. The importance of this process has declined over the years and today a few % of the world's pulp is made by sulfite pulping. In the sulfite process the wood chips are cooked at elevated temperature and pressure in a mixture of sulfurous acid and bisulfite. Originally calcium bisulfite have also been used, but later magnesium, sodium or ammonium bisulfite has also been used. The main advantages of the sulfite process are high brightness of the pulp and easy bleachability [11] (although in many applications no further bleaching is necessary). In addition the chemicals used are inexpensive. In the sulfite process a large portion of the hemicellulose is removed from cellulose, which is an advantage if the cellulose is further chemically modified. The main disadvantage with sulfite pulping is that cellulose fibres typically are weaker than those produced by the alkaline Kraft process. In addition, it is only applicable to a limited number of wood species. However, in Asia the sulfite process is used to process biomass from different annual plants as raw material.

Approximately, 1.2 million tonnes of lignosulfonate are produced annually. This accounts for around 10% of the lignin in the biomass pulped annually worldwide. In many plants operating with the Kraft process the produced lignin is burnt, both to recover chemicals for the process and for its caloric value. Thus, even if the amount of Kraft lignin produced annually is significantly larger than the amount of lignosulfonate, it is only technically available after precipitation by acidification followed by filtration. Advanced commercial applications (i.e. LignoBoost) are rare. Moreover, large amounts of wood are processed to fibres in other processes where cellulose and lignin are not separated.

Lignin, which binds bundles of cellulose fibres in wood and annual plants, is assumed to have a three-dimensional structure, but the structure has not been determined exactly. It is believed that the biosynthesis involves radical polymerization of phenyl propanoid molecules [12–14], but its exact pathway has not been determined. The phenyl propanoid monomers are linked together with ether bonds primarily between aromatic hydroxyl groups and the phenyl side chain. During the sulfite process the α-0-4 ether bonds are broken and sulfonic acid groups are introduced [11]. Introduction of sulfonic acid groups in the lignin structure renders the lignin fragments soluble in water, hence the term lignosulfonate. A tentative illustration of this process is given in Figure 14.2. (Water-soluble lignins can also be manufactured from Kraft lignin and in this process the sulfur-free Kraft lignin is sulfomethylated. After sulfomethylation, the sulfonated Kraft lignin is soluble in water.) The insoluble cellulose is separated from lignosulfonate by filtration. Typically, the spent sulfite liquor contains sugars originating from hemicellulose and inorganic salts in addition to lignosulfonate.

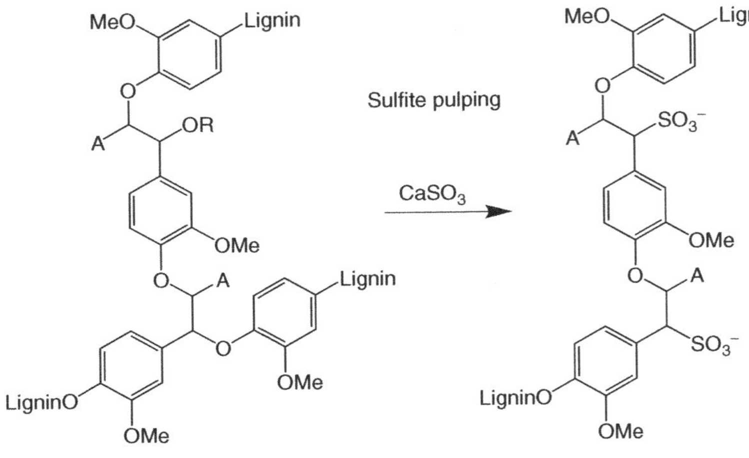

Figure 14.2 *A tentative description of the sulfite pulping process. Lignin in the structure represents the sites where the lignin structure propagates.*

Further refinement of the spent sulfite liqueurs is possible and common. Typically, the pH must be increased to reduce corrosion rates on process equipment and ease the handling. Fermentation of hexoses into bioethanol or pentoses into yeast is feasible. Alternatively, they can be converted to acids in different chemical processes. Lignosulfonates have a broad molecular weight distribution, but with the use of ultrafiltration lignosulfonates with narrower molecular weight distributions may be produced.

The lignosulfonate counterion is determined by the bisulfite salt used together with sulfurous acid during pulping of the wood chips. It is possible via different ion-exchange processes to change the counterion.

14.4 Environmental Issues

About 2 kg of atmospheric CO_2 is required for the biosynthesis of 1 kg of lignin. The harvesting of wood, transport to the pulp mills and pulping process all require energy and fossil fuels. In a thorough analysis [15], Østfoldforskning determined that 1.3 kg CO_2 is released to produce 1 kg dry lignosulfonate. This number includes contributions from the entire production, starting with the transportation of wood to the plant mill, processing and shipping the product to the customer. Parts of the energy for the pulping process might be derived from wood waste, and thus be CO_2 neutral.

One of the major applications for lignosulfonates is its use as a plasticizing or superplasticizing agent in concrete. In this application, lignosulfonates serve as a dispersant that reduces the amount of water and cement required in concrete. Incorporation of lignosulfonate in the concrete matrix removes the lignosulfonate from the natural carbon cycle. Since lignin originates from plants that produce it by uptake of CO_2 from the air, one may argue that the use of lignin in concrete applications acts as a carbon sink.

In typical lignosulfonate applications, like dust binder, soil conditioning agent and dispersant for pesticides and micronutrients within agriculture, lignosulfonate is released into nature. In other applications like feed pellets and in dyestuff formulations, lignosulfonates will end up in sewage or waste water treatment plants. For these applications it is of interest to know the impact of lignosulfonate released into different environments.

Lignin has no regular structure, which suggests that its degradation is not a simple reversal of its biosynthesis. Rather, the degradation process involves several changes in the molecular structure [16]. The term 'enzymatically catalysed burning' was coined to illustrate this difference [17]. Since degradation of lignin will not yield sugars, which is the case for many other natural polymers, its degradation process is more complex. Different investigations have revealed that various extracellular enzymes primarily present in soil are capable of degrading lignosulfonates [18–23]. The major difference between lignin and lignosulfonate is the sulfonic acid groups introduced during chemical pulping. It is reasonable that after removal or degradation of these groups the remaining lignin skeleton degrades according to normal pathways. Hence, the sulfur removal from lignosulfonate has been the topic of many investigations. Abernathy [18] did not find any liberation of sulfur from lignosulfonates in a study that lasted 45 days. Bollen [24] found that cultures of mixed soil bacteria were best suited for lignosulfonate degradation, approximately half of the sulfur present being liberated as sulfate within approximately a year. In another study which utilized radio-labelled ^{35}S, around 4% of the sulfonic acid groups disappeared within seven days [25], while another study [26] found a loss of 27% of the sulfur within 17 weeks. These findings suggest that several microorganisms are capable of digesting the sulfonic acid groups, although the exact rate probably varies with the ambient conditions and the types of microorganisms present.

The general mechanism for degradation of lignosulfonates in nature appears to start with various oxidative enzymes that generate free radicals in the polymeric structure. Lignin and lignosulfonates are bulky rigid molecules and it is reasonable that the enzymatic attack takes place in an extracellular process. Alternatively, radicals required for degradation of the lignin/lignosulfonate skeleton may be generated by light. Light irradiation degrades lignosulfonates [27–29] in particular, the content of methoxyl groups is reduced [25]. In the presence of organic material [19, 20, 30–32] like glucose, light-induced lignosulfonate degradation gives low molecular weight compounds rather than further polymerization of lignosulfonate fragments. In the course of this degradation, around 15–20% of the organic carbon [18, 21, 22, 25, 26] is lost as CO_2. While undegraded lignin and lignosulfonate might be inaccessible for bacteria and microorganisms, breakdown products appear to be available. This was demonstrated in a study [27] of degradation of Marasperse CB (a commercially available lignosulfonate) in the presence of light, where an increased growth of microorganisms was observed in the course of degradation. In a related study [28, 29], sunlight was found to mineralize lignin in seven days. This study also found an increased biomass production in the presence of lignin degradation products.

It is difficult to describe a detailed degradation pathway for lignosulfonates for several reasons. Firstly, the chemical structure of lignosulfonates is complex, which complicates a mechanistic understanding of the degradation process. Secondly, lignin and lignosulfonates are not metabolized as a fuel source, but rather the breakdown products appear to be utilized in the biosynthesis of biomass, which perhaps broadens the scope of possible reaction pathways and interesting products. However, natural mechanisms exist that degrade both lignin and lignosulfonates.

14.5 Lignosulfonates as Stabilizers for Emulsions and Suspoemulsions

The use of lignosulfonates to stabilize emulsions is well-known [1]. Lignosulfonates impart stability towards temperature gradients and electrolytes, but it is difficult to create emulsions using lignosulfonate alone as the interfacial agent. One likely reason for this is the poor solubility in most organic solvents. Use of high-shear mixing followed by high-pressure homogenization is a process where one can prepare oil-in-water lignosulfonate-based emulsions without additional emulsifiers. Below, we will illustrate this concept in more detail.

As discussed at the start of this chapter, lignosulfonates are randomly branched polymers and their shape is believed to be ellipsoidal (see Figure 14.1). The random character of lignosulfonate differentiates it from the typical surfactant structure with a defined hydrophilic and hydrophobic part. However, as is demonstrated in Figure 14.3, lignosulfonate does lower the surface tension in a similar fashion as conventional surfactants. Due to its heterogeneous nature, there is no well-defined critical micelle concentration. The viscosity of dense lignosulfonate solutions is high [5] and an interfacial lignosulfonate film is rigid and provides good film stability in emulsified systems, even in highly concentrated emulsions and in foams. One way to demonstrate this is to look at the compressibility of lignosulfonate films (Figure 14.3). The sample named 'lignosulfonate 1' forms a stable film that cannot be compressed further while 'lignosulfonate 2' can be compressed. In this example, 'lignosulfonate 1' has a lower degree of sulfonation than 'lignosulfonate 2'. A lower degree of sulfonation also implies a more significant hydrophobic character, meaning that 'lignosulfonate 1' molecules at the air–water interface may form hydrophobic associations to form a transient network that aids in the stabilization of the film. It also shows that even if the properties of lignosulfonates to a large degree are determined by the pulping conditions at the mill, post-processing of the spent sulfite liquor can modify lignosulfonate properties, for instance by modifying the degree of sulfonation or fractionation by molecular weight.

The resulting droplet size distribution of a specific emulsion recipe is inversely proportional to the applied energy in addition to how efficiently the surface-active agent lowers the interfacial tension. On an industrial scale, high-shear mixing and high-pressure homogenization are the most common methods utilized. Given the bulky nature of lignosulfonate molecules, more energy than required for typical emulsifiers is necessary to form an emulsion. In Figure 14.4 the average droplet size of an emulsion is shown as a function of mixing time in a high shear mixer. At three stages in the mixing process, a sample is withdrawn and further processed in a high-pressure homogenizer. Using pressures in the range of 100–200 bars, droplets in the micrometre and submicrometre range are easily obtained.

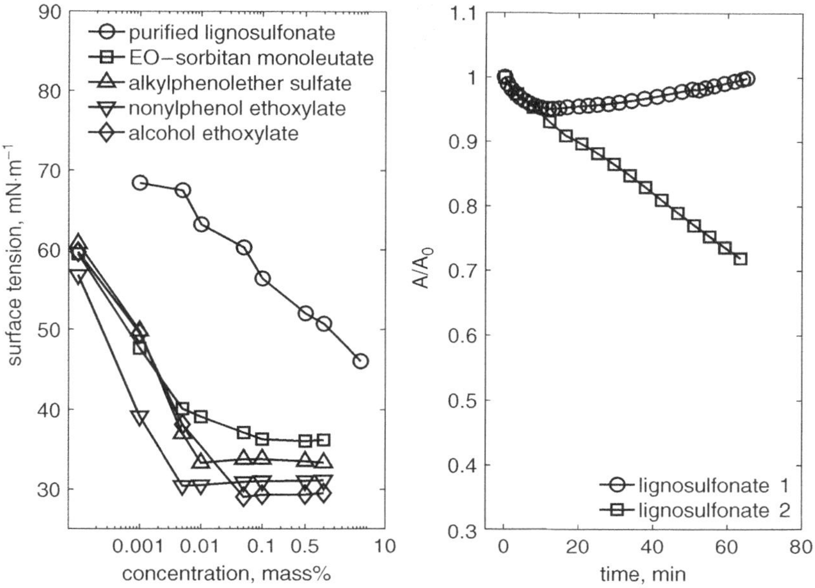

Figure 14.3 *(a) Surface tension of lignosulfonate as a function of concentration compared to typical surfactants. Lignosulfonates do not have a critical micelle concentration and hence surface tension versus concentration does not level out. (b) The ratio between initial area, A_0, and area as a function of time, A, when a steady force is compressing the layer. Lignosulfonate 1 gives a rigid layer which appears highly stable. The layer formed with lignosulfonate 2 is less stable and some deformation takes place.*

(b) Reprinted from *Colloids and Surfaces A: Physicochemical and Engineering Aspects*, **182** (1–3), 199–218, S. A. Gundersen, M. H. Ese and J. Sjöblom (2001), Langmuir surface and interface films of lignosulfonates and Kraft lignins in the presence of electrolytes and asphaltenes: Correlation to emulsion stability. Copyright 2001 with permission from Elsevier.

Emulsions are thermodynamically unstable systems and will eventually phase separate. In particular for industrial emulsions, the shelf-life is important and many methods have been established to increase the emulsion stability. One way is to add viscosity modifiers that raise the viscosity of the continuous phase. This slows down the diffusion of emulsion droplets and reduces the probability for droplet coalescence. Lignosulfonates are compatible with typical viscosity modifiers like clays, gums, starches and cellulose derivatives. Another commonly encountered destabilization mechanism is Ostwald ripening. Ostwald ripening describes the process where large droplets grow at the expense of smaller droplets. Physically, this process is related to the higher interfacial energy of small droplets, which have a higher curvature than large droplets. Hence, Ostwald ripening is a transportation process where material is transported from small to larger droplets. One successful method to inhibit Ostwald ripening is the addition of a nonmobile component to the dispersed phase [33, 34]. If mass transport of the dispersed phase occurs, the concentration of the nonmobile component in the dispersed phase increases along with its chemical potential. This is an unfavourable thermodynamic process and is an efficient method to prevent Ostwald ripening. In Figure 14.5 this concept is demonstrated.

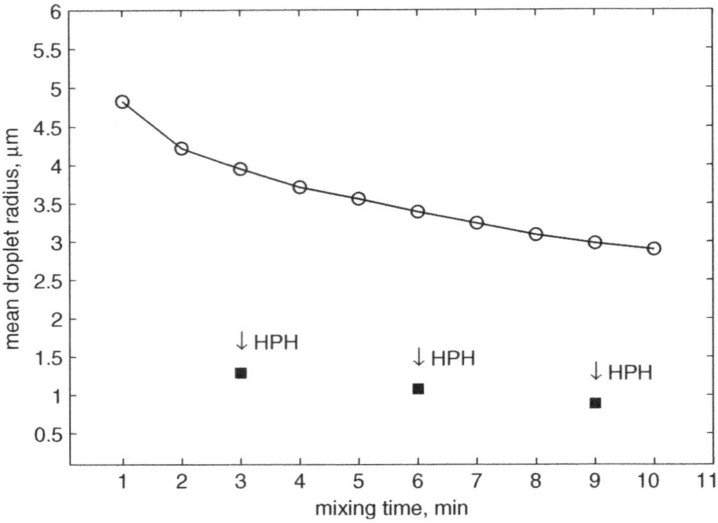

Figure 14.4 *The circles describe the reduction in average droplet radius as a function of mixing time in a high-shear mixer. At three times during high-shear mixing, a pre-emulsion may be further homogenized in a high-pressure homogenizer, which reduces the average droplet radius further (squares). In this example the emulsion is composed of equal parts of water and soya bean oil and 4 mass % lignosulfonate as stabilizer.*

Without a nonmobile component the average droplet size increases fairly rapidly, as displayed in Figure 14.5(a). In Figure 14.5(b) the evolution of the average droplet size with time is shown for the same emulsion as shown in Figure 14.5(a), but 6 vol % of an organic polymer [35] (polyisobutylene-butene) with no solubility in water is added to the oil phase (Solvesso 200); thus the droplet radius remains constant for prolonged periods of time.

As an alternative to solvents to dissolve, for instance, pesticides within agrochemical emulsions, vegetable oils constitute an interesting option. Vegetable oils have lower solubility in water than typical solvents, which means that Ostwald ripening is less significant; creaming, coalescence and flocculation are more important destabilization mechanisms. Additional advantages with vegetable oils are lower evaporation rates and easier handling. Lignosulfonates are compatible with vegetable oils, as demonstrated in Figure 14.6, where stable emulsions are obtained. These emulsions were formulated without viscosity modifiers so a dense layer of oil droplets formed at the top of samples between measurements. The emulsions were stirred gently to homogenize them before measurement of the droplet size distribution. Since the mean droplet size remains constant this suggests that lignosulfonates provides a robust stabilization mechanism, even in concentrated emulsions at elevated temperatures (40 °C).

Suspoemulsions constitute an advanced formulation comprising both an emulsion and a suspension concentrate. Two immediate examples can illustrate this concept: oilwell fluids typically contain oil, water and clay/silica particles. The characteristics of this fluid must be controlled to facilitate pumping through pipelines and this can be achieved

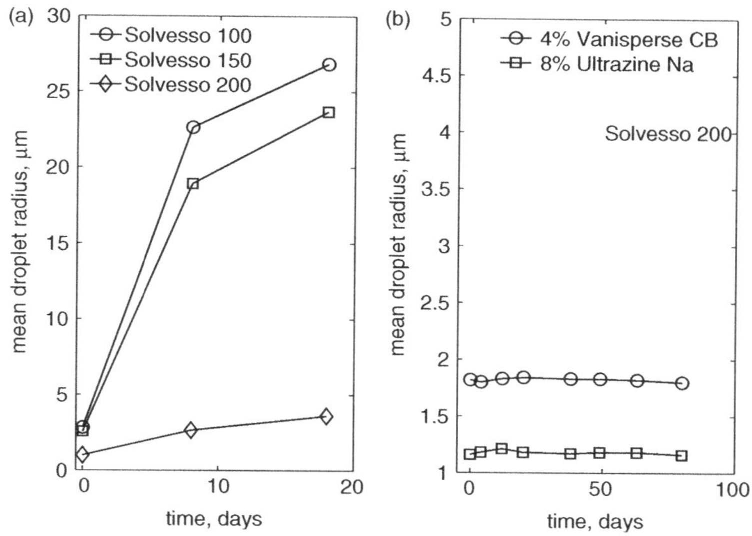

Figure 14.5 (a) The increase in average droplet radius as a function of time of storage at 40 °C. Ostwald ripening takes place at different rates in the three systems. In (b) the average droplet radius is followed for 80 days without significant variation. In this system, 6% of a nonmobile component is added to the oil (Solvesso 200) and this effectively inhibits Ostwald ripening.

via suspoemulsions [36, 37]. One of the important issues is to prevent separation of the three phases during transportation. Agrochemical formulations [38–40] containing active ingredients may be formulated as suspoemulsions. In this case, more than one active ingredient may be included in the formulation. If the two active ingredients are poorly compatible different solvents might be required to make a stable formulation. One can conceive different stabilization mechanisms in a suspoemulsion. In one case emulsion droplets and suspended particles act as individual particles. Alternatively, the oil wets the particles and the resulting solution is an emulsion with particles dissolved in the oil phase. Another possibility is that the particles enrich at the interface of oil droplets and stabilize the system [41].

Lignosulfonates are excellent stabilization agents in emulsions when they are enriched at an interface. A wide variety of particulate substrates can be distributed as suspensions using lignosulfonate as the stabilization agent. Using this as a basis, a suspoemulsion based upon a well-known pesticide, Chlorothalonil, was formulated and its stability evaluated at elevated temperatures. The stability of this model suspoemulsion, stored at 40 °C, was evaluated by rheological measurements. Since typical rheological parameters change in response to storage, suggesting that the suspoemulsion ages. Among typical ageing processes, sedimentation of the suspended particles can lead to poor resuspensibility and deteriorate the performance. In Figure 14.7 the amplitude sweeps at increasing shear stress and flow curve measurements obtained immediately after mixing, two and three weeks after mixing is shown. The complex viscosity obtained from the amplitude sweeps reveals a slight increase upon storage. An apparent yield stress can be estimated

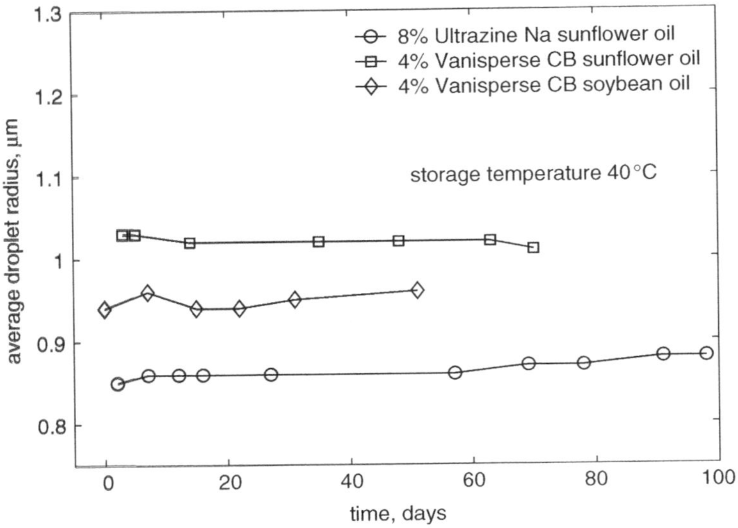

Figure 14.6 Lignosulfonates stabilize vegetable oil emulsions, here demonstrated with soya bean oil and sunflower oil. No viscosity modifier was added so a creaming layer formed upon storage. Before measurement of the droplet size the emulsions were gently stirred for homogenization. No coalescence or flocculation was observed and the droplet size remained constant.

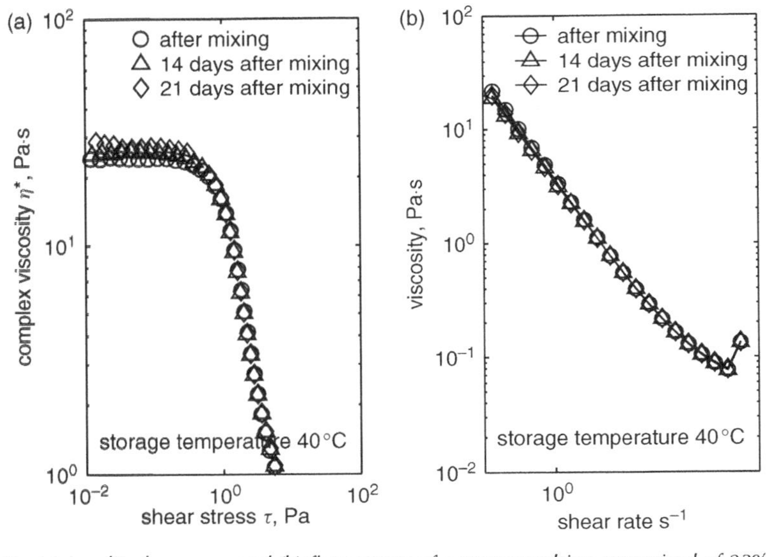

Figure 14.7 (a) Amplitude sweeps and (b) flow curves of a suspoemulsion comprised of 23% sunflower oil, 23% Chlorothalonil, 4% lignosulfonate and biocide, wetting agent and xanthan gum for stabilizaton. The plateau value of the complex viscosity increases slightly with time, but the yield stress remains similar for the duration of the trial. From measurements of the shear viscosity, no changes are discerned in the flow curves (b).

from the shear stress where the complex viscosity drops. It remains constant around 10 Pa for all times. There is no change in the flow curve. If sedimentation occurs in the formulation the viscosity of the continuous phase would decrease with time until all the suspended particles had precipitated. Aggregation of emulsion droplets would encapsulate water in the aggregates and decrease the amount of water available as a continuous phase. This would give an increase in the suspoemulsion viscosity. The lack of changes in the rheological experiments suggests that the suspoemulsion is stable over a period of three weeks.

14.6 Superplasticizers for Concrete

In the cement and concrete industry, dispersants are often referred to as 'water reducers' because addition of a dispersant reduces the amount of water required to produce a concrete suspension of a given viscosity. The strength of the concrete increases as the water content is reduced. Roughly 50% of the annual production of lignosulfonates is used as admixtures in concrete [1]. There are many parameters that are important in the formulation of concrete; in this discussion we will focus on slump and slump retention. Slump [42] is an easy method to evaluate the fluidity of concrete. It makes use of a standardized cone that is filled with concrete. After a prescribed time, the cone is removed and the difference in height between the cone and the concrete cone is measured and referred to as slump. A high slump value indicates that a concrete formulation is fluid; thus the slump is inversely proportional to the yield value. Slump retention relates to the time required for the concrete formulation to solidify. As concrete solidification proceeds, the slump value of the concrete diminishes and at some point the concrete cone is strong enough to support itself. At some construction sites concrete is delivered in concrete trucks. It is vital that it is possible to pump and pour the concrete into formworks at the construction site when the truck arrives. This is achieved by addition of additives like lignosulfonate.

Standard lignosulfonates are moderately efficient concrete dispersants. They are often mixed with other more efficient dispersants, referred to as superplasticizers, to give a concrete formulation its required characteristics. Recent development work has resulted in a lignosulfonate quality that has superplasticizing properties. One of the benefits of this lignosulfonate quality, referred to as Solus 5, is high slump retention. The slump retention is higher than normally found for other superplasticizers, meaning that a smaller amount of Solus 5 must be added at the workplace to produce a high initial slump since the slump values decrease more slowly. This concept [43] is illustrated in Figure 14.8. In Figure 14.8(a), the slump values of two concrete mixes are shown as a function of time after mixing, where the dosage of superplasticizer is adjusted to give the same initial slump in both mixes. The concrete mix formulated with Solus 5 maintains a high slump value in the entire measurement interval. The slump values for the concrete mix with the conventional naftalene sulfonate-based superplasticizer loose the slump values faster. The concrete mix formulated with Solus 5 has a higher slump than required, suggesting that the dosage may be reduced. In Figure 14.8(b) the compressive strengths of the

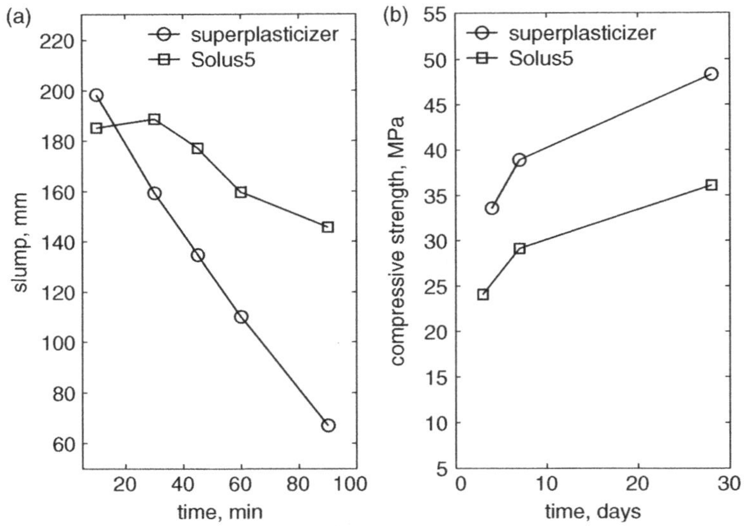

Figure 14.8 (a) Slump values for concrete mixes formulated with a conventional naftalene sulfonate-based superplasticizer and Solus 5. The concrete mixes are formulated to give approximately the same initial slump value. Solus 5 maintains a high slump value for a longer time. (b) The evolution in compressive strength for the same concrete mixes. The naftalene sulfonate-based superplasticizer gives a higher strength than Solus 5. However, since the concrete mix formulated with Solus 5 has a higher slump, the amount of water in the mix can be reduced, giving strengths in the same range as the naftalene sulfonate-based superplaticizer.

two concrete mixes are compared. The concrete formulated with Solus 5 has a lower compressive strength, but since the slump values are higher it is possible to reduce the water content in the formulation and match the strength of the concrete mix formulated with naftalene sulfonate.

14.7 Summary

Since Gargulak and Lebo's review nine years ago significant development has taken place in the lignosulfonate industry. Lignosulfonates are produced from a renewable resource, are degradable and nontoxic. In applications like concrete where lignosulfonate becomes immobilized in the concrete matrix, one may argue that it acts like a CO_2 sink, removing CO_2 from the environment. The applicability of lignosulfonates in different complex formulations demonstrates that it is a versatile dispersant with significant robustness against high levels of electrolyte and temperature variations. In addition to becoming a global industry with global actors, dedicated efforts to research and development have led to the realization of specialized lignosulfonate qualities that are used in advanced formulations. In several applications, lignosulfonates now constitute a real alternative to synthetic chemicals based on oil or gas.

Acknowledgements

Bjørn Arild Jensen is thanked for doing the experimental work on the agrochemical emulsions and suspoemulsions.

This contribution is dedicated to the memory of Roar Pihlstrøm.

References

1. Gargulak, J. D. and Lebo, S. E. (2000) Commercial use of lignin-based materials, in *Lignin; Historical, Biological, and Materials Perspectives* (eds W. G. Glasser, R. A. Northey and T. P. Schultz), American Chemical Society, Washington, DC.
2. Lebo, S. E., Braaten, S. M., Fredheim, G. E. *et al.* (2008) Recent advances in the characterisation of lignosulfonates, in *Characterization of Lignocellulosic Materials* (ed. T. Q. Hu), Wiley-Blackwell.
3. Goring, D. A. I. (1962) The physical chemistry of lignin. *Pure Appl. Chem.*, **5**, 233–254.
4. Myrvold, B. O. (2008) A new model for the structure of lignosulphonates. Part 1. Behaviour in dilute solutions. *Ind. Crops Prod.*, **27**, 214–219.
5. Vainio, U., Lauten, R. A. and Serimaa, R. (2008) Structural and rheological properties of aqueous lignosulfonate solutions. *Langmuir*, **24**, 7735–7743.
6. Browning, W. C. (1957) Flow properties of lignosulfonates. *Ind. Eng. Chem.*, **49**, 1401.
7. Goring, D. A. I., Vuong, R., Gancet, C. and Chanzy, H. (1979) The flatness of lignosulfonate macromolecules as demonstrated by electron microscopy. *J. Appl. Polym. Sci.*, **24**, 931–936.
8. Ekeberg, D., Gretland, K. S., Gustaffson, J. *et al.* (2006) Characterisation of lignosulphonates and Kraft lignin by hydrophobic interaction chromatography. *Anal. Chim. Acta*, **565**, 121–128.
9. Moacanin, J., Felicetta, V. F., Haller, W. and McCarthy, J. L. (1955) Lignin. VI. Molecular weights of lignin sulfonates by light scattering. *J. Am. Chem. Soc.*, **77**, 3470–3475.
10. Fredheim, G. E., Bråten, S. M. and Christensen, B. E. (2002) The molecular weight determination of lignosulfonates by size-exclusion chromatography and multi-angle laser light scattering. *J. Chromatogr. A*, **942**, 191–199.
11. Bryce, J. R. G. (1980) Sulfite pulping, in *Pulp and Paper, Chemistry and Chemical Technology* (ed. J. P. Casey), Wiley, New York.
12. Ralph, J., Lundquist, K., Brunow, G. *et al.* (2004) Lignins: natural polymers from oxidative coupling of 4-hydroxyphenyl-propanoids. *Phytochem. Rev.*, **3**, 29–60.
13. Davin, L. B. and Lewis, N. G. (2005) Lignin primary structures and dirigent sites. *Curr. Opin. Biotechnol.*, **16**, 407–415.
14. Ralph, J., Brunow, G., Harris, P. J. *et al.* (2008) Lignification: are lignins biosynthesized via simple combinatorial chemistry or via proteinaceous control and template replication?, in *Recent Advances in Polyphenol Research*, vol. 1 (eds F. Daayf and V. Lattanzio), Wiley-Blackwell.
15. Modahl, I. S., Brekke, A. and Raadal, H. L. (2009) *Livsløpanalyse av cellulose, etanol, lignin og vanillin frå Borregaard*, Østfoldforskning, Sarpsborg, ISBN 978-82-7520-607-5/82-7520-607-3.
16. Salkinoja-Salonen, M. and Sundman, V. (1980) Regulation and genetics of the biodegradation of lignin derivatives in pulp mill effluents, in *Lignin Biodegradation*, vol. 2 (ed. T. K. Kirk), CRC Press.

17. Kirk, T. K. and Farrell, R. L. (1987) Enzymatic 'combustion': the microbial degradation of lignin. *Annu. Rev. Microbiol.*, **41**, 465–505.
18. Abernathy, A. R. (1963) Microbial degradation of lignin sulfonate, PhD Dissertation, University of North Carolina.
19. Hiroi, T. and Eriksson, K.-E. (1976) Microbial degradation of lignin. Part 1. Influence of cellulose on the degradation of lignins by the white rot fungus. *Sven. Papperstidn.*, **79**, 157–161.
20. Hiroi, T., Eriksson, K.-E. and Stenlund, B. (1976) Microbial degradation of lignin. Part 2. Influence of cellulose upon the degradation of calcium lignosulfonate of various molecular sizes by the white rot fungus *Pleurotus ostreatus*. *Sven. Papperstidn.*, **79**, 162–166.
21. Pandila, M. M. (1973) Micro-organisms associated with microbiological degradation of lignosulphonates: a review of literature. *Pulp Pap. Mag. Can.*, **74**, 80–84.
22. Brauns, F. E. and Brauns, D. A. (1960) *The Chemistry of Lignin*, Academic Press, London.
23. Johjima, T., Itoh, N., Kabuto, M. *et al.* (1999) Direct interaction of lignin and lignin peroxidase from *Phanerochaete chrysosporium*. *Proc. Natl Acad. Sci. USA*, **96**, 1989–1994.
24. Bollen, W. B. (1942) Decomposition of waste sulfite liquor in soil. *Pac. Pulp Pap. Ind.*, **16**, 30–34.
25. Haider, K. and Kladivko, E. J. (1980) Transformation of ^{14}C and ^{35}S labelled lignosulfonates during soil incubation. *Soil Biol. Biochem.*, **12**, 275–279.
26. Lundquist, K., Kirk, T. K. and Conners, W. J. (1977) Fungal degradation of kraft lignin and lignin sulfonates prepared from synthetic ^{14}C-lignins. *Arch. Microbiol.*, **112**, 291–296.
27. Klein, D. A., Rockhill, R. C., Eldridge, J. P. and Park, J. E. (1970) Fungal and bacterial responses to photooxided lignin sulfonate substrates. *Tappi*, **53**, 1469–1472.
28. Salonen, K. and Vähätalo, A. (1994) Photochemical mineralisation of dissolved organic matter in Lake Skjervatjern. *Environ. Int.*, **20**, 307–312.
29. Vähätalo, A. V., Salonen, K., Salkinoja-Salonen, M. and Hatakka, A. (1999) Photochemical mineralization of synthetic lignin in lake water indicates enhanced turnover of aromatic organic matter under solar radiation. *Biodegradation*, **10**, 415–420.
30. Hütterman, A., Gebauer, M., Volger, C. and Rösger, C. (1977) Polymerisation und abbau van natrium-lignosulfonate durch *Fomes annosus* (Fr.) cooke. *Holzforschung*, **31**, 83–89.
31. Jong, B.-H., Kim, Y.-S. and Mun, S.-P. (1995) Degradation characteristics of lignosulfonate by laccase and Mn-peroxidase. *J. Korean Wood Sci. Technol.*, **23**, 27–32.
32. Selin, J.-F., Sundman, V. and Räihä, M. (1975) Utilization and polymerisation of lignosulfonates by wood-rotting fungi. *Arch. Microbiol.*, **103**, 63–70.
33. Buscall, R., Davis, S. S. and Potts, D. C. (1979) The effect of long-chain alkanes on the stability of oil-in-water emulsions. The significance of Ostwald ripening. *Colloid Polym. Sci.*, **257**, 636–644.
34. Higuchi, W. I. and Misra, J. (1961) Physical degradation of emulsions via the molecular diffusion route and the possible prevention thereof. *J. Pharm. Sci.*, **51**, 459–466.
35. Mulqueen, P. (2003) Recent advances in agrochemical formulation. *Adv. Colloid Interface Sci.*, **106**, 83–107.
36. Yan, Y., Pal, R. and Masliyah, J. (1991) Rheology of oil in water emulsions with added kaolinite clay. *Ind. Eng. Chem. Res.*, **30**, 1931–1936.
37. Pal, R. and Masliyah, J. (1990) Rheology of oil in water emulsions with added solid. *Can. J. Chem. Eng.*, **68**, 24–28.
38. Faers, M. A. and Pontzen, R. (2008) Factors influencing the association between active ingredient and adjuvant in the leaf deposit of adjuvant-containing suspoemulsion formulations. *Pest Manag. Sci.*, **64**, 820–833.
39. Chen, C.-Y. R. (2006) Fertilizer-compatible composition, 11/041167[US 2006/0166939 A1]. January 21, 2005.

40. Gundersen, S. A., Pihlstrøm, R. and Lauten, R. A. (2005) Lignin-based stabilisers for emulsion and suspo-emulsion formulations, in Book of Abstracts (2) for Poster Sessions, 11th IUPAC International Congress of Pesticide Chemistry, Kobe, Japan, 2005, p. 167.
41. Midmore, B. R. (1998) Preparation of a novel silica-stabilized oil/water emulsion. *Colloids Surf. A*, **132**, 257–265.
42. Coussot, P. (2005) Practical rheometrical techniques, in *Rheometry of Pastes, Suspensions and Granular Materials*, Chapter 7, Wiley Blackwell, New Jersey.
43. Reknes, K., Jha, P., Sharma, S. and Chuah, P. (2008) A lignosulfonate superplasticizer for concrete, in ACI-India Conference in Sustainable Development of Concrete, Ratnagiri, India, 2008.

15
Dispersion Stabilizers Based on Inulin

Tharwat Tadros[1] **and Bart Levecke**[2]

[1] *Consultant, Wokingham, UK*

[2] *Beneo Bio Based Chemicals, a Division of Beneo-Remy NV, Leuven-Wijgmaal, Belgium*

15.1 Introduction

In recent years, there has been considerable interest in surfactants from renewable resources. One of the most commonly used types are those derived from mono- and polysaccharides, which are synthesized by reaction with the multifunctional hydroxyl groups. The technical problem is one of joining a hydrophobic group (such as an alkyl chain) to the multihydroxyl structure. Several surfactants have been made, e.g. esterification of sucrose with fatty acids or fatty glycerides, to produce sucrose esters [1, 2]. The most interesting sugar surfactants are the alkyl polyglucosides (APGs), which are produced by reaction of a fatty alcohol (which is derived from vegetable oils) directly with glucose [1, 2]. The properties of APG surfactants depend on the alkyl chain length and average degree of polymerization (DP) (which range from 1.1 to 3.0).

Although the above sugar surfactants found many applications, particularly in cosmetics and personal care products, they are seldom very effective in stabilization of disperse systems against flocculation and/or coalescence. This is due to the reversible nature of adsorption of these molecules at the solid/liquid or liquid/liquid interfaces. For that reason we have developed a polymeric surface-active molecule based on inulin (which is extracted from chicory roots). Inulin is a polydisperse polysaccharide consisting mainly, if not exclusively, of $\beta(2 \rightarrow 1)$ fructosyl fructose units with normally, but not necessarily, one glucose unit at the reducing end. The extracted inulin is fractionated to obtain

Surfactants from Renewable Resources Edited by Mikael Kjellin and Ingegärd Johansson

the high molecular weight fraction (with a DP > 23) inulin. To produce the amphipathic graft copolymer, the chain is modified by introduction of alkyl groups (C_4–C_{18}) on the polyfructose backbone [3, 4]. The first molecule in this series is obtained by random grafting of long-chain inulin with several C_{12} alkyl chains, hydrophobically modified inulin (HMI).

In this chapter we will start with a section on the raw materials used to produce HMI, the possible production methods of this product and its safety. The second section will give a short description of the solution properties of long-chain inulin and HMI. This is followed by a section on the interfacial properties of HMI at the air/liquid, liquid/liquid and solid/liquid interfaces. Particular attention will be given in describing the effectiveness of HMI as a stabilizer for various disperse systems, e.g. emulsions, nanoemulsions and latexes. The application of HMI in the formulation of emulsions, latex dispersions and nano-emulsions will be described in subsequent sections.

15.1.1 Raw Materials – Inulin

Inulin, the reserve polysaccharide of many plants, is a polydisperse carbohydrate material consisting mainly of $\beta(2 \rightarrow 1)$ fructosyl–fructose links [5]. A starting glucose moiety can be present but is not necessary. The DP of inulin depends on the type of plant from which it is extracted, on the weather conditions during the growth and on the physiological age of the plant [6]. The most important sources of inulin are *Cichorium intybus* (chicory), *Dahlia pinuata* Cav. (dahlia) and *Helianthus tuberosus* (Jerusalem artichoke). From these three plants, chicory is the favoured one. The inulin extracted from that plant has a DP that varies from 2 to 70 fructose units. The molecular structure of inulin is shown in Figure 15.1.

Generally speaking, three types of inulin products can be distinguished and are commercially available as white powders. First of all is the native inulin with a mean DP of

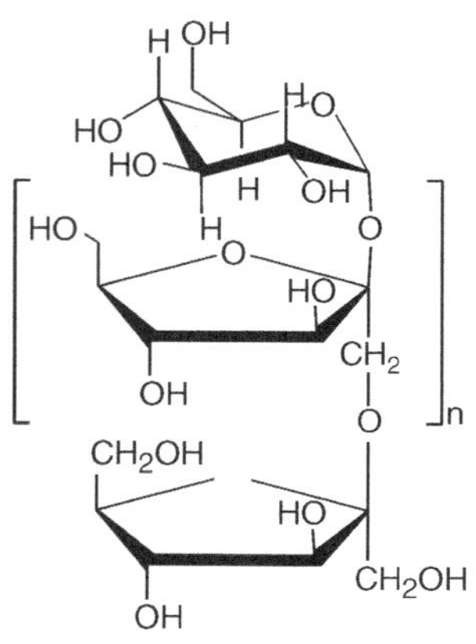

(GF$_n$)

Figure 15.1 Molecular structure of inulin.

about 10–12, containing glucose, fructose, sucrose and small oligosaccharides as well as high-DP fructose polymers. Secondly, the long-chain inulin (INUTEC®N25) with a mean DP greater than 23, and in which chains smaller than 10 units are absent [6], is obtained by physical fractionation of the native inulin, eliminating smaller oligosaccharides and monomers that are naturally present. Thirdly, the complementary product of the latter, containing no molecules with a DP larger than 10, can be obtained by a controlled enzymatic hydrolysis process. The above-mentioned types of inulin can be used to produce inulin-based surfactants.

In the industrial production of inulin, one can clearly distinguish two steps. The first one, which is virtually similar to the sugar beet process, includes the extraction and a primary purification step that results in a raw impure syrup. The second step is the refining phase, which results in a commercial end product with a purity greater than 99.5%. This is achieved by using ion exchange resins for demineralization and active carbon for decolorization, ending with a spray-drying process to obtain a storable, microbiological stable and commercial end product.

Inulin is widely applied in food products. It can be used for either its nutritional advantages or its technological properties, though it is often applied to offer a double benefit: an improved organoleptic quality and a better-balanced nutritional composition.

15.1.2 Inulin-Based Surfactants

As an answer to the growing demand for products made from 'renewable resources' and with (modified) starch as a reference in mind, a lot of these starches have been devoted to the chemical modification of inulin [3]. Hydrophobization of inulin, e.g. grafting of alkyl chains on to the inulin backbone and as such obtaining an inulin derivative showing the wanted amphiphilic character, can be done in several ways: by esterification, by etherification or by carbamoylation. The reactions are usually performed in solvents like pyridine, dimethylformamide or dimethylsulfoxide, using catalysts like, for example, sodium acetate, potassium carbonate or triethylamine.

Esterification can be carried out using conventional reactions with acid chloride or with acid anhydrides. The etherification of inulin can be done using alkylepoxides as reactant. Carbamoylation reactions using alkyl isocyanate can be performed, obtaining inulin carbamates.

Different HMI surfactants can be obtained by changing three parameters: the DP of the inulin backbone, the length of the alkyl chain and the degree of substitution. INUTEC®SP1, commercially available from BENEO Bio Based Chemicals, consist of an inulin backbone with a DP greater than 23 on which several C_{12} alkyl chains are randomly grafted.

15.1.3 Product Safety

With the ever-increasing pressure on environmentally unfriendly and/or toxicologically unfavourable chemicals, this biopolymer-based surface-active molecule shows a good combination of interesting functional properties with a positive safety profile. Due to the large carbohydrate part in the molecule, its biodegradability is high and its toxicity is low. The product is completely biocompatible and safe, with no primary skin irritation, oral toxicity, sensitization, mutagenicity or aquatic toxicity.

15.2 Solution Properties of Long-Chain Inulin and Hydrophobically Modified Inulin (HMI)

Studying the solution properties of the graft copolymer HMI is key to understanding the conformation of the polymer in solution, its aggregation behaviour and its subsequent adsorption at various interfaces. As mentioned above, the molecule consists of a hydrophilic backbone (linear polyfructose with DP $>$ 23), to be referred to as the stabilizing chain A (see below) and several hydrophobic alkyl chains to be referred to as the 'anchor' chains B (see below). In aqueous media the hydrophilic A chain is highly soluble and strongly hydrated by water molecules (hydrogen bonding between the OH groups of polyfructose and water molecules). In contrast the B chains are highly insoluble in the medium. The presence of sequences with very different solvent affinities have a consequence on the conformation of the isolated polymer chains. This results in the formation of aggregates involving several polymer molecules, even in dilute solutions [7], as illustrated below using light scattering techniques.

It is perhaps useful to discuss the solubility of the A chain, long-chain inulin, in water as a function of temperature [8]. Long-chain inulin forms a clear solution in water at concentrations $\leq$1 wt% in the temperature range 20–100 °C. However, at concentrations $>$1% phase separation occurs forming a solid + liquid phase and a clear solution is obtained above a certain temperature that depends on the long-chain inulin concentration. In other words, the inulin–water system shows an upper consolute temperature boundary (UCST). The lower consolute temperature boundary is higher than 100 °C. Addition of electrolytes can cause a shift in the UCST with $CaCl_2$ showing a shift to lower temperatures whereas $MgSO_4$ and Na_2SO_4 show a shift to higher temperatures. Measurement of the cloud point of 1, 2, 3 and 4% inulin, both in water and various concentrations (0.5–4.0 mol·dm^{-3}) of $CaCl_2$, $MgSO_4$ and Na_2SO_4, showed a value greater than 100 °C, indicating the strong hydration of the polymer chains both in water and high electrolyte concentrations [8]. These results have a direct implication on the stabilization mechanism produced when using HMI in disperse systems.

As mentioned above, HMI is expected to show aggregation in very dilute solutions. This is illustrated in Figure 15.2, which shows the variation of the intensity of scattered light as a function of HMI concentration [9]. The results clearly show a rapid increase in light scattering intensity above a critical concentration, which is identified as the critical association concentration (CAC). The latter value is equal to 5×10^{-6} mol/dm^3, which is low and typical for graft copolymers that associate in very dilute solutions, as discussed above.

Information about the size and morphology of aggregates can be obtained by measuring the light scattering at the CAC as a function of scattering angle. Unfortunately, the results did not show the expected linear relationship between log of the scattering intensity versus the square of scattering vector [8] and hence no information could be obtained on the aggregate geometry using the light scattering technique.

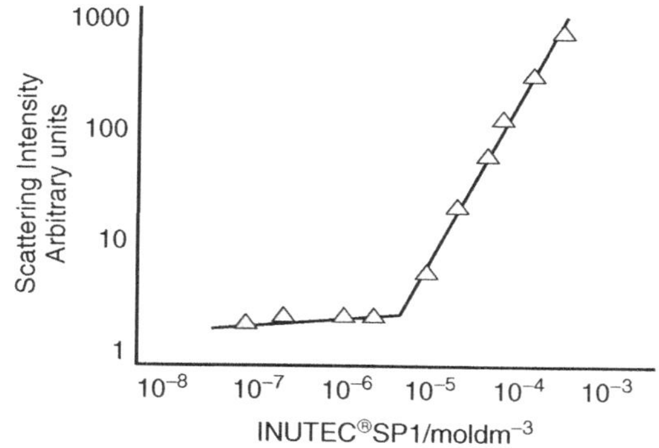

Figure 15.2 Light scattering intensity as a function of HMI concentration.

15.3 Interfacial Aspects of HMI at Various Interfaces

Figure 15.3 shows the variation of surface tension γ with log HMI concentration. The results clearly show that γ decreases linearly with an increase in HMI concentration till a critical concentration is reached above which γ remains virtually constant with a further increase in polymer concentration. The break point at the critical concentration can be identified with the CAC. The latter was found to be 6.6×10^{-7} mol/dm^3, which is significantly lower than the value obtained using light scattering. It is well known that the CAC value depends on the technique used for its measurement [9].

Figure 15.4 shows the variation of interfacial tension (at the isoparaffinic oil/water, or O/W, interface) with HMI concentration [10] at two different NaCl concentrations, namely 2.0×10^{-4} and 5.0×10^{-1} mol/dm^3. As with the results at the air/water (A/W) interface, γ at the O/W interface also decreases with an increase in HMI concentration

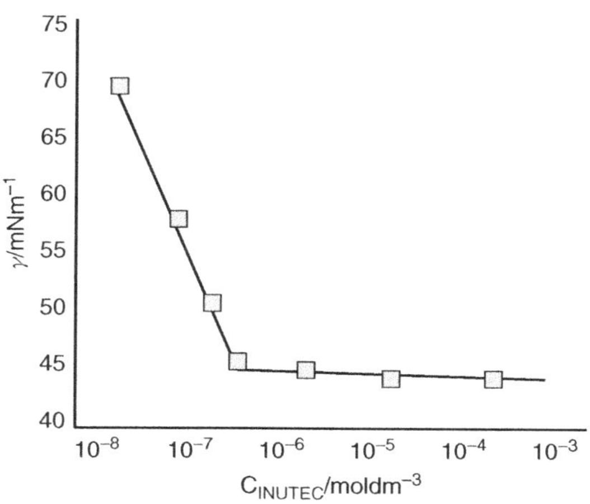

Figure 15.3 $\gamma - \log C_{HMI}$ curve at the air/water interface.

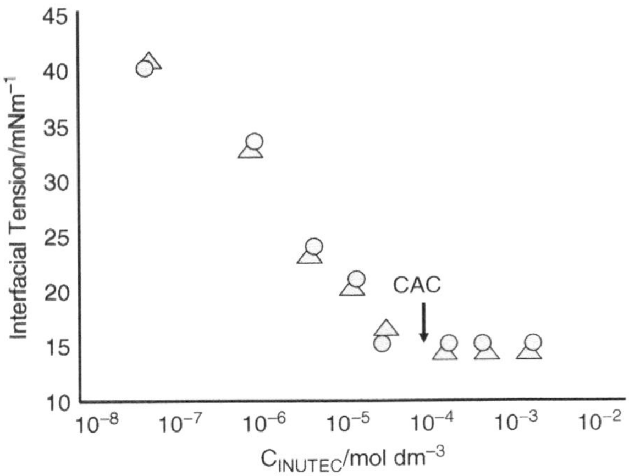

Figure 15.4 *$\gamma - \log C_{HMI}$ curve at the oil/water interface; Δ, 2×10^{-4} mol/dm^3 NaCl; o, 5×10^{-1} mol/dm^3 NaCl.*

till a CAC is reached, after which γ remains virtually constant with a further increase in polymer concentration. The $\gamma - \log C$ curve is similar at the two NaCl concentrations. The CAC obtained in this case is 5×10^{-5} mol/dm^3, which is significantly higher than the value obtained at the A/W interface. This reflects the different adsorption characteristics at the two interfaces. The penetration of the alkyl chains in the oil phase seems to affect the association behaviour in the bulk solution.

The adsorption of HMI on solid particles was investigated [11] using two different latex dispersions, namely polystyrene (PS) and polymethylmethacrylate (PMMA). Both lattices have a narrow particle size distribution with PS having a diameter of 321 nm and polydispersity index of 0.03 and PMMA having a diameter of 273 nm and polydispersity index of 0.05. The results are shown in Figure 15.5, which shows the amount of adsorption Γ in μmol/m^2 versus HMI concentration (μmol/dm^3).

The results of Figure 15.5 show that the adsorption isotherms of INUTEC®SP1 on PS and PMMA are of the high affinity type that is characteristic for polymer adsorption with multipoint attachment (with several alkyl chains on the latex surface). The plateau adsorption value is higher for PMMA when compared to PS. This is probably due to the higher adsorption energy of INUTEC®SP1 on PMMA latex.

15.4 Emulsions Stabilized Using HMI

Emulsions of Isopar M/water and cyclomethicone/water were prepared using HMI. The 50/50 v/v O/W emulsions were prepared and the emulsifier concentration was varied from 0.25 to 2.0% w/v based on the oil phase. The 0.5% w/v emulsifier was sufficient for stabilization of these 50/50 v/v emulsions [12].

The emulsions were stored at room temperature and 50 °C and optical micrographs were taken at intervals of time (for one year) in order to check the stability. Emulsions prepared in water were very stable, showing no change in droplet size distribution over

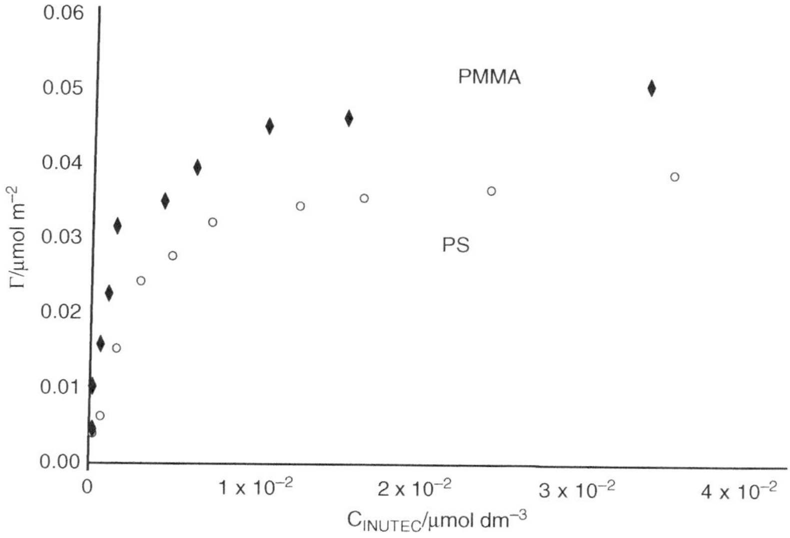

Figure 15.5 *Adsorption isotherms of HMI on PS and PMMA latexes.*

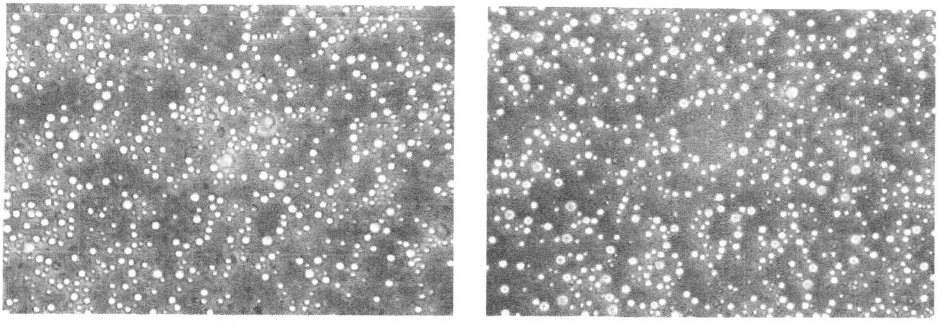

Figure 15.6 *Optical micrographs of O/W emulsions stabilized with HMI stored at 50 °C for (a) 1.5 weeks and (b) 14 weeks.*

more than a year period, which indicated the absence of coalescence. Any weak flocculation that occurred was reversible and the emulsion could be redispersed by gentle shaking. Figure 15.6 shows an optical micrograph for a dilute 50/50 v/v emulsion that was stored for 1.5 and 14 weeks at 50 °C.

No change in droplet size was observed after storage for more than one year at 50 °C, indicating the absence of coalescence. The same result was obtained when using different oils. Emulsions were also stable against coalescence in the presence of high electrolyte concentrations (up to 4 mol/dm^3 or ~25% NaCl).

The above stability in high electrolyte concentrations is not observed with polymeric surfactants based on polyethylene oxide (PEO). The high stability observed using HMI is related to its strong hydration both in water and in electrolyte solutions. The hydration of inulin (the backbone of HMI) could be assessed using cloud point measurements. A comparison was also made with PEO with two molecular weights, namely 4000 and 20 000.

Solutions of PEO 4000 and 20 000 showed a systematic decrease of cloud point with an increase in NaCl or $MgSO_4$ concentration. In contrast, inulin showed no cloud point up to 4 mol/dm^3 NaCl and up to 1 mol/dm^3 $MgSO_4$.

The above results explain the difference between PEO and inulin. With PEO, the chains show dehydration when the NaCl concentration is increased above 2.0 or 0.5 mol/dm^3 $MgSO_4$. The inulin chains remain hydrated at much higher electrolyte concentrations. It seems that the linear polyfructose chains remain strongly hydrated at high temperature and high electrolyte concentrations.

The high emulsion stability obtained when using HMI can be accounted for by the following factors:

1. The multipoint attachment of the polymer by several alkyl chains that are grafted on the backbone.
2. The strong hydration of the polyfructose 'loops' both in water and high electrolyte concentrations (the Flory–Huggins interaction parameter χ remains below 0.5 under these conditions).
3. The high volume fraction (concentration) of the loops at the interface.
4. Enhanced steric stabilization; this is the case with multipoint attachment, which produces strong elastic interaction.

Evidence for the high stability of the liquid film between emulsion droplets when using HMI was obtained by Exerowa *et al.* [10] using disjoining pressure measurements. This is illustrated in Figure 15.7, which shows a plot of disjoining pressure versus separation

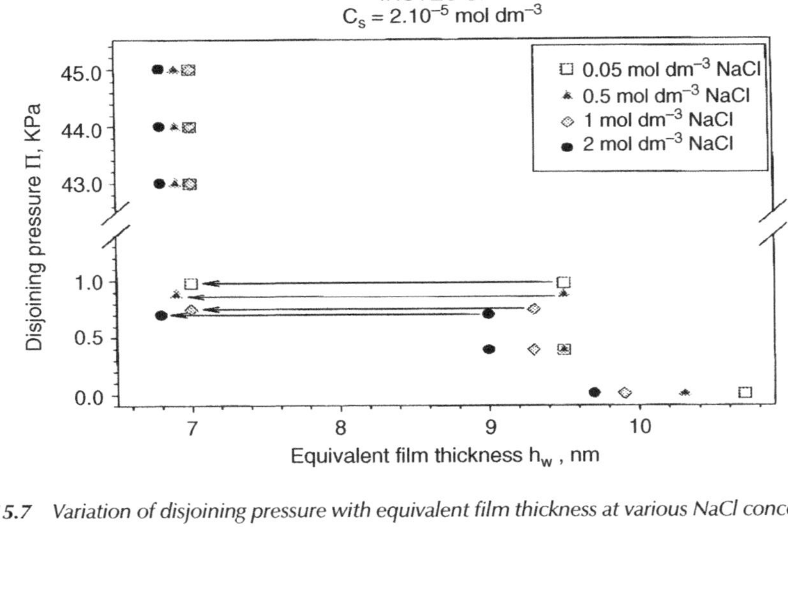

Figure 15.7 *Variation of disjoining pressure with equivalent film thickness at various NaCl concentrations.*

distance between two emulsion droplets at various electrolyte concentrations. The results show that by increasing the capillary pressure a stable Newton black film (NBF) is obtained at a film thickness of ~7 nm.

The lack of rupture of the NBF up to the highest pressure applied, namely 4.5×10^4 Pa, clearly indicates the high stability of the liquid film in the presence of high NaCl concentrations (up to 2 mol/dm^3). This result is consistent with the high emulsion stability obtained at high electrolyte concentrations and high temperature. Emulsions of Isopar M in water are very stable under such conditions and this could be accounted for by the high stability of the NBF. The droplet size of 50 : 50 O/W emulsions prepared using 2% HMI is in the region of 1–10 μm. This corresponds to a capillary pressure of $\sim 3 \times 10^4$ Pa for the 1 μm drops and $\sim 3 \times 10^3$ Pa for the 10 μm drops. These capillary pressures are lower than those to which the NBF has been subjected to, which clearly indicates the high stability obtained against coalescence in these emulsions.

15.5 Emulsion Polymerization Using HMI

Recently, the graft copolymer of HMI has been used in emulsion polymerization of styrene, methyl methacrylate, butyl acrylate and several other monomers [7]. All lattices were prepared by emulsion polymerization using potassium persulfate as initiator. The z-average particle size was determined by photon correlation spectroscopy (PCS) and electron micrographs were also taken.

Emulsion polymerization of styrene or methylmethacrylate showed an optimum weight ratio of HMI/monomer of 0.0033 for PS and 0.001 for PMMA particles. The initiator/monomer ratio was kept constant at 0.00125. The monomer conversion was higher than 85% in all cases. Latex dispersions of PS reaching 50% and of PMMA reaching 40% could be obtained using such low concentrations of HMI. Figure 15.8 shows the variation of particle diameter with monomer concentration.

The stability of the latexes was determined by finding the critical coagulation concentration (CCC) using $CaCl_2$. The CCC was low (0.0175 – 0.05mol/dm^3) but this was higher than that for the latex prepared without surfactant. Post-addition of HMI resulted in a large increase in the CCC, as is illustrated in Figure 15.9, which shows log W – log C curves (where W is the ratio between the fast flocculation rate constant to the slow flocculation rate constant, referred to as the stability ratio) at various additions of HMI.

As with the emulsions, the high stability of the latex when using HMI is due to the strong adsorption of the polymeric surfactant on the latex particles and formation of strongly hydrated loops and tails of polyfructose that provide effective steric stabilization. Evidence for the strong repulsion produced when using HMI was obtained from atomic force microscopy investigations [11], whereby the force between hydrophobic glass spheres and hydrophobic glass plate, both containing an adsorbed layer of HMI, was measured as a function of distance of separation both in water and in the presence of various Na_2SO_4 concentrations. The results are shown in Figures 15.10 and 15.11.

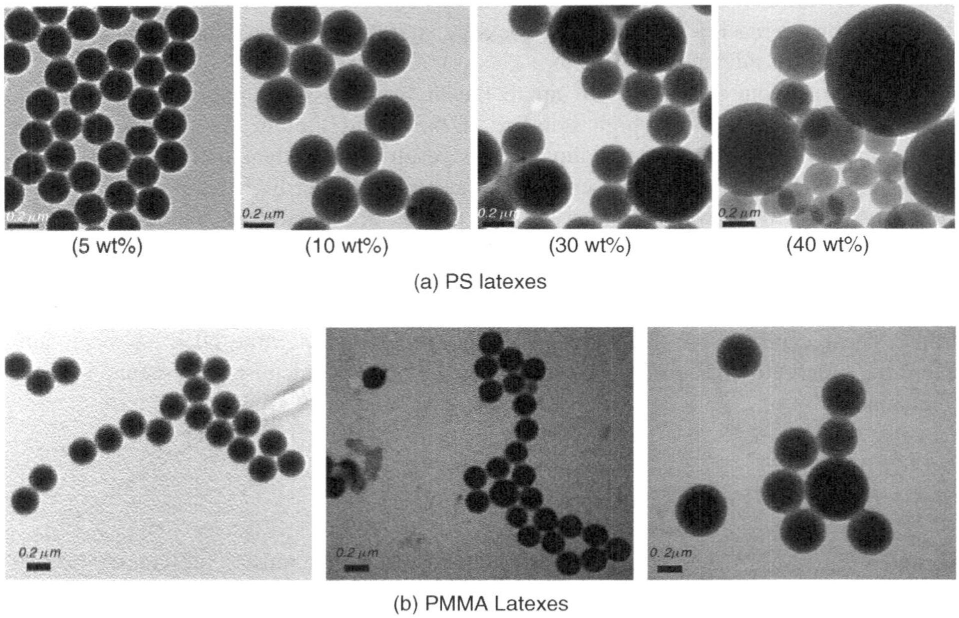

Figure 15.8 *Electron micrographs of the latexes.*

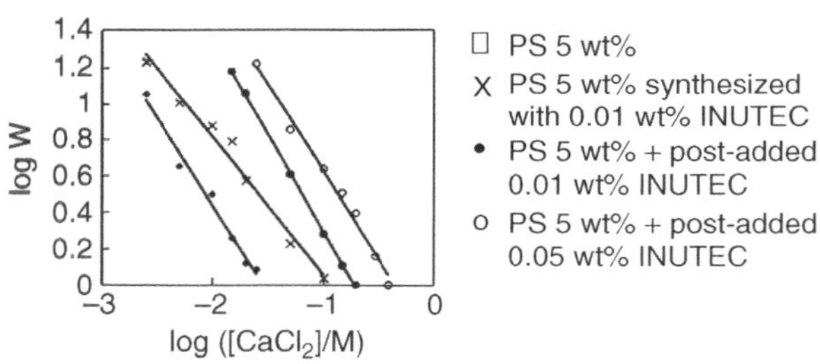

Figure 15.9 *Influence of post-addition of HMI on the latex stability.*

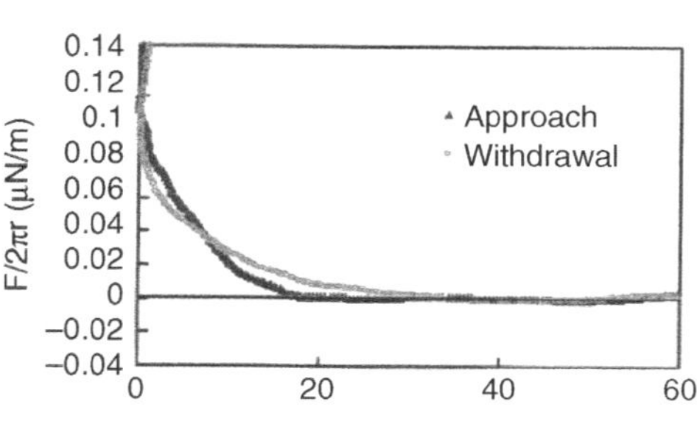

Figure 15.10 *Force–distance (in nm) curves between hydrophobized glass surfaces containing adsorbed HMI in water.*

Printed and bound by CPI Group (UK) Ltd, Croydon, CR0 4YY
16/09/2026
14963475-0001

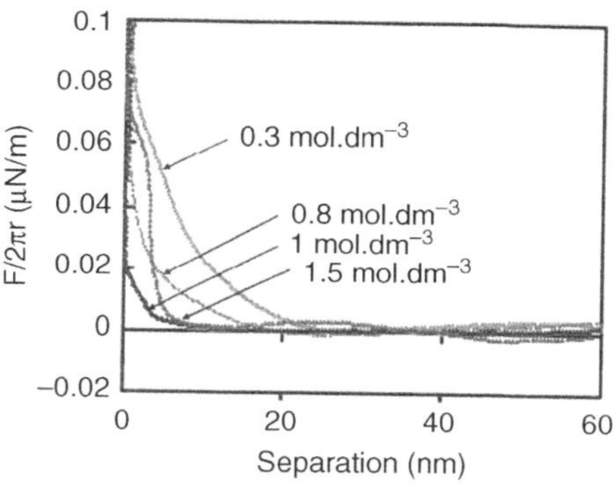

Figure 15.11 Force–distance curves for hydrophobized glass surfaces containing adsorbed HMI at various Na_2SO_4 concentrations.

15.6 Use of HMI for Preparation and Stabilization of Nanoemulsions

Nanoemulsions are systems that cover the size range 20–200 nm [13, 14]. They can be transparent, translucent or turbid depending on the droplet radius and refractive index difference between the droplets and the continuous phase. This can be understood from consideration of the dependence of light scattering (turbidity) on the above two factors. For droplets with a radius that is less than 1/20 of the wavelength of the light, the turbidity τ is given by the following equation:

$$\tau = KN_o V^2 \tag{15.1}$$

where K is an optical constant that is related to the difference in refractive index between the droplets n_p and the medium n_o and N_o is the number of droplets each with a volume V.

It is clear from Equation (15.1) that τ decreases with a decrease in K that is smaller $(n_p - n_o)$ than a decrease in N_o and a decrease in V. Thus to produce a transparent nanoemulsion one has to decrease the difference between the refractive index of the droplets and the medium (i.e. try to match the two refractive indices). If such matching is not possible then one has to reduce the droplet size (by high-pressure homogenization) to values below 50 nm. It is also necessary to use a nanoemulsion with a low oil volume fraction (in the region of 0.2).

Nanoemulsions are only kinetically stable. They have to be distinguished from microemulsions (that cover the size range 5–50 nm), which are mostly transparent and thermodynamically stable. The long-term physical stability of nanoemulsions (with no apparent flocculation or coalescence) make them unique and they are sometimes referred to as ‘approaching thermodynamic stability’. The inherently high colloid stability of nanoemulsions can be well understood from a consideration of their steric stabilization (when using nonionic surfactants and/or polymers) and how this is affected by the ratio of the adsorbed layer thickness to the droplet radius [15].

Unless adequately prepared (to control the droplet size distribution) and stabilized against Ostwald ripening (that occurs when the oil has some finite solubility in the continuous medium), nanoemulsions may show an increase in the droplet size and an initially transparent system may become turbid on storage.

The attraction of nanoemulsions for application in personal care and cosmetics as well as in health care is due to the following advantages:

1. The very small droplet size causes a large reduction in the gravity force and the Brownian motion may be sufficient to overcome gravity. This means that no creaming or sedimentation occurs on storage.
2. The small droplet size also prevents any flocculation of the droplets. Weak flocculation is prevented and this enables the system to remain dispersed with no separation.
3. The small droplets also prevent their coalescence, since these droplets are nondeformable and hence surface fluctuations are prevented. In addition, the significant surfactant film thickness (relative to the droplet radius) prevents any thinning or disruption of the liquid film between the droplets.

The production of small droplets (submicrometre) requires application of high energy. The process of emulsification is generally inefficient. Simple calculations show that the mechanical energy required for emulsification exceeds the interfacial energy by several orders of magnitude. For example, to produce a nanoemulsion at $\phi = 0.1$ with an average radius R of 200 nm, using a surfactant that gives an interfacial tension $\gamma = 10$ mN/m, the net increase in surface free energy is $A\gamma = 3\phi\gamma/R = 1.5 \times 10^4$ J/m^3. The mechanical energy required in a homogenizer is 1.5×10^7 J/m^3, i.e. an efficiency of 0.1%. The rest of the energy (99.9%) is dissipated as heat.

The intensity of the process or the effectiveness in making small droplets is often governed by the net power density ($\varepsilon(t)$):

$$p = \varepsilon(t)\,\mathrm{d}t \tag{15.2}$$

where t is the time during which emulsification occurs.

The break-up of droplets will only occur at high ε values, which means that the energy dissipated at low ε levels is wasted. Batch processes are generally less efficient than continuous processes. This shows why with a stirrer in a large vessel most of the energy applied at low intensity is dissipated as heat. In a homogenizer, p is simply equal to the homogenizer pressure.

Several procedures may be applied to enhance the efficiency of emulsification when producing nanoemulsions:

1. One should optimize the efficiency of agitation by increasing ε and decreasing dissipation time.
2. The nanoemulsion is preferably prepared at a high volume faction of the disperse phase and diluted afterwards. However, very high ϕ values may result in coalescence during emulsification.
3. Add more surfactant, whereby creating a smaller γ_{eff} and possibly diminishing recoalescence.
4. Use surfactant mixture that gives more reduction in γ than the individual components.

5. If possible dissolve the surfactant in the dispersed phase rather than in the continuous phase – this often leads to smaller droplets.
6. It may be useful to emulsify in steps of increasing intensity, particularly with nanoemulsions having a highly viscous dispersed phase.

The high kinetic stability of nanoemulsions can be explained from consideration of the energy–distance curves for sterically stabilized dispersions shown in Figure 15.12. It can be seen from Figure 15.12 that the depth of the minimum decreases with increasing δ/R (adsorbed layer thickness/radius). With nanoemulsions having a radius in the region of 50 nm and an adsorbed layer thickness of, say, 10 nm, the value of δ/R is 0.2. This high value (when compared with the situation with macroemulsions where δ/R is at least an order of magnitude lower) results in a very shallow minimum (which could be less than kT). This situation results in very high stability with no flocculation (weak or strong). In addition, the very small size of the droplets and the dense adsorbed layers ensures a lack of deformation of the interface, lack of thinning and disruption of the liquid film between the droplets, and hence coalescence is also prevented.

One of the main problems with nanoemulsions is Ostwald ripening, which results from the difference in solubility between small and large droplets. The difference in chemical potential of dispersed phase droplets between different-sized droplets, as given by Lord Kelvin [16], is

$$s(r) = s(\infty)\exp\left(\frac{2\gamma V_{\mathrm{m}}}{rRT}\right) \tag{15.3}$$

where $s(r)$ is the solubility surrounding a particle of radius r, $s(\infty)$ is the bulk phase solubility and V_{m} is the molar volume of the dispersed phase. The quantity $(2\gamma V_{\mathrm{m}}/RT)$ is termed the characteristic length. It has an order of ~1 nm or less, indicating that the difference in solubility of a 1 μm droplet is of the order of 0.1% or less.

Theoretically, Ostwald ripening should lead to condensation of all droplets into a single drop (i.e. phase separation). This does not occur in practice since the rate of growth decreases with increase of droplet size. For two droplets of radii r_1 and r_2

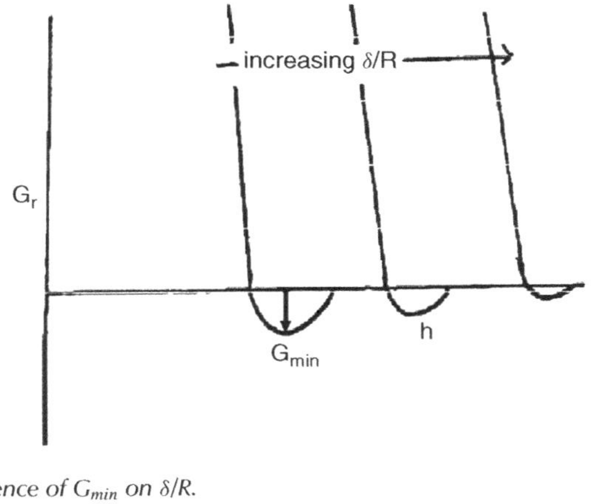

Figure 15.12 Dependence of G_{min} on δ/R.

(where $r_1 < r_2$),

$$\frac{RT}{V_m} \ln \left[\frac{s(r_1)}{s(r_2)} \right] = 2\gamma \left(\frac{1}{r_1} - \frac{1}{r_2} \right) \tag{15.4}$$

Equation (15.4) shows that the larger the difference between r_1 and r_2, the higher the rate of Ostwald ripening. Ostwald ripening can be quantitatively assessed from plots of the cube of the radius versus time t [17–19]:

$$r^3 = \frac{8}{9} \left(\frac{s(\infty)\gamma V_m D}{\rho RT} \right) \tag{15.5}$$

where D is the diffusion coefficient of the disperse phase in the continuous phase and ρ is its density.

Ostwald ripening can be reduced by incorporation of a second component that is insoluble in the continuous phase (e.g. squalane). In this case significant partitioning between different droplets occurs, with the component having low solubility in the continuous phase expected to be concentrated in the smaller droplets. During Ostwald ripening in a two-component disperse phase system, equilibrium is established when the difference in chemical potential between different sized droplets (which results from curvature effects) is balanced by the difference in chemical potential resulting from partitioning of the two components. If the secondary component has zero solubility in the continuous phase, the size distribution will not deviate from the initial one (the growth rate is equal to zero). In the case of limited solubility of the secondary component, the distribution is the same as governed by Equation (15.5); i.e. a mixture growth rate is obtained that is still lower than that of the more soluble component.

Another method for reducing Ostwald ripening depends on modification of the interfacial film at the O/W interface. According to Equation (15.5), reduction in γ results in reduction of Ostwald ripening. However, this alone is not sufficient since one has to reduce γ by several orders of magnitude. Walstra [20, 21] suggested that by using surfactants that are strongly adsorbed at the O/W interface (i.e. polymeric surfactants) and that do not desorb during ripening, the rate could be significantly reduced. An increase in the surface dilational modulus ε and decrease in γ would be observed for the shrinking drops. The difference in ε between the droplets would balance the difference in capillary pressure (i.e. curvature effects).

To achieve the above effect it is useful to use A–B–A block copolymers that are soluble in the oil phase and insoluble in the continuous phase. A strongly adsorbed polymeric surfactant that has limited solubility in the aqueous phase can also be used, e.g. HMI. This is illustrated in Figure 15.13, which shows plots of r^3 versus time for 20% v/v silicone oil-in-water emulsions at two concentrations of HMI (1.6%, top curve, and 2.4%, bottom curve) [22]. The concentration of HMI is much lower than that required when using nonionic surfactants.

The rate of Ostwald ripening is 1.1×10^{-29} and 2.4×10^{-30} m^3/s at 1.6 and 2.4% HMI, respectively. These rates are ~3 orders of magnitude lower than those obtained using a nonionic surfactant. Addition of 5% glycerol was found to decrease the rate of Ostwald ripening in some nanoemulsions, which may be due to the lower oil solubility in the water–glycerol mixture.

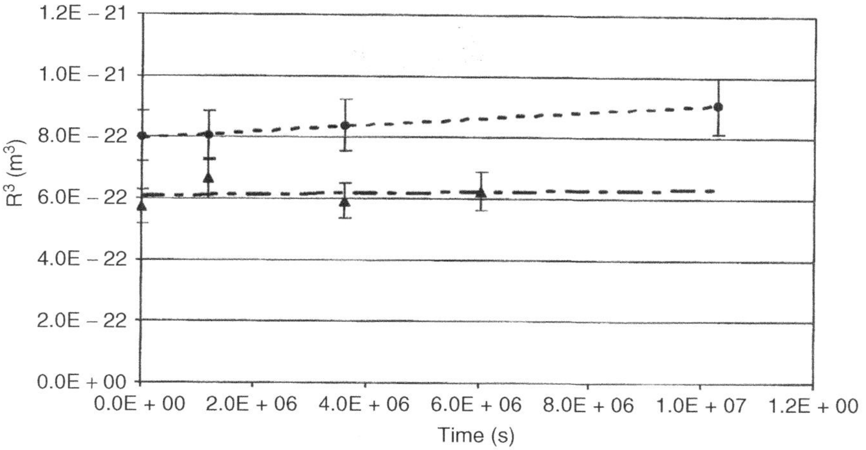

Figure 15.13 *Silicone oil in water nanoemulsions stabilized with HMI (top curve: 1.6% w/w HMI; bottom curve: 2.4% w/w HMI).*

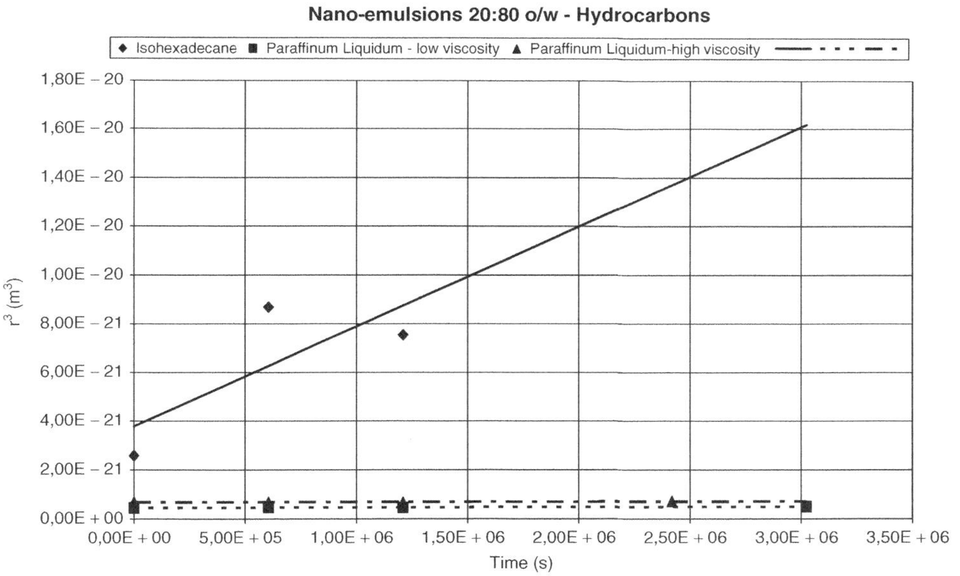

Figure 15.14 *r^3 versus t for nanoemulsions based on hydrocarbon oils.*

Various nanoemulsions with hydrocarbon oils of different solubility were prepared using HMI. Figure 15.14 shows plots of r^3 versus t for nanoemulsions of the hydrocarbon oils that were stored at 50 °C. It can be seen that both paraffinum liquidum with low and high viscosities give almost a zero slope, indicating the absence of Ostwald ripening in this case. This is not surprising since both oils have very low solubility and the HMI strongly adsorbs at the interface, giving high elasticity that reduces both Ostwald ripening and coalescence.

With the more soluble hydrocarbon oils, namely isohexadecane, there is an increase in r^3 with time, giving a rate of Ostwald ripening of 4.1×10^{-27} m^3/s. The rate for this oil is almost 3 orders of magnitude lower than that obtained with a nonionic surfactant, namely laureth-4 (C_{12} alkyl chain with 4 moles of ethylene oxide). This clearly shows the effectiveness of HMI in reducing Ostwald ripening. This reduction can be attributed to the enhancement of the Gibbs dilational elasticity, which results from the multipoint attachment of the polymeric surfactant with several alkyl groups to the oil droplets. This results in a reduction of the molecular diffusion of the oil from the smaller to the larger droplets.

References

1. Tadros, T. F. (2005) *Applied Surfactants*, Wiley-VCH Verlag GmbH, Weinheim, Germany.
2. Holmberg, K., Jonsson, B., Kronberg, B. and Lindman, B. (2003) *Surfactants and Polymers in Aqueous Solutions*, John Wiley & Sons, Ltd, Chichester.
3. Stevens, C. V., Meriggi, A., Peristerpoulou, M. *et al.* (2001) Polymeric surfactants based on inulin, a polysaccharide extracted from chicory. 1. Synthesis and interfacial properties. *Biomacromolecules*, **2** (4), 1256–1259.
4. Hirst, E. L., McGilvary, D. I. and Percival, E. G. (1950) Studies on fructosans. Part I. Inulin from dahlia tubers. *J. Chem. Soc.*, 1297–1302.
5. Watherhouse, A. L. and Chatterton, N. J. (1993) Glossary of fructan terms, in *Science and Technology of Fructans* (eds M. Suzuki and N. J. Chatterton), CRC Press, Boca Raton, Florida, pp. 2–7.
6. De Leenheer, L. (1996) Production and use of inulin: industrial reality with a promising future, in *Carbohydrates as Organic Raw Materials III* (eds H. van Bekkum, H. Roper and A. L. J. Voragen), Wiley-VCH Verlag GmbH, Weinheim, Germany, pp. 67–92.
7. Nestor, J., Esquena, J., Solans, C. *et al.* (2005) Emulsion polymerization of styrene and methyl methacrylate using a hydrophobically modified inulin and comparison with other surfactants. *Langmuir*, **21** (11), 4837–4841.
8. Nestor, J. (2005) Study of the Stabilization of Colloidal Dispersions by New Sugar-Derivativ Polymeric Surfactant, PhD Thesis, University of Barcelona.
9. Hunter, R. J. (1993) *Foundations of Colloid Science*, vols 1 and 2, Oxford University Press, Oxford.
10. Exerowa, D., Gotchev, G., Kolarev, T. *et al.* (2007) Interaction forces in thin liquid films stabilized by hydrophobically modified inulin polymeric surfactant. 2. Emulsion films. *Langmuir*, **23** (4), 1711–1715.
11. Nestor, J., Esquena, J., Solans, C. *et al.* (2007) Interaction forces between particles stabilized by a hydrophobically modified inulin surfactant. *J. Colloid Interface Sci.*, **311** (2), 430–437.
12. Tadros, Th. F., Vandamme, A., Levecke, B. *et al.* (2004) Stabilization of emulsions using polymeric surfactants based on inulin. *Adv. Colloid Interface Sci.*, **108–109**, 207–226.
13. Nakajima, H., Tomomossa, S. and Okabe, M. (1993) Proceedings of the First Emulsion Conference, Paris, 1993.
14. Nakajima, H. (1997) Microemulsions in cosmetics, in *Industrial Applications of Microemulsions* (eds C. Solans and H. Konieda), Marcel Dekker, New York, pp. 175–197.

15. Tadros, T., Izquierdo, P., Esquena, J. and Solans, C. (2004) Formation and stability of nano-emulsions. *Adv. Colloid Interface Sci.*, **108–109**, 303–318.
16. Thompson, W. (Lord Kelvin) (1871) *Philos. Mag.*, **42**, 448.
17. Lifshitz, I. M. and Slesov, V. V. (1959) *Sov. Phys.*, **35**, 331.
18. Wagner, C. (1935) *Z. Electrochem.*, **35**, 581.
19. Kabalnov, A. S. and Schukin, E. D. (1992) Ostwald ripening theory: applications to fluoro-carbon emulsion stability. *Adv. Colloid Interface Sci.*, **38**, 69–97.
20. Walstra, P. (1983) Formation of emulsions, in *Encyclopedia of Emulsion Technology* (ed. P. Becher), Marcel Dekker, New York, pp. 57–127.
21. Walstra, P. and Smoulders, P. E. A. (1998) Emulsion formation, in *Modern Aspects of Emulsion Science* (ed. B. P. Binks), The Royal Society of Chemistry, Cambridge, pp. 56–99.
22. Tadros, T., Vandekerckhove, E., Vandamme, A. *et al.* (2005) A hydrophobically modified inulin surfactant for preparing and stabilizing nanoemulsions. *Cosmet. Toiletries*, **120** (2), 45.

Index

Italic entries indicate references to a figure, **bold** entries indicate references to a table.

Surfactants from Renewable Resources Edited by Mikael Kjellin and Ingegärd Johansson

Printed and bound in the UK by
CPI Antony Rowe, Eastbourne